Chlamydia pneumoniae Infection and Disease

INFECTIOUS AGENTS AND PATHOGENESIS

Series Editors: Mauro Bendinelli, *University of Pisa*
Herman Friedman, *University of South Florida College of Medicine*

Recent volumes in this series:

CHLAMYDIA PNEUMONIAE
Infection and Disease
Edited by Herman Friedman, Yoshimasa Yamamoto, and Mauro Bendinelli

DNA TUMOR VIRUSES
Oncogenic Mechanisms
Edited by Giuseppe Barbanti-Brodano, Mauro Bendinelli, and Herman Friedman

ENTERIC INFECTIONS AND IMMUNITY
Edited by Lois J. Paradise, Mauro Bendinelli, and Herman Friedman

HELICOBACTER PYLORI INFECTION AND IMMUNITY
Edited by Yoshimasa Yamamoto, Herman Friedman, and Paul S. Hoffman

HERPESVIRUSES AND IMMUNITY
Edited by Peter G. Medveczky, Herman Friedman, and Mauro Bendinelli

HUMAN RETROVIRAL INFECTIONS
Immunological and Therapeutic Control
Edited by Kenneth E. Ugen, Mauro Bendinelli, and Herman Friedman

MICROORGANISMS AND AUTOIMMUNE DISEASES
Edited by Herman Friedman, Noel R. Rose, and Mauro Bendinelli

OPPORTUNISTIC INTRACELLULAR BACTERIA AND IMMUNITY
Edited by Lois J. Paradise, Herman Friedman, and Mauro Bendinelli

PULMONARY INFECTIONS AND IMMUNITY
Edited by Herman Chmel, Mauro Bendinelli, and Herman Friedman

RAPID DETECTION OF INFECTIOUS AGENTS
Edited by Steven Specter, Mauro Bendinelli, and Herman Friedman

RICKETTSIAL INFECTION AND IMMUNITY
Edited by Burt Anderson, Herman Friedman, and Mauro Bendinelli

STAPHYLOCOCCUS AUREUS INFECTION AND DISEASE
Edited by Allen L. Honeyman, Herman Friedman, and Mauro Bendinelli

---

A Continuation Order Plan is available for this series. A continuation order will bring delivery of each new volume immediately upon publication. Volumes are billed only upon actual shipment. For further information please contact the publisher.

Chlamydia pneumoniae Infection and Disease

Edited by

Herman Friedman
University of South Florida College of Medicine
Tampa, Florida

Yoshimasa Yamamoto
Osaka University
Osaka, Japan

and

Mauro Bendinelli
University of Pisa
Pisa, Italy

Kluwer Academic / Plenum Publishers
New York, Boston, Dordrecht, London, Moscow

Library of Congress Cataloging-in-Publication Data

Chlamydia pneumoniae: infection and disease/edited by Herman Friedman, Yoshimasa Yamamoto, and Mauro Bendinelli.
p. ; cm. — (Infectious agents and pathogenesis)
Includes bibliographical references and index.
ISBN 0-306-48487-0
1. Chlamydophila pneumoniae infections. 2. Chlamydophila pneumoniae I. Friedman, Herman, 1931– II. Yamamoto, Yoshimasa. III. Bendinelli, Mauro. IV. Series.
[DNLM: 1. Chlamydophila Infections. 2. Chlamydophila pneumoniae. WC 600 C5435 2004]
RC124.5.C44 2004
616.9′235—dc22

2004044175

ISBN: 0-306-48487-0
eISBN: 0-306-48741-1

ISSN: 1075-1289

233 Spring Street, New York, NY 10013

http://www.wkap.nl/

10 9 8 7 6 5 4 3 2 1

A C.I.P. record for this book is available from the Library of Congress

Printed in the United States of America

Contributors

LUIGI ALLEGRA • Istituto di Tisiologia e Malattie dell'Apparato Respiratorio, Universita degli Studi di Milano, Pad. Litta, IRCCS Ospedale Maggiore di Milano, I-20122, Milan, Italy

DENAH M. APPELT • Department of Biomedical Sciences, Philadelphia College of Osteopathic Medicine, Philadelphia, Pennsylvania

BRIAN J. BALIN • Departments of Pathology, Microbiology, and Immunology, and Philadelphia College of Osteopathic Medicine, Philadelphia, Pennsylvania 19131

ROBERT J. BELLAND • Rocky Mountain Labs NIAID/NIH, Hamilton, Montana

SVEND BIRKELUND • Department of Medical Microbiology and Immunology, University of Aarhus, Aarhus, Denmark

FRANCESCO BLASI • Istituto di Tisiologia e Malattie dell'Apparato Respiratorio, Universita degli Studi di Milano, Pad. Litta, IRCCS Ospedale Maggiore di Milano, I-20122, Milan, Italy

GERALD I. BYRNE • Department of Molecular Sciences, University of Tennessee, Health Science Center, Memphis, Tennessee

GUNNA CHRISTIANSEN • Department of Medical Microbiology and Immunology, University of Aarhus, Aarhus, Denmark

ROBERTO COSENTINI • Department of Emergency Medicine, IRCCS Ospedale, Maggiore di Milano, Universita degli Studi di Milano, 1-20122 Milan, Italy

IGNATIUS W. FONG • St. Michael's Hospital, Division of Infectious Diseases, Department of Medicine, University of Toronto, Toronto, Ontario M5B 1W8, Canada

JENS GIEFFERS • Medical University of Luebeck, Luebeck, Germany

HÅKAN G. GNARPE • Institute of Medical Sciences, University of Uppsala, Uppsala, Sweden

JUDY A. GNARPE • Institute of Medical Microbiology and Immunology, University of Alberta, Alberta, Edmonton, Canada

SANDEEP GUPTA • Department of Cardiology, Whipps Cross University Hospital, Leytonstone, London E11 1NR, United Kingdom

DAVID L. HAHN • Arcand Park Clinic Division of Dean Medical Center, Madison, Wisconsin

MARGARET R. HAMMERSCHLAG • Department of Pediatrics, SUNY Downstate Medical Center, Brooklyn, New York

CHRISTINE J. HAMMOND • Departments of Pathology, Microbiology, and Immunology, Philadelphia College of Osteopathic Medicine, Philadelphia, Pennsylvania 19131

ALAN P. HUDSON • Wayne State University School of Medicine, Detroit, Michigan

C. SCOTT LITTLE • Departments of Pathology, Microbiology and Immunology, Philadelphia College of Osteopathic Medicine, Philadelphia, Pennsylvania 19131

ANGELA MACINTYRE • Department of Biomedical Sciences, Philadelphia College of Osteopathic Medicine, Philadelphia, Pennsylvania 19131

AKIRA MATSUMOTO • Department of Urology, Okayama University, Okayama, Japan

TOSHIHARU MATSUSHIMA • Division of Respiratory Diseases, Department of Medicine, Kawasaki Medical School, Kurashiki City, Okayama 701-0192, Japan

NAOYUKI MIYASHITA • Division of Respiratory Diseases, Department of Medicine, Kawasaki Medical School, Kurashiki City, Okayama 701-0192, Japan

JOSEPH B. MUHLESTEIN • Division of Cardiology, LDS Hospital, Salt Lake City, Utah

JOSEPH NGEH • Department of Clinical Gerantology, Radcliffe Infirmary, Oxford OX2 6HE, United Kingdom

KAZENOBU OUCHI • Department of Pediatrics, Kawasaki Medical School, Kurashiki, Okayama, Japan

SCOT P. OUELLETTE • Department of Molecular Sciences, University of Tennessee Health Science Center, Memphis, Tennessee

ANTONIO ROTHFUCHS • Microbiology and Tumorbiology Center, Karolinska Institute, Stockholm, Sweden

MARTIN E. ROTTENBERG • Microbiology and Tumorbiology Center, Karolinska Institute, Stockholm, Sweden

H. RALPH SCHUMACHER • University of Pennsylvania School of Medicine and D.V.A. Medical Center, Philadelphia, Pennsylvania

SUBRAMANIAM SRIRAM • Department of Neurology, Multiple Sclerosis Research Laboratory, Vanderbilt School of Medicine, 1222H Vanderbilt Stallworth Rehabilitation Hospital, Nashville, Tennessee

CHARLES W. STRATTON • Departments of Pathology, Multiple Sclerosis Research Laboratory, Vanderbilt School of Medicine,1222H Vanderbilt Stallworth Rehabilitation Hospital, Nashville, Tennessee

HELJÄ-MARJA SURCEL • National Public Health Institute, Department of Microbiology, Oulu, Finland

PAOLO TARSIA • Department of Emergency Medicine, IRCCS Ospedale Maggiore di Milano, Universita degli Studi di Milano, 1-20122 Milan, Italy

BRIAN VANDAHL • Loke Diagnostics ApS, Science Park, Aarhus, Denmark
JUDITH A. WHITTUM-HUDSON • Wayne State University School of Medicine, Detroit, Michigan
HANS WIGZELL • Microbiology and Tumorbiology Center, Karolinska Institute, Stockholm, Sweden
YOSHIMASA YAMAMOTO • Division of Molecular Microbiology, Department of Basic Laboratory Sciences, School of Allied Health Sciences, Osaka University, Osaka, Japan

Preface to the Series

The mechanisms of disease production by infectious agents are presently the focus of an unprecedented flowering of studies. The field has undoubtedly received impetus from the considerable advances recently made in the understanding of the structure, biochemistry, and biology of viruses, bacteria, fungi, and other parasites. Another contributing factor is our improved knowledge of immune responses and other adaptive or constitutive mechanisms by which hosts react to infection. Furthermore, recombinant DNA technology, monoclonal antibodies, and other newer methodologies have provided the technical tools for examining questions previously considered too complex to be successfully tackled. The most important incentive of all is probably the regenerated idea that infection might be the initiating event in many clinical entities presently classified as idiopathic or of uncertain origin.

Infectious pathogenesis research holds great promise. As more information is uncovered, it is becoming increasingly apparent that our present knowledge of the pathogenic potential of infectious agents is often limited to the most noticeable effects, which sometimes represent only the tip of the iceberg. For example, it is now well appreciated that pathologic processes caused by infectious agents may emerge clinically after an incubation of decades and may result from genetic, immunologic, and other indirect routes more than from the infecting agent in itself. Thus, there is a general expectation that continued investigation will lead to the isolation of new agents of infection, the identification of hitherto unsuspected etiologic correlations, and, eventually, more effective approaches to prevention and therapy.

Studies on the mechanisms of disease caused by infectious agents demand a breadth of understanding across may specialized areas, as well as much cooperation between clinicians and experimentalists. The series *Infectious Agents and Pathogenesis* is intended not only to document the state of the art in this fascinating and challenging field but also to help lay bridges among diverse areas and people.

Mauro Bendinelli
Herman Friedman

Introduction

This volume on the topic *Chlamydia pneumoniae* infection and disease is appropriate, we believe, for our continuing series on infectious agents and pathogenesis. This book concerns both basic and clinical concepts and implications of infection by the ubiquitous, opportunistic intracellular bacterium *Chlamydia pneumoniae* (Cpn), which was recognized only in the last few decades as a cause of significant morbidity and disease in humans. Infections by Chlamydia are now believed to cause chronic diseases, including serious sequelae. *C. pneumoniae* infection was first recognized as a cause of acute upper respiratory tract infection, especially pneumonia and bronchitis in young individuals. However, nearly all adults have evidence of being exposed to this organism, since relatively high serum antibody levels are present in the blood of most individuals, even those with no evidence of previous clinical disease. Thus, there is now much interest in the likelihood that chronic sequelae developing after either acute or, most likely, asymptomatic infections may result in chronic inflammatory diseases, including coronary heart disease, such as atherosclerosis, and possibly some neurologic autoimmune disease like multiple sclerosis and, even, Alzheimer's disease.

This volume presents up-to-date reviews of Chlamydia infection and disease by acknowledged experts on the basis of basic science and clinical studies. The first chapter of this volume is a general overview of the expanding knowledge that this ubiquitous intracellular bacterium may be associated with many diseases besides pulmonary infections. Drs. Oulette and Byrne from the University of Tennessee review the basic biology of this organism, including information concerning different forms of Chlamydia, such as the EB and RB bodies. Dr. Miyashita and colleagues from Japan describe the morphology of these bacteria. The cellular and molecular biology of this obligate intracellular bacterium are discussed by Dr. Christiansen and colleagues from Sweden. An overview of the immune response to Chlamydia, including involvement of innate and adoptive immune defenses, is discussed by Dr. Surcel and colleagues from Finland. Dr. M. Rottenberg and associates from Sweden describe current information concerning development of vaccines against this bacterium. Dr. Margaret Hammerschlag from New York presents a review of the efficacy of antibiotic treatment for *C. pneumoniae* infection, while Dr. Yamamoto from

Japan describes newer laboratory techniques, especially PCR methods, for rapid detection of Chlamydia in blood.

A review by Drs. Nge and Gupta from England details the possible role of this microorganism in cardiovascular disease pathogenesis. Dr. Byrne from Tennessee gives an introduction into current concepts that *C. pneumoniae* is associated with important chronic diseases, especially atherosclerosis. Animal models of Cpn infection, especially as related to atherosclerosis, are discussed by Dr. Fong from Canada. Dr. Oulette and colleagues from Tennessee then describe some of the known factors associated with this bacterium in the pathogenesis of cardiovascular disease. Dr. Muhlestein from Utah describes some of the clinical evidence aimed at determining whether antimicrobial treatment of Chlamydia infection has a significant effect on coronary artery disease. Drs. Harken and Judy Gnarpe from Canada present information concerning myocarditis associated with Cpn infection.

Involvement of this bacterium with other chronic diseases, including reactive arthritis, as well as the possibility that infection with Cpn is associated with diseases of the nervous system, including late onset of Alzheimer's disease and multiple sclerosis, is discussed. Drs. Stratton and Sriram from Vanderbilt University present a summary of their study, indicating that Cpn infection is associated with multiple sclerosis, an association which has been controversial but has also been confirmed by a number of other investigators. Dr. Balin and colleagues from Philadelphia describe histologic evidence suggesting that *C. pneumoniae* may be associated with the pathogenesis of Alzheimer's disease, at least late disease. Dr. Whittin-Hudson and colleagues from Detroit discuss the possible involvement of this bacterium in arthritis. Dr. David Hahn from Madison, WI, discusses the role of *C. pneumoniae* in asthma. Dr. Ouchi from Japan discusses the classic situation of respiratory infections caused by *C. pneumoniae* from the viewpoint of pediatric patients. Dr. Blasi and colleagues from Italy also discuss information concerning pneumonia caused by this organism.

It is widely acknowledged that research and clinical studies concerning infection by this pathogen will have numerous beneficial results for both the medical and the scientific communities. The editors of this volume, as well as authors of individual chapters, are encouraged by recent developments concerning advances in knowledge about this organism and its association not only with acute infections such as respiratory diseases but also chronic inflammatory diseases like atherosclerosis and autoimmunity, especially CNS diseases. We believe the emerging interest in this important opportunistic intracellular bacterium is of value not only for clinicians but also for laboratory investigators and biomedical scientists, as well as students and health professionals in general. The editors wish to express their thanks and gratitude to Ms. Ilona Friedman, who again served as an outstanding editorial assistant for this volume, as for all the books in this series.

Yoshimasa Yamamoto
Herman Friedman
Mauro Bendinelli

Contents

1. *Chlamydia pneumoniae*: Prospects and Predictions for an Emerging Pathogen 1
SCOT P. OUELLETTE and GERALD I. BYRNE

1. Overview 1
2. Insights: Basic Biology 2
3. Insights: Respiratory Infections 4
4. Insights: Cardiovascular Disease 5
5. Comments on Various Chronic Diseases 6
6. Concluding Remarks 7
References 7

2. Morphology of *Chlamydia pneumoniae* 11
NAOYUKI MIYASHITA and AKIRA MATSUMOTO

1. Introduction 11
2. Chlamydial Developmental Cycle 11
3. Morphology of EB 15
4. Structure of the Outer Membrane 16
5. Surface Projections and Related Structures 20
6. Persistent Infection and Aberrant Forms 23
7. Summary 25
References 26

3. Cell and Molecular Biology of *Chlamydia pneumoniae* 29
GUNNA CHRISTIANSEN, BRIAN VANDAHL, and SVEND BIRKELUND

1. Introduction 29
2. Ultrastructure of *Chlamydia* EB and RB 30
3. Components Responsible for the Structural Appearance 30

4. Physical Localization of the Components 31
5. Chlamydia Genomes 32
6. The *C. pneumoniae* Proteome 33
7. The Pmp Gene Family 35
8. Pmp Structure 36
9. Type III Secretion System 36
10. Secreted Proteins 37
11. Vesicle Trafficking 39
12. Host Attack on *Chlamydia* 39
13. Perspectives 40
References 40

4. Antibiotic Susceptibility and Treatment of *Chlamydia pneumoniae* Infections 45
MARGARET R. HAMMERSCHLAG

1. Introduction 45
2. *In Vitro* Susceptibility of *C. pneumoniae* 45
2.1. Is There Antimicrobial Resistance in *C. pneumoniae*? 47
2.2. Treatment of Persistent *C. pneumoniae* Infection *In Vitro* 49
3. Treatment of *C. pneumoniae* Respiratory Infections 49
4. Relevance for Cardiac Clinical Treatment Intervention Studies 52
References 53

5. Pneumonia Caused by *Chlamydia pneumoniae* 57
FRANCESCO BLASI, ROBERTO COSENTINI, PAOLO TARSIA, and LUIGI ALLEGRA

1. Introduction 57
2. Epidemiology 58
3. CAP in Pediatric Patients 59
4. CAP in Adults 60
5. Infection in Immunocompromised Subjects 62
6. Conclusions 62
References 63

6. Chlamydia Detection in Blood 65
YOSHIMASA YAMAMOTO

1. Introduction 65
2. PCR—Methodological Aspects 66
2.1. Contamination 66
2.2. DNA Extraction 67
2.3. Target Genes 67
2.4. PCR Protocols 68
2.5. Detection of PCR Products 70

3. Detection of Chlamydia in Blood by PCR 70
3.1. *C. pneumoniae* DNA Extraction 71
3.2. PCR Method 71
3.3. Detection of *C. pneumoniae* in Blood 74
3.4. Location of *C. pneumoniae* in Blood 75
3.5. Association of Blood *C. pneumoniae* with Cardiovascular Disease 75
4. Chlamydia Antigen Detection by Immunostaining 76
5. Conclusions 77
References 77

7. *Chlamydia pneumoniae* Infection and Diseases: Imunity to *Chlamydia pneumoniae* 81
HELJÄ-MARJA SURCEL
1. Introduction 81
2. Innate Immune Mechanisms 82
3. The Role of Macrophages in *C. pneumoniae* Immunity 83
4. Acquired Immunity 84
4.1. Effector Mechanisms of Protective Immunity 84
4.2. Cell-Mediated Immunity in Humans 86
4.3. Immune Escape 88
5. *C. pneumoniae*–Specific Antibodies 88
6. Conclusion 90
References 90

8. Vaccines Against *Chlamydia pneumoniae*: Can They Be Made? 99
MARTIN E. ROTTENBERG, ANTONIO GIGLIOTTI ROTHFUCHS, and HANS WIGZELL
1. Introduction 99
2. Do We Need an Anti–*C. pneumoniae* Vaccine? 100
3. *C. pneumoniae* Vaccine Development 100
3.1. Adaptive Immune Responses 101
3.2. Escape Mechanisms 103
4. Antigens 104
5. Adjuvants 107
5.1. Activation of Innate Immunity 107
5.2. Adjuvants and Delivery Systems 108
6. Conclusions 108
References 109

9. *Chlamydia pneumoniae* and Atherosclerosis—an Overview of the Association 113
JOSEPH NGEH and SANDEEP GUPTA

1. Introduction 113
1.1. Historical Perspectives on the Association between Infection and Atherosclerosis 113
1.2. Microbiology and Epidemiology of *C. pneumoniae* 115
2. Evidence Linking *C. pneumoniae* and Atherosclerosis 116
2.1. Seroepidemiological Observations 116
2.2. Pathological Specimen Findings 117
2.3. Animal Experimental Models 119
2.4. Molecular and Immunological Mechanisms 121
2.5. Antibiotic Treatment Studies 123
3. *C. pneumoniae* Atherosclerotic Risk Factors and Koch's Postulates 130
4. Concluding Comments 131
References 132

10. The Biology of *Chlamydia pneumoniae* in Cardiovascular Disease Pathogenesis 135
SCOT P. OUELLETTE, ROBERT J. BELLAND, JENS GIEFFERS, and GERALD I. BYRNE
1. Introduction ....
2. Chlamydial Life Cycle and Persistence 137
3. *C. pneumoniae* in the Pathogenesis of Cardiovascular Disease 139
3.1. *C. pneumoniae* and Host Cell Signaling 139
3.2. Chlamydiae–Macrophage Interactions 142
4. *C. pneumoniae* Strain Heterogeneity in Cardiovascular Disease 144
4.1. The *tyrP* Polymorphism and SNPs 144
4.2. Host Tropism and Polymorphic Outer Membrane Proteins 147
5. Concluding Remarks 148
References 149

11. Animal Models of *Chlamydia pneumoniae* Infection and Atherosclerosis 153
IGNATIUS W. FONG
1. Introduction 153
1.1. Animal Models Used in Atherosclerotic Research 154
1.2. *C. pneumoniae* Lung Infection Model 156
2. Murine Models of *C. pneumoniae*–Related Atherosclerosis 157
3. Rabbit Models of *C. pneumoniae* Infection and Atherosclerosis 159
4. Effect of Antibiotics on *C. pneumoniae* Animal-Induced Atherosclerosis 163
5. Summary and Future Directions 166
5.1. Future Directions 167
References 168

12. Antiinfective Trials for the Treatment of *Chlamydia pneumoniae* in Coronary Artery Disease 173
JOSEPH B. MUHLESTEIN

1. Introduction 173
2. Animal Studies of *Chlamydia pneumoniae*, Atherosclerosis and Antiinfective Therapy 174
3. Observational Studies of the Effect of Antimicrobials on the Course of Coronary Artery Disease 175
4. Randomized Secondary Prevention Trials for Clinical Peripheral Vascular Disease 176
5. Randomized Secondary Prevention Trials for Clinical Coronary Artery Disease 178
6. Ongoing Clinical Antiinfective Trials 183
7. Potential Concerns 184
References 185

13. *Chlamydia pneumoniae* and Myocarditis 187
HÅKAN G. GNARPE and JUDY A. GNARPE

1. Introduction 187
2. Myocarditis 188
3. Endocarditis Cases 190
4. Dilated Cardiomyopathy 192
5. Animal Models of Myocarditis 192
6. Possible Mechanisms 193
7. Summary 194
References 194

14. *Chlamydia pneumoniae* as a Candidate Pathogen in Multiple Sclerosis 199
CHARLES W. STRATTON and SUBRAMANIAM SRIRAM

1. Introduction 199
2. Autoimmune Basis for MS 199
3. MS as an Autoimmune Disease: Unresolved Issues 200
4. Infection as an Alternate Hypothesis for MS 201
5. Candidate Pathogens 203
5.1. Human Herpes Virus-6 (HHV-6) 204
5.2. Retroviruses 204
5.3. *Chlamydia pneumoniae* 205
6. Mechanism of *C. pneumoniae*–Mediated Accentuation of Autoimmune Disease 205
7. Conclusions 207
References 208

15. *Chlamydia pneumoniae* in the Pathogenesis of Alzheimer's Disease........ 211
BRIAN J. BALIN, CHRISTINE J. HAMMOND, C. SCOTT LITTLE, ANGELA MACINTYRE, and DENAH M. APPELT

1. Introduction 211
2. Classification and Characteristics of Alzheimer's Disease 212
3. Risk Factors in Alzheimer's Disease 212
4. Association of *C. pneumoniae* and Alzheimer's Disease 213
5. Entry of *C. pneumoniae* into the Nervous System 214
6. *C. pneumoniae* and Neuroinflammation 216
7. Consequences of *C. pneumoniae* Infection *In Vitro* 218
8. Animal Model for Sporadic Alzheimer'S Disease 219
9. Summary 221
10. Future Directions 222
References 223

16. *Chlamydia pneumoniae* and Inflammatory Arthritis........ 227
JUDITH A. WHITTUM-HUDSON, H. RALPH SCHUMACHER, and ALAN P. HUDSON

1. Introduction 227
2. Reactive Arthritis 228
3. *Chlamydia trachomatis*, Persistence, and Reactive Arthritis 229
4. *Chlamydia pneumoniae* and Arthritis 231
5. *Chlamydia pneumoniae* and Cytokines 233
6. Summary 235
References 236

17. Role of *Chlamydia pneumoniae* as an Inducer of Asthma........ 239
DAVID L. HAHN

1. Introduction 239
   1.1. Definition of Induction 239
   1.2. Definition of Asthma 240
   1.3. Importance of Asthma 241
   1.4. Current Asthma Treatments Are Palliative, Not Curative 241
2. *Chlamydia pneumoniae* as a Promoter of Asthma Severity 242
   2.1. *Chlamydia pneumoniae*–Asthma Serologic Associations 242
   2.2. Results of Antibiotic Treatment Directed against Chlamydial Infection in Asthma 243
3. *Chlamydia pneumoniae* as an Asthma Initiator 248
   3.1. Asthma Is Often Associated with Preceding Respiratory Illnesses 248
   3.2. New-Onset Asthma after Acute Cpn Infection 248
   3.3. Infection Should Be Added to the List of Possible Causes for Worldwide Increases in Asthma 249

4. Summary: *Chlamydia pneumoniae* as an Asthma Inducer 249
5. *Chlamydia pneumoniae* and Lung Remodeling 251
5.1. Infection May Also Be Related to Airway Remodeling in Asthma and COPD 251
5.2. Cpn Produces Cytokines Linked to Asthma and Lung Remodeling 252
5.3. Summary: *Chlamydia pneumoniae* and Lung Remodeling 253
6. Conclusion: *Chlamydia pneumoniae*, Chronic Nonspecific Lung Disease (CNSLD) and the "Dutch Hypothesis" 253
References 254

18. Respiratory Tract Infections Caused by *C. pneumoniae* in Pediatric Patients 263
KAZUNOBU OUCHI

1. Introduction 263
2. Respiratory Disease Associated with *Chlamydia pneumoniae* 263
2.1. Carrier Status 263
2.2. Upper Respiratory Tract Infection 264
2.3. Lower Respiratory Tract Infection 265
3. Diagnosis 267
4. Treatment 269
5. Conclusion 270
References 271

Index 275

1

Chlamydia pneumoniae: Prospects and Predictions for an Emerging Pathogen

SCOT. P. OUELLETTE and GERALD. I. BYRNE

1. OVERVIEW

Chlamydial diseases have long been recognized as a significant cause of morbidity in humans. Although acute disease is rarely fatal, chronic infections with their associated sequelae have an enormous impact both on the economy and on the quality of life of those who suffer from these conditions. For *Chlamydia trachomatis*, acute infections include conjunctivitis and the sexually transmitted diseases; chronic infections can lead to trachoma, tubal factor infertility, pelvic inflammatory disease, and reactive arthritis.[1] The identification and speciation of *Chlamydia pneumoniae* about 15 years ago and the recognition that a large percentage of the world's population has been exposed to this organism[2] have led many investigators to inquire into certain diseases with unknown etiology, some of which were not thought to have an infectious component. For *Chlamydia pneumoniae*, acute infections are localized to the airways, with pneumonia and bronchitis being the most common disease conditions.[3] Chronic sequelae, developing from acute or asymptomatic infections, cannot be definitively attributed to *C. pneumoniae* at this time, although evidence is accumulating to suggest that *C. pneumoniae* infection may also lead to debilitating (asthma) and even fatal (heart disease) conditions. Table I shows acute diseases caused by *C. pneumoniae* and the chronic diseases with which it has been associated. Possible pathological associations are also listed.

C. pneumoniae, like other members of the genus, has the characteristic biphasic life cycle between the infectious, metabolically inert elementary body (EB) and the noninfectious, metabolically active reticulate body (RB). EBs are

SCOT P. OUELLETTE and GERALD I. BYRNE • Department of Molecular Sciences, University of Tennessee Health Science Center, Memphis, Tennessee.

TABLE I
***C. pneumoniae* Diseases and Putative Chronic Sequelae**

Acute disease	Chronic disease	Possible chlamydia-associated pathology	Epidemiology[a]	Refs.
Community-acquired pneumonia	Atherosclerosis	Macrophage/endothelial dysfunction, chronic inflammatory response, LDL oxidation	Serology, IHC, PCR, EM, culture	4, 5, 6, 7 (see 8 for review)
Bronchitis Pharyngitis	Asthma	Ciliary dysfunction, inflammatory and IgE responses, bronchial hyperresponsiveness	Serology, PCR, MIF, culture, treatment	9, 10 (see 11 for review)
Other respiratory tract diseases	Reactive arthritis	Chronic inflammatory response, HLA type association	RT/PCR	12, 13
	Alzheimer's (late onset)	Chronic inflammation, glial cell dysfunction	RT/PCR, EM	14
	Multiple sclerosis	Chronic inflammation	PCR	15

[a]IHC: immunohistochemistry; MIF: microimmunofluorescence; EM: electron microscopy; RT/PCR: reverse transcriptase/polymerase chain reaction.

internalized into a pathogen-modified phagosome that avoids fusion with lysosomes. This is referred to as an inclusion. The ability of *C. pneumoniae* to enter a non-productive growth state, often termed persistence, further highlights the similarities between *C. trachomatis* and *C. pneumoniae* (see ref. 16 for review). However, as is apparent in the chapters of this text, it is quite clear that *C. pneumoniae* also has unique characteristics that contribute to its disease-evoking potential.

2. INSIGHTS: BASIC BIOLOGY

Understanding the basic biology of the bacteria is a critical first step to developing better treatment strategies. Miyashita *et al.* have provided an ultrastructural assessment of the *C. pneumoniae* EB and RB, with particular focus on the surface projections that have been documented in other *Chlamydia.* In spite of the oft-seen pear-shaped EBs of *C. pneumoniae,* the authors show that at the electron microscopic level the basic morphology remains the same between all isolates examined. The surface projections span the chlamydial membranes to effectively connect the bacterial cytoplasm with its surrounding environment. RBs also appear to associate with the inclusion membrane in such a way as to allow direct connection with the host cytoplasm. These observations are intriguing and lead to speculation about possible roles for type III secretion and parasite-mediated host cell disruption. Type III secretion is a common feature among Gram-negative pathogens that allows the organism to directly inject effector molecules into the host cell cytoplasm that modulate host cell function in ways that are advantageous to the bacteria.[17]

Christiansen *et al.* have reviewed the cell and molecular biology of this organism. *Chlamydia* are obligate intracellular parasites and consequently have several obstacles to overcome in establishing a successful growth phase. The mechanism of attachment to host epithelial cells remains unknown, but surface components that may play a role include the major outer membrane protein, cysteine-rich outer membrane proteins, and other polymorphic membrane proteins.[18] The broad host cell tropism that is characteristic of *C. pneumoniae* likely has its origins in the outer membrane complex. Modification of the phagosome into which the bacteria enters is an important survival mechanism as it prevents fusion with the lysosome. The chlamydial proteins on the inclusion membrane responsible for inhibiting this process are not definitively known. Chlamydiae secrete inclusion membrane proteins (Incs),[19] which are inserted into the membrane and may help prevent lysosome fusion via a type III dependent process. Belland *et al.* have recently identified a hypothetical gene that is transcribed very early in the course of infection and may also serve to prevent phagosome maturation.[20] Other areas of interest where transcriptional profiling may be useful include how the organism directs differentiation from EBs to RBs and vice versa. Such techniques must be used because there are no methods for genetic transfer, which makes members of the genus *Chlamydia* some of the most difficult organisms to study. Developing a tractable genetic system to allow for targeted gene disruption would be a great asset for chlamydial research. Unfortunately, such a system is difficult to envision. The EB has a tightly cross-linked outer membrane and its chromosome is packaged in histone-like proteins. Thus, introducing DNA vectors into EBs may prove difficult, and DNA may not be recombined into the chromosome. Targeting the RB for transformation also has problems. Once inside the cell, there are three membranes separating the organism's cytoplasm from the extracellular environment: the host cell membrane, the inclusion membrane, and the bacterial cell wall. Furthermore, RBs are fragile, so harsh transformation methods such as electroporation would not likely have success.

Immune responses to *Chlamydia* have traditionally been difficult to dissect. As an intracellular pathogen, immune responses to *C. pneumoniae* will likely involve TH1 responses. However, there will be significant differences from *C. trachomatis* in terms of innate defenses because of the anatomical and physiological differences between the respiratory and genitourinary tract systems. For *C. trachomatis,* there is a clear role for TH1 responses. CD4+ T cells stimulate CD8+ responses to secrete interferon gamma and to lyse infected cells (see ref. 21 for review). The role of B cells and antibody remains uncertain; however, there does appear to be a synergistic effect between B and CD4+ T cells since depleting both leaves mice highly susceptible to reinfection.[22] H.-M. Surcel provides an overview of immune responses to *C. pneumoniae* from the innate to the specific. The epithelial barrier of the respiratory system is the first significant defense against inhaled pathogens, but *C. pneumoniae* efficiently attaches to and is internalized by these cells. The secretion of proinflammatory cytokines and recruitment of phagocytes is effected by epithelia. However, macrophages may be used to disseminate infection through the circulation. This may also be true for T cells; thus the organism seems to exploit the defenses used against it. Understanding both the primary and secondary immunological responses to this organism is critical for the successful design of vaccines. Rottenberg *et al.* have approached the problem of vaccine design with the aid of the recent sequencing of the organism. Vaccine candidates can be identified and studied in appropriate mouse models. However, a significant immunological question that must be addressed is the propensity for *C. pneumoniae* to enter a nonproductive growth state. Any successful vaccine strategy must be able to prevent and eradicate such cases.

3. INSIGHTS: RESPIRATORY INFECTIONS

Upper and lower respiratory tract disease is the most common manifestation of symptomatic *C. pneumoniae* infection. K. Ouchi has discussed the role of this organism in respiratory diseases of pediatric patients. Acute infections, including atypical pneumonia, in young children are perhaps more common than originally thought and can induce wheezing in children with reactive airway disease. Chronic infections in otherwise healthy children make it difficult to determine carrier status. Proper diagnosis and treatment methods remain an understudied area that requires more input to ensure eradication. Blasi *et al.* complement this analysis by discussing the role of community acquired pneumonia (CAP) caused by the organism, particularly as it relates to coinfections and to immunocompromised status. *C. pneumoniae* is commonly found in CAP patients and can lead to more serious disease sequelae when left untreated in the presence of other bacterial pathogens. More studies are required to assess the infection rate in immunocompromised patients. A recurring theme is that these infections are difficult to diagnose and treat.

The association between *C. pneumoniae* and asthma has gained much credence in recent years. D. Hahn has outlined a role for this organism as an inducer of asthma. Although there is some disagreement in the medical profession vis á vis diagnosing asthma versus chronic bronchitis or emphysema, the clinical manifestations such as wheezing, coughing, and shortness of breath are endpoints that can be useful in determining lung dysfunction. An intriguing theory is that chlamydial infection leads to the establishment of nonatopic asthma in some individuals (see ref. 11 for review). Perhaps the most convincing data in support of this is the amelioration of symptoms in patients treated with antibiotics effective against *Chlamydia.* Clinical studies have demonstrated a more rapid decline in lung function in patients with *C. pneumoniae* seroreactivity. Because of the increasing numbers of reactive airway diseases in recent decades, the possibility of reversing this trend with antibiotics is an important outcome.

4. INSIGHTS: CARDIOVASCULAR DISEASE

Chronic infections with *C. pneumoniae* and the sequelae with which it has been associated could arguably place this bacterium among the most important pathogens in the developed world. The most significant disease state connected to the organism is atherosclerosis and cardiovascular disease.[(23)] Ngeh and Gupta provide an overview of this association citing the large body of epidemiological studies. This field is highly controversial due in large part to the lack of standardized methods between laboratories; for example, there is a variety of "home brew" methods currently used for DNA extraction and for serological testing. Also, antibiotic treatment studies that have shown inconclusive data may need further refinement to better identify those patients who would have the greatest benefit. *Chlamydia pneumoniae* as an agent of myocarditis is discussed by Gnarpe and Gnarpe. In spite of the association of this bacterium with atherosclerosis, there have been relatively few documented cases of myo-, perimyo-, or endocarditis attributed to it. Strain differences and host genetic factors may be controlling factors in this disease state.

Before a causal connection can be made between *C. pneumoniae* and atherosclerosis, pathogenic mechanisms must be defined to support a role for this organism in the disease. Ouellette *et al.* discuss in some detail potential mechanisms for *C. pneumoniae* to exacerbate or promote lesion development and instability. Chlamydial virulence determinants include its Hsp60 and lipopolysaccharide (LPS). Both of these molecules are capable of inducing changes in macrophage biology consistent with proatherogenic events.[(24)] Furthermore, the organism is capable of inducing changes in a variety of cell types at the site of lesions. Induction of signaling pathways results in the inappropriate transcriptional activation of key genes involved in atherogenesis. With the recent sequencing of the genome, other genes may be identified as virulence determinants. Because atherosclerosis is a complicated and slow-developing disease, further studies will be required to assess a role for this bacteria and its pathogenic mechanisms.

Establishing a causal link between *C. pneumoniae* and atherosclerosis would be greatly enhanced by the use of an appropriate animal model. I. Fong has summarized the status of these studies. Although there are some dissenting data, most experimental models of atherosclerosis in hypercholesterolemic mice have shown that infection with *C. pneumoniae* can enhance the disease. However, infection alone is incapable of inducing the pathology, but, consistent with *in vitro* data, infection will lead to inflammation and endothelial dysfunction. In rabbits, similar results can be obtained, but early lesions can be induced in normal animals although the appearance of such does not precisely resemble the human disease. Interestingly, antibiotic treatment of animals can have beneficial effects on atherosclerosis when treated early after infection but not at later times.

The burgeoning interest in *C. pneumoniae* and atherosclerosis has prompted the establishment of clinical trials to test the efficacy of antichlamydial drugs on cardiac events. J. Muhlestein gives an overview of the status of these studies.

Although clinical results have been ambiguous, there are several factors that may influence the experimental outcome. Only those with significant serological responses to the organism may benefit from therapy: Mahdi *et al.* recently showed a correlation between elevated IgG responses to chlamydial Hsp60 and cardiovascular disease.[(25)] The type of antibiotic used may also have a significant effect. M. Hammerschlag discusses antibiotic treatment options for chlamydial infections, and such points must be taken in to consideration. In animal models azithromycin and rifampin elicited the greatest eradication, but drugs that specifically target a nonproductively growing chlamydia may result in greater eradication rates. Finally, because atherosclerosis is a multifactorial disease, combination therapies with both an antibiotic (e.g. azithromycin) and a cholesterol-lowering agent (statins) may provide the best analysis for this association.

5. COMMENTS ON VARIOUS CHRONIC DISEASES

Although *C. pneumoniae* is commonly associated with atherosclerosis and asthma, there are several other chronic diseases that may be attributed to the organism. Hudson *et al.* challenge the accepted definition of reactive arthritis. Reactive arthritis has traditionally been defined as a sterile immune-mediated pathology to bacterial antigens in the joint after infections in the gastrointestinal or urogenital tract. PCR diagnostics have detected a variety of organisms in the joint. However, data exist to show that *C. trachomatis* not only localizes to the joint but is metabolically active. Because *C. pneumoniae* has the ability to disseminate through the bloodstream, it is not surprising that it too has been localized to synovial tissue. Further work is necessary to elucidate a pathogenic role for the organism in this disease.

Perhaps some of the more controversial associations with *C. pneumoniae* are those to nervous system diseases, including late-onset Alzheimer's and multiple sclerosis. These diseases are every bit as complicated as atherosclerosis and asthma and could quite likely contain pathogen-mediated components. However, it remains equally plausible to posit that the presence of the organism in the nervous system may be due to its propensity to disseminate through the circulation rather than as a directed disease-contributing agent. Balin *et al.* discuss the potential for *C. pneumoniae* to be a factor in late-onset Alzheimer's. The classical risk factors for this disease include atherosclerosis, and *C. pneumoniae* dissemination to the brain may occur following head trauma or by infected monocytes crossing the blood–brain barrier. The bacterium has been detected in glial cells within the brain of affected patients and may promote dysfunctional activity in these cell types, which would result in inflammation. In a mouse model of the disease, increased plaque pathology after infection is consistent with that seen in Alzheimer's. Stratton and Sriram discuss the possibility of *C. pneumoniae* as a contributing factor in the etiology of multiple sclerosis. Although there is a strong genetic predisposition for this disease, infectious agents may add to the heterogeneity of disease phenotypes. A common theme for these diseases is the ability of *C. pneumoniae* to apparently persist for long durations within cell types and elicit low-level inflammatory responses.

6. CONCLUDING REMARKS

Advancing the field of *C. pneumoniae* research will have numerous beneficial effects for both the medical and scientific communities. By developing sensitive and specific diagnostic tests for *C. pneumoniae*, infected patients can be reliably identified and appropriately treated. These tests must also be standardized as there are many laboratories using varied techniques (e.g. DNA isolation). This is particularly important in chronic diseases, such as atherosclerosis and asthma, with a potential infectious etiology where only a subset of patients may benefit from antibiotic therapy. Treatment, therefore, is crucial to reduce the impact of such diseases on society. New antimicrobials or antibiotic regimens must be developed to better combat persistently growing organisms. This is likely the state in which *C. pneumoniae* can be found in chronic disease sites. The impact of acute respiratory diseases must not be overlooked, however, as this is one means for the bacteria to enter the host and disseminate to extrapulmonary locations. Vaccines would be the best strategy for reducing morbidity if they could provide long-lasting immunity. The successful design of a vaccine should enlist both B and T cells. Finally, basic research on the biology of *C. pneumoniae* will aid all of these areas by providing new targets for diagnostics, antimicrobials, and vaccines.

With the recent sequencing of three strains of *C. pneumoniae*,[26–28] we are now acquiring the necessary information to facilitate these studies. Combining genomic and proteomic analyses, in an effort to link genes with functions, will likely yield the most fruitful results. This ubiquitous bacterium is clearly an important and complicated emerging pathogen.

ACKNOWLEDGMENTS. Work in GIB's laboratory is supported in part by NIH grants HL71735 and AI42790.

REFERENCES

1. Schachter, J., 1978, Chlamydial infections, *N. Engl. J. Med.* **298:**428–435.
2. Grayston, J. T., 1992, Infections caused by *Chlamydia pneumoniae* strain TWAR, *Clin. Infect. Dis.* **15:**757–761.
3. Kalayoglu, M., Hahn, D., and Byrne, G. I., 1998, *Chlamydia* infection and pneumonia, in: Opportunistic Intracellular Bacteria and Immunity (L. Paradise and H. Friedman, ed.), Plenum, New York, pp. 233–253.
4. Danesh, J., Whincup, P., and Peto, R., 1997, Chronic infections and coronary heart Disease: Is There a Link? *Lancet* **350:**430–436.
5. Kuo, C. C., Shor A., Campbell, L. A., Fukushi, H., Patton, D. L., and Grayston, J. T., 1993, Demonstration of *Chlamydia pneumoniae* in atherosclerotic lesions of coronary arteries, *J. Infect. Dis.* **167:**841–849.

6. Maass, M., Bartels, C., Kruger, S., Krause, E., Engel, P. M. and Dalhoff, K., 1998. Endovascular presence of *Chalmydia pneumoniae* DNA is a generalized phenomenon in atherosclerotic vascular disease, *Atherosclerosis* **140:**25S–30S.
7. Apfalter, P., Loidl, M., Nadrchal, R., Makristathis, A., Rotter, M., Bergmann, M., Palterauer, P., and Hirschl, A. M., 2000. Isolation and continuous growth of *Chlamydia pneumoniae* from arterectomy specimens, *Eur. J. Clin. Microbiol. Infect. Dis.* **19:**305–308.
8. Kalayoglu, M. V., Libby, P., and Byrne, G. I., 2002, *Chlamydia pneumoniae* as an emerging risk factor in cardiovascular disease, *JAMA* **288:**2724–2731.
9. Grayston, J. T., Kuo, C. C., Wang, S. P., and Altman, J., 1986, A new *Chlamydia psittaci* strain, TWAR, isolated in acute respiratory tract infections, *N. Engl. J. Med.* **315:**161–168.
10. Normann, E., Gnarpe, J., Gnarpe, H., and Wettergren, B., 1998, *Chlamydia pneumoniae* in children with acute respiratory tract infections, *Acta Paediatr.* **87:**23–27.
11. Hahn, D., 1999, *Chlamydia pneumoniae,* asthma, and COPD: What is the evidence? Ann. Allergy Asthma Immunol. **83:**271–291.
12. Schumacher, H. R. Jr., Gerard, H. C., Arayssi, T. K., Pando, J. A., Branigan, P. J., Saaibi, D. L., and Hudson, A. P., 1999, Lower prevalence of *Chlamydia pneumoniae* DNA compared with *Chlamydia trachomatis* DNA in synovial tissue of arthritis patients. *Arthritis Rheumatol.* **42:**1889–1993.
13. Gerard, H. C., Schumacher, H. R., El-Gabalawy, H., Goldbach-Mansky, R., and Hudson, A. P., 2000, *Chlamydia pneumoniae* present in the human synovium are viable and metabolically active, *Microb. Pathog.* **29:**17–24.
14. Balin, B. J., Gerard, H. C., Arking, E. J., Appelt, D. M., Branigan, P. J., Abrams, J. T., Whittum-Hudson, J. A., and Hudson, A. P., 1998, Identification and localization of *Chlamydia pneumoniae* in the Alzheimer's brain, *Med. Microbiol. Immunol. (Berlin)* **187:**23–42.
15. Layh-Schmitt, G., Bendl, C., Hildt, U., Dong-Si, T., Juttler, E., Schnitzler, P., Grond-Ginsbach, C., and Grau, A. J., 2000, Evidence for infection with *Chlamydia pneumoniae* in a subgroup of patients with multiple sclerosis, *Ann. Neurol.* **47:**652–655.
16. Beatty, W. L., Morrison, R. P., and Byrne, G. I., 1994, Persistent chlamydiae: From cell culture to a paradigm for chlamydial pathogenesis, *Microbiol. Rev.* **58:**686–699.
17. Cornelis, G. R., 2002, The *Yersinia* Ysc-Yop 'Type III' Weaponry, *Nat. Rev. Mol. Cell Biol.* **3:**742–752.
18. Rocha, E. P. C., Pradillon, O., Bui, H., Sayada, C., and Denamur, E., 2002, A new family of highly variable proteins in the *Chlamydophila pneumoniae* genome, *Nucleic Acids Res.* **30:**4351–4360.
19. Subtil, A., Parsot, C., and Dautry-Varsat, A., 2001, Secretion of predicted Inc proteins of *Chlamydia pneumoniae* by a heterologous type III machinery, *Mol. Microbiol.* **39:**792–800.
20. Belland, R. J., Zhong, G., Crane, D. D., Hogan, D., Sturdevant, D., Sharma, J., Beaty, W. L., and Caldwell, H. D., 2003, Genomic transcriptional profiling of the developmental cycle of *Chlamydia trachomatis, Proc. Nat. Acad. Sci. U.S.A.* **100:** 8478–8483.
21. Loomis, W. P., and Stranbach, M. N., 2002, T cell responses to *Chlamydia trachomatis, Curr. Opin. Microbiol.* **5:**87–91.
22. Morrison, S. G., Su, H., Caldwell, H. D., and Morrison, R. P., 2000, Immunity to murine *Chlamydia trachomatis* genital tract reinfection involves B cells and CD4(+) T cells but not CD8(+) T cells, *Infect. Immun.* **68:**6979–6987.
23. Saikku, P., Leinonen, M., Mattila, K., Ekman, M.-R., Nieminen, M. S., Mäkelä, P. H., Huttunen, J. K., and Valtonen, V., 1988, Serological evidence of an association of a novel *Chlamydia,* TWAR, with chronic coronary heart disease and acute myocardial infarction, *Lancet* **2:**983–986.
24. Kalayoglu, M. V., Indrawati, Morrison, R. P., Morrison, S. G., Yuan, Y., and Byrne, G. I., 2000, Chlamydial virulence determinants in atherogenesis: The role of chlamydial lipopolysaccharide and heat shock protein 60 in macrophage–lipoprotein interactions, *J. Infect. Dis.* **181:**S483–S489.
25. Mahdi, O. S. M., Horne, B. D., Mullen, K., Muhiestein, J. B., and Byrne, G. I., 2002, Serum immunoglobulin G antibodies to chlamydial heat shock protein 60 but not to human and bacterial homologs are associated with coronary artery disease, *Circulation* **106:**1659–1663.

26. Kalman, S., Mitchell, W., Marathe, R., Lammel, C., Fan, J., Hyman, R. W., Olinger, L., Grimwood, J., Davis, R. W., and Stephens, R. S., 1999, Comparative genomes of *Chlamydia pneumoniae* and *C. trachomatis*, *Nat. Genet.* **21**:385–389.
27. Read, T. D., Brunham, R. C., Shen, C., Gill, S. R., Heidelberg, J. F., White, O., Hickey, E. K., Peterson, J., Utterback, T., Berry, K., Bass, S., Linher, K., Weidman, J., Khouri, H., Craven, B., Bowman, C., Dodson, R., Gwinn, M., Nelson, W., DeBoy, R., Kolonay, J., McClarty, G., Salzberg, S. L., Eisen, J., and Fraser, C. M., 2000, Genome sequences of *Chlamydia trachomatis* MoPn and *Chlamydia pneumoniae* AR39, *Nucleic Acids Res.* **28**:1397–1406.
28. Shirai, M., Hirakawa, H., Kimoto, M., Tabuchi, M., Kishi, F., Ouchi, K., Shiba, T., Ishii, K., Hattori, M., Kuhara, S., and Nakazawa, T., 2000, Comparison of whole genome sequences of *Chlamydia pneumoniae* J138 from Japan and CWL029 from USA, *Nucleic Acids Res.* **28**:2311–2314.

2

Morphology of *Chlamydia pneumoniae*

NAOYUKI MIYASHITA and AKIRA MATSUMOTO

1. INTRODUCTION

Chlamydiae, obligate intracellular parasites require host cell biosynthetic machinery for several metabolic functions. All chlamydiae multiply through a common, unique developmental cycle in which there are two morphologically and functionally distinct forms: one is the infectious elementary body (EB) and the other is the reproductive reticulate body (RB). Morphologically, EBs have a high density and are small in size (0.3 to 0.35 μm in diameter), whereas RBs consist of homogeneous internal material and are large in size (0.5 to 2.0 μm in diameter). RBs are metabolically active and reproductive, but noninfectious. In contrast, EBs are infectious but metabolically inactive, suggesting their adaptation to an extracellular environment. This unique developmental cycle occurs in a membrane-bound cytoplasmic vacuole, termed inclusion. In this chapter, we discuss the morphology of *Chlamydia pneumoniae* as revealed by transmission and scanning electron microscopy.

2. CHLAMYDIAL DEVELOPMENTAL CYCLE

The central core of the developmental cycle is the alternating and complementary nature of its distinct developmental forms. Contact with the host cell and entry of the EB are the first steps in a complicated interaction between the infecting chlamydiae and the invaded host cell. A simplified drawing of the basic developmental cycle of *C. pneumoniae* is presented in Fig. 1, and

NAOYUKI MIYASHITA • Division of Respiratory Diseases, Department of Medicine, Kawasaki Medical School, Kurashiki City, Okayama, Japan. AKIRA MATSUMOTO • Department of Urology, Okayama University, Okayama, Japan.

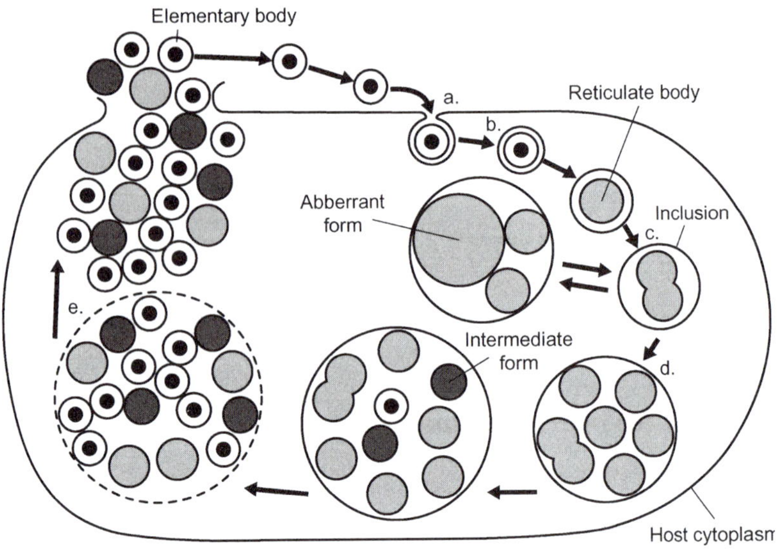

FIGURE 1. A line drawing of a generalized *Chlamydia pneumoniae* developmental cycle. Infection begins when an infectious but metabolically inactive EB comes in contact with a host cell (a) and is endocytozed (b). The phagocytic vacuole migrates toward the Golgi apparatus, and the EB differentiates into a noninfectious but metabolically active RB (c). RB division ensues and the inclusion increases in size (d). RBs then begin to reorganize back into EBs, and the inclusion grows until it occupies the entire cytoplasm of the infected cell (e). The inclusion lyses, the host cell lyses, and EBs are freed to infect another cell.

electron microscopic images of the *in vitro* developmental prospect are presented in Fig. 2. The initial events in the *C. pneumoniae* infection are the attachment of the EB to the host cell by electrostatic interaction, and successive association of the EB to the host cell receptor, followed by internalization into the host cytoplasm by induction of active phagocytosis. The phagocytic vesicle containing the EB migrates toward the Golgi area. In *in vivo C. pneumoniae* infection, the EB exhibit affinity for epithelial cells of the mucosal membrane in the respiratory tract. When the EB is alive, it prevents lysosomal fusion and permits replication in the phagosome (inclusion). At about 8 h after infection, the EB begins to undergo profound changes in its morphology to form a metabolically active RB. Then the resulting RB multiplies by binary fission (Fig. 2a,b). At around 36 h after infection, some RBs begin to decrease in size and continue to reorganize back into EBs through transitional, intermediate forms until 60 h after infection (Fig. 2d,e). Most RBs continue to multiply until the host cell cytoplasm is almost filled by the inclusion, which expands in size concurrently with increase of the chlamydial population. The developmental cycle ceases at 72 to 80 h and the host cell dies, releasing EBs (Fig. 2f) that infect new host cells.[1,2] However, little is known about the mechanism of the bursting of the host cell to release progeny organisms. The EB formation observed by electron microscopy coincides well with the infectivity kinetics in a one-step growth curve (Fig. 3), in which the infectivity rapidly increases from 36 h and reaches a maximum level during the next 60 h after infection.[3]

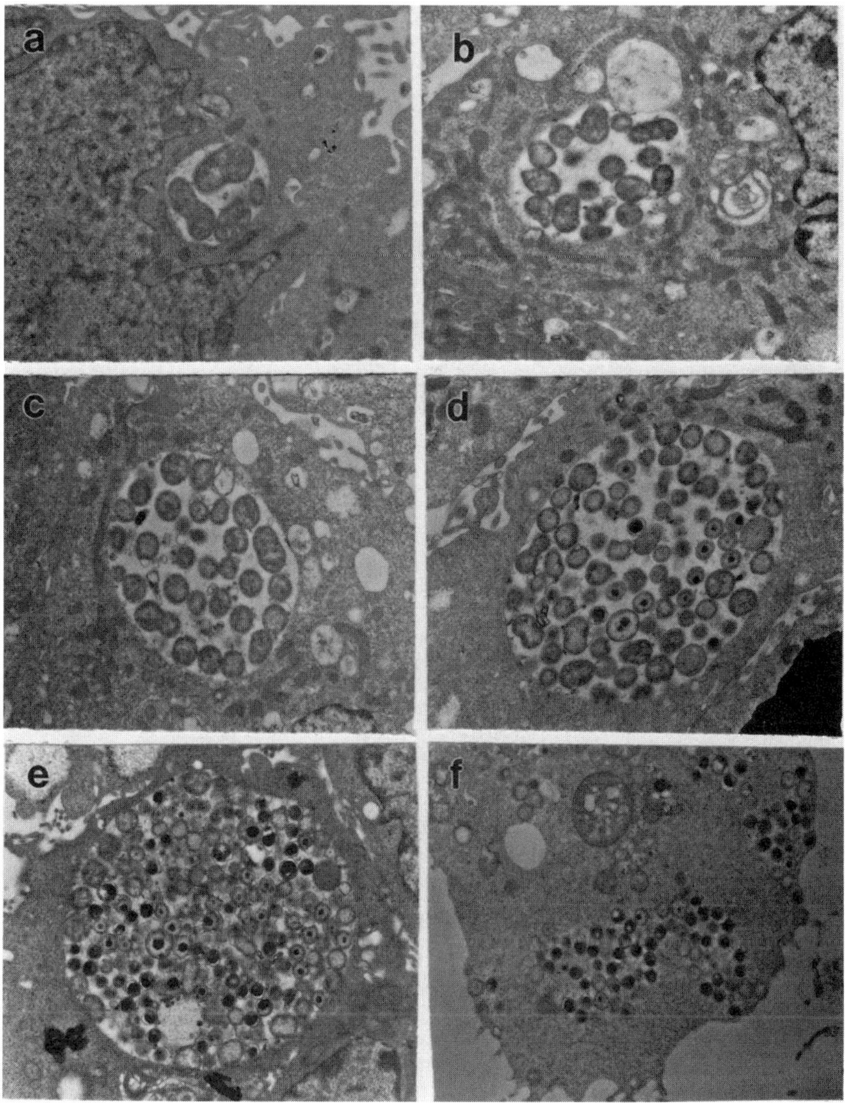

FIGURE 2. Electron micrographs of the sequential changes of inclusion development in the *Chlamydia pneumoniae* KKpn-15 strain in HEp-2 cells. (a) 12 h, (b) 24 h, (c) 36 h, (d) 48 h, (e) 60 h, and (f) 80 h postinoculation.

The properties of the inclusion morphology of Chlamydia are summarized in Table I. The inclusions of *C. pneumoniae* are positively stained by immunostaining with a genus-specific monoclonal antibody, but not with iodine. The inclusion morphology in susceptible cells, such as HEp-2 cells, closely resembles that of *Chlamydia psittaci* and *Chlamydia pecorum* inclusions, which are oval in shape and are stained densely by Giemsa staining. In contrast to *Chlamydia*

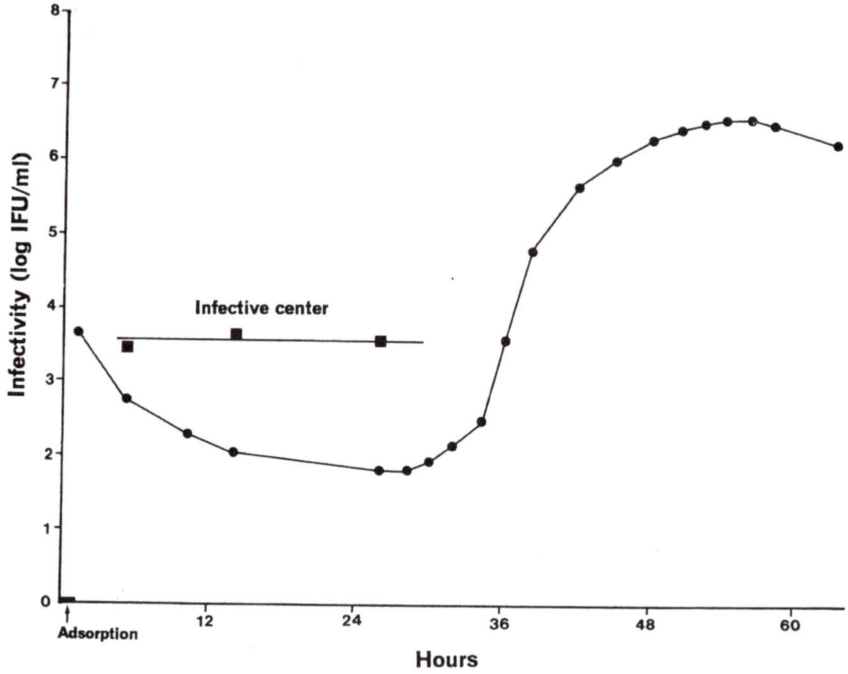

FIGURE 3. A one-step growth curve of the *Chlamydia pneumoniae* KKpn-15 strain in HEp-2 cells. The infectivity increased sharply from 36 h after infection and the growth rate of *C. pneumoniae* is apparently slower than that of other Chlamydia species.

trachomatis inclusion, *C. pneumoniae* inclusion does not compress the host nucleus even though it fills the host cytoplasm.

Chlamydiae are thought to be "energy parasites" because they obtain ATP from their host cells.[4] Positive evidence of this hypothesis was provided by direct demonstration of use of the host for chlamydial biosynthesis in host-free RBs.[5] However, recent sequence analyses of genomic DNA have identified additional genes that may allow chlamydiae to generate at least a minimal amount of their own ATP,[6,7] ATP–ADP translocases, which exchange ATP and ADP, are essential for chlamydial growth. In this regard, the inclusion/mitochondria association

TABLE I
Differentiation of Inclusion and EB Morphology

	C. pneumoniae	*C. trachomatis*	*C. psittaci*	*C. pecorum*
Inclusion morphology	Oval	Oval	Variable	Variable
Intrainclusion space	Dense	Vacuolar	Dense	Dense
Glycogen in inclusions	–	+	–	–
Compression of the host nucleus	–	+	–	–
Mitochondrial association	–	–	+	+
EB morphology	Pear-shaped, round	Round	Round	Round

in a cell infected with *C. psittaci* is quite interesting. This association is so tight that mitochondria remain associated with the inclusion membrane even after the isolation of inclusions from infected cells.[8] However, mitochondria do not associate with inclusions containing *C. trachomatis* and *C. pneumoniae*.[8,9] Little is known about the mechanism of the association, although many chlamydial proteins, termed Inc proteins, have been found in the inclusion membrane of cells infected with *C. psittaci*, *C. trachomatis*, and *C. pneumoniae*.[10]

3. MORPHOLOGY OF EB

The morphology of EBs has been proposed as one of the criteria for distinguishing *C. pneumoniae* from other chlamydial species.[11,12] The EBs of the *C. pneumoniae* TW-183 and AR (TWAR) strains have a wide periplasmic space limited by a wavy outer membrane, creating "pear-shaped" profiles in thin sections (Fig. 4a,b). In contrast, the EBs of the KKpn strains, which were isolated in Kawasaki Medical School Hospital and number from 1 to 18, were found to be round in shape with a narrow periplasmic space (Fig. 4c).[9,13–15] This morphological difference between the TWAR and KKpn strains was also distinct when purified EBs of both strains were air-dried and then shadowcast (Fig. 5). KKpn EBs have a round "fried egg" appearance that is indistinguishable from the EBs' of other chlamydiae, such as *C. trachomatis*, *C. psittaci*, and *C. pecorum*, but TWAR EBs have a pear-shaped morphology with a wide, flat outer membrane, perhaps the result of air-drying (Fig. 5).[16] Moreover, Carter *et al.*[17] found the EB morphology of the *C. pneumoniae* IOL-207 strain to also be round in thin sections. Similarly, Popov *et al.*[18] found the EBs of the *C. pneumoniae* Kajaani-6 strain to be round. These facts, together with the results obtained from our examinations, strongly suggest that among all the *C. pneumoniae* strains the morphology of pear-shaped EBs is uncommon. Thus it is likely that the "pear-shape" is not valid as a morphological criterion for the differentiation of *C. pneumoniae* species.[9]

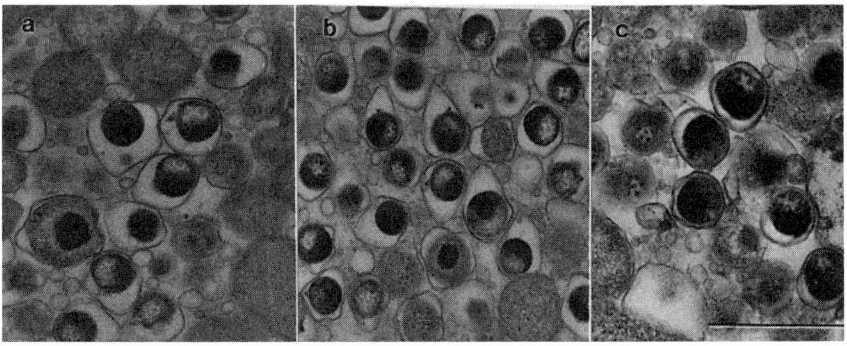

FIGURE 4. Thin sections of *Chlamydia pneumoniae* strains in HeLa 229 cells at 60 h after infection. (a) TW-183 strain, (b) AR-39 strain, (c) KKpn-15 strain. In the KKpn-15, which was isolated in Kawasaki Medical School Hospital, EBs have a narrow periplasmic space and are round in shape, whereas TW-183 and AR-39 EBs are enclosed by a wavy outer membrane and are pear-shaped in profile. Bars indicate 500 nm.

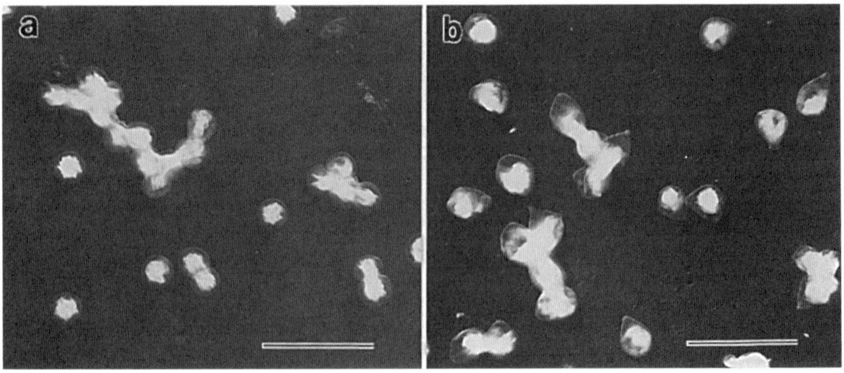

FIGURE 5. Shadowcast of purified EBs of the *Chlamydia pneumoniae* TW-183 (a) and KKpn-15 (b) strains. The EBs were air-dried on a smooth-surfaced agar plate, transferred to a specimen grid by the pseudoreplica method with collodion, and then shadowcast with Pt–Pd alloy. Bars indicate 1 μm.

Kuo *et al.*[19] studied the entry of TWAR organisms into HeLa cells by electron microscopy and suggested that the binding mechanism of *C. pneumoniae* TWAR EBs may differ from those of *C. trachomatis* and *C. psittaci*, which are bound to host cells by a stretch of EB surface in contact with the host cell plasma membrane. However, we could find no differences in the physicochemical and antigenic properties of the pear-shaped and round-shaped EBs in the *C. pneumoniae* species.[13] Therefore, the biological function of the unique wavy outer membrane of the TWAR EBs is still unclear.

4. STRUCTURE OF THE OUTER MEMBRANE

EBs of all chlamydial species are stable against mechanical agitation, such as that from ultrasonic, freeze-thawing, or osmotic shock. To improve this property, EBs can be recovered from infected cells at an extremely high level of purity.[16,20] Peptidoglycan (PG) is an ubiquitous cell wall component that is responsible for the rigidity of free-living bacteria. However, no positive evidence for the presence of PG and its precursor has been documented by biochemical analyses of chlamydial organisms,[21] whereas genome sequencing on *C. trachomatis* and *C. pneumoniae* has revealed virtually complete sets of genes encoding the machinery for PG production. The rigidity of *C. psittaci*, instead, is retained with several kinds of cysteine-rich proteins linked by disulfide bondings that construct a rigid network in the EB outer membrane.[22] This may be supported by our biochemical analysis of *C. pneumoniae*, which showed the presence of ^{35}S-labeled proteins at 60, 40, and 9 kDa molecular masses.

By electron microscopy of the purified EB outer membrane of the *C. psittaci* california 10 (meningopneumonitis) strain, Manire[23] first demonstrated regularly arranged, hexagonal macromolecular structures approximately 100 Å in diameter located on the inner surface of the outer membrane and suggested

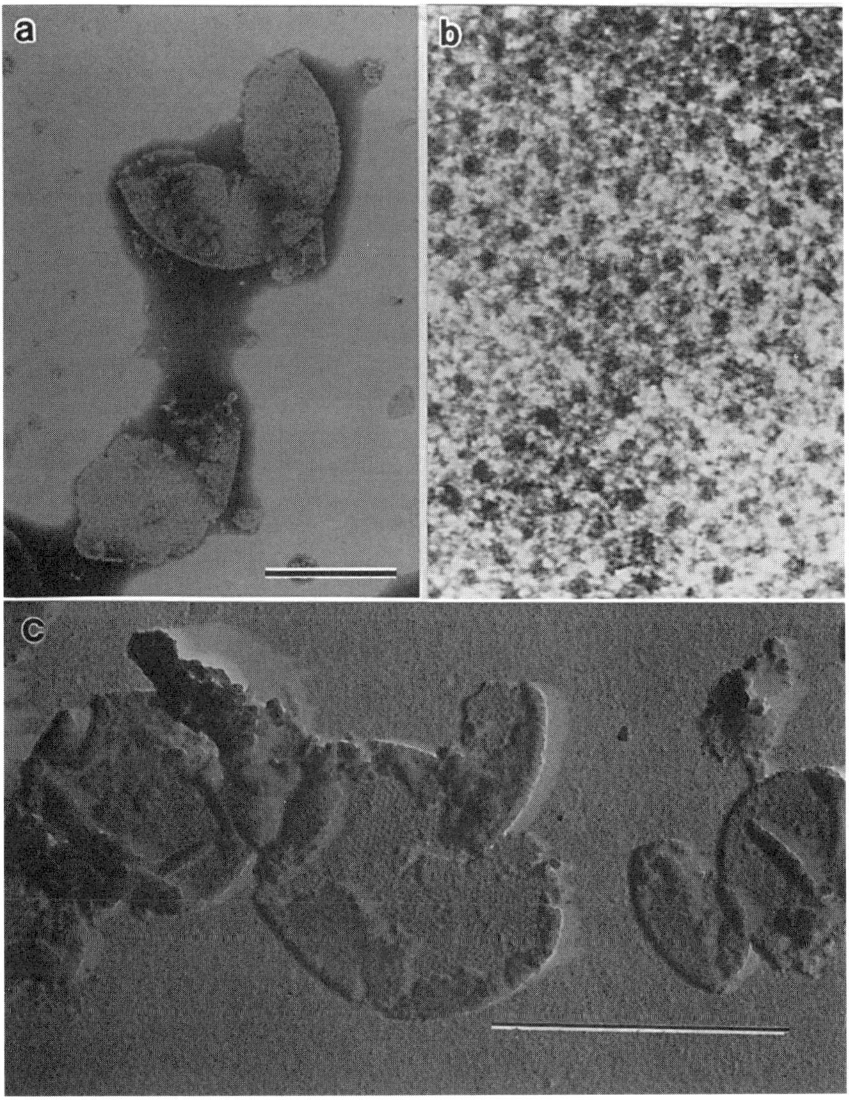

FIGURE 6. Negatively stained (a) and shadowcast images (c) of EB outer membranes isolated from purified EBs of the *Chlamydia pneumoniae* KKpn-15 strain. (b) Higher magnification obtained from the micrograph shown in (a). The hexagonal structure seen throughout the outer membrane (a) is seen only on the inner face (c). Bars indicate 200 nm.

that these structures are responsible for the rigidity of the EB outer membranes against mechanical agitation. An identical structure, termed hexagonally arrayed structure (HAS), was partially seen only on the partially exposed inner surface of the outer membrane of the *C. pneumoniae* EB when prepared by the shadowcasting technique (Fig. 6).[9] However, the HAS was evident all over the

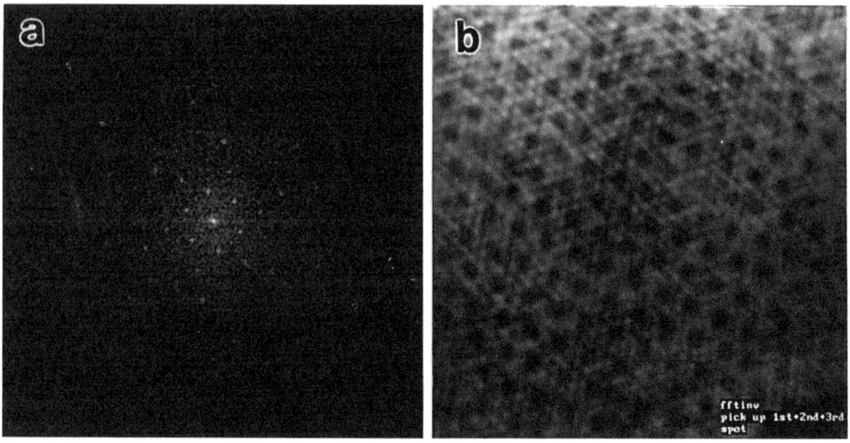

FIGURE 7. Image processing of the hexagonal structures in the outer membrane of *Chlamydia pneumoniae* KKpn-15 EB. (a) Diffraction spots obtained by Fourier transformation from the original micrograph shown in Fig. 6b; (b) Image reconstituted by inverse Fourier transformation from the six spots of the diffraction shown in Fig 7a.

outer membrane when stained negatively because of penetration of the staining solution into the outer membranes, indicating that the whole inner layer of EB outer membrane is composed of HAS (Fig. 6). To determine the periodicity of the HAS, the computer Fourier transformation was applied to outer membranes stained negatively (Fig. 7). The result obtained demonstrated that the EB outer membrane was composed of three structural units with dimensions of 176, 90 and 50 Å respectively. When an image was reconstituted using only spots showing 176 Å periodicity by the inverse Fourier transformation, a distinct HAS image was obtained (Fig. 7b), suggesting strongly that the periodicity of the HAS in the *C. pneumoniae* EB outer membrane is 176 Å. Using such a procedure, the HAS dimensions of other chlamydiae can be determined. Our results are summarized in Table II. The imperceptible differences in the values listed are likely to be meaningless if the dimensional aberration caused by a mechanical aberration in the electron microscope is considered in the comparison of HAS periodicity. This led us to conclude that the basic macromolecular structure, HAS, in the inner layer of the EB outer membrane is common to all members of the genus Chlamydia. Applying the EnvA–EnvB network in the *C. psittaci* outer membrane[22] to the HAS architecture is an attractive possibility, but no clear evidence showing correlation of that rigid network and HAS in the outer membrane of *C. pneumoniae* EB has been found. Although a structural unit with 50 Å periodicity appeared to be similar to one with 60 Å periodicity that was considered to be composed of fine particles consisting of the outermost surface layer of the *C. psittaci* EB outer membrane,[1] little is known about this unit, or one with 90 Å periodicity, in the outer membrane of *C. pneumoniae.*

TABLE II
Comparison of Periodicity in Chlamydial Species

Species	Strain	Mean periodicity ± SD (Å)[a]
C. pneumoniae		
Round shape	KKpn-1	176 ± 7
	KKpn-15	179 ± 8
	KKpn-16	178 ± 8
	IOL-207	177 ± 7
	Kajaani-6	178 ± 6
	YK-41	176 ± 8
Pear shape	TW-183	173 ± 7
	AR-39	174 ± 8
	AR-388	173 ± 8
C. trachomatis	D/UW-3/Cx	177 ± 8
	L2/434/Bu	176 ± 7
C. psittaci	Frt-Hu/Cal 10	175 ± 8

[a] Periodicity measured from center to center by Fourier transform.

C. psittaci RB outer membranes can also be recovered in high purity from purified RBs.[24] The outer membranes are normally composed of fine particles about 50 Å in diameter. Some RB outer membranes showing HAS identical to that of the EB outer membrane were seen in a limited area.[25] Such partial HAS were only occasionally encountered in negative-stained preparations. However, these partial HAS progressively disappeared during multiplication, indicating that the HAS in outer membranes are carried over from the infected EB, but are not newly synthesized in RBs. Although no evidence has been reported on the *C. pneumoniae* RB outer membrane, its morphology seems to be identical with that of the *C. psittaci* RB outer membrane on the basis of such physicochemical properties as its fragility to mechanical agitation and protein composition.

The diameter of EBs in thin sections is approximately 0.3 μm, smaller than that of RBs, which range from 0.5 to, occasionally, more than 1 μm in diameter. This means that the area of each RB outer membrane should be diminished by less than one-third to form the EB outer membrane at the final stage of multiplication. Although little is known about the mechanism of outer membrane reorganization, it has been speculated that there are two mechanisms: one is reduction due to the formation of a rigid network by the disulfide bonding connecting cysteine-rich proteins and the other is exclusion of excess outer membrane from the RBs. Figure 8 shows an inclusion containing chlamydial bodies of a KKpn-8-1 clone that was recently established by the plaque cloning technique.[26] Many small vesicles, ranging from 50 to 100 nm in diameter, budding from the surface of RBs and floating in the inclusion space, are clearly seen. From these observations, it may be possible to speculate that RBs decrease their size by pinching off their excess outer membrane to form EBs and that although small vesicle formation occurs slowly in this *C. pneumoniae* clone, it

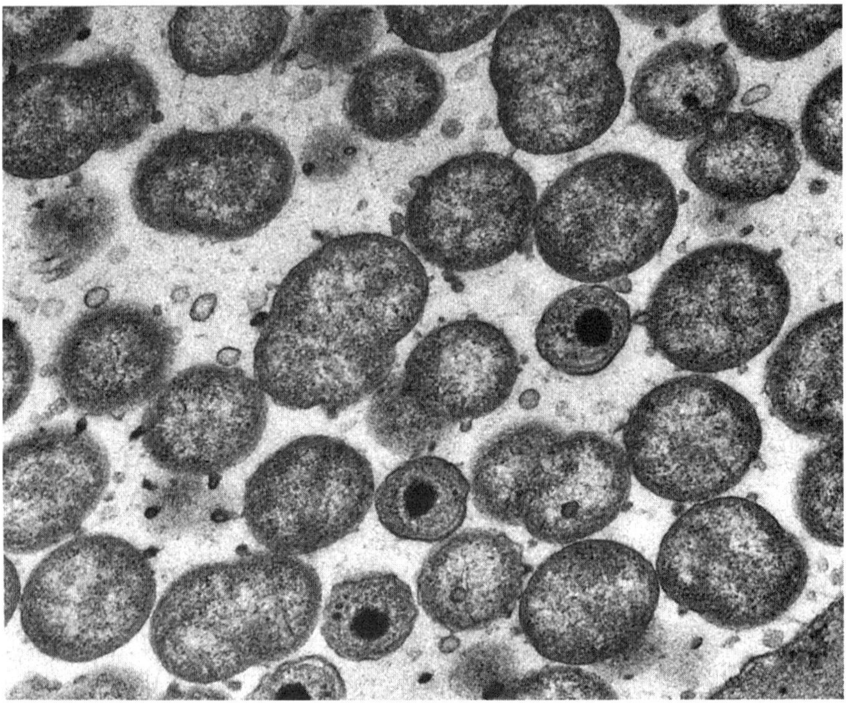

FIGURE 8. Inclusion of the cloned *Chlamydia pneumoniae* KKpn-8-1 strain. Many small vesicles, ranging from 50 to 100 nm in diameter, budding from the surface of RBs and floating in the inclusion space are clearly seen.

normally occurs so rapidly that budding of the RB outer membrane is hard to observe in ordinary strains of chlamydiae. In addition, we speculate that the outer membrane of the pear-shaped EB is formed prior to extrusion of the excess of RB outer membrane by budding, resulting in the formation of a baggy outer membrane in each EB.

5. SURFACE PROJECTIONS AND RELATED STRUCTURES

Despite the difference in the shape of the *C. pneumoniae* EBs, their basic morphology is identical. Each EB has a dense nucleus in an eccentric region and cytoplasm that contains ribosomes, moderately dense particles, and amorphous material. These components are tightly enclosed within a cytoplasmic membrane so that the cytoplasm is seen to be round in shape in both the round and pear-shaped EBs of *C. pneumoniae*, and this results in the formation of a "cytoplasmic body" and a wide, less dense periplasmic space (Fig. 4). The periplasmic space in the round-shaped EBs of *C. pneumoniae* is seen to be wider than that in the EBs of *C. trachomatis* and *C. psittaci*. Interestingly, the cytoplasmic body tends to be associated closely with the outer membrane at a site far from the nucleus. This image strongly suggests that the association of the cytoplasmic body and

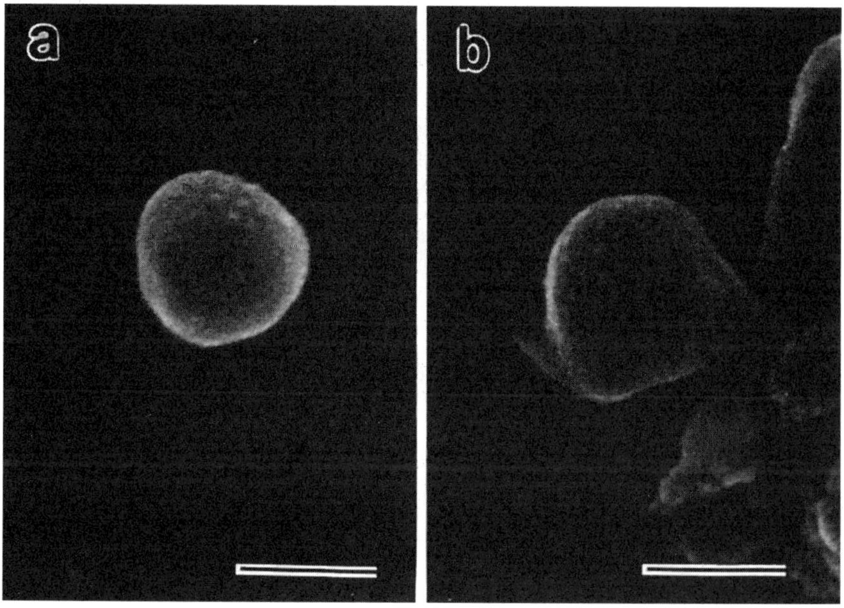

FIGURE 9. Scanning electron micrographs of EBs of *Chlamydia pneumoniae* KKpn-15 (a) and AR-39 (b). Both micrographs show hexagonally arrayed projections in a limited area of the surface. Bars indicate 300 nm.

outer membrane is mediated by surface projections, but these projections have not been visualized because of inadequate opacity in thin sections prepared by ordinary procedures. This supposition is supported by evidence that the projections are located in a group on a limited surface area (Fig. 9) and that one end of each projection is fitted into the cytoplasmic membrane, while the other end protrudes beyond the outer membrane.[1]

By the freeze-replica technique, *in situ* inclusions at the late stage of multiplication are visualized as convex and concave faces (Fig. 10). Button structures or craters have been noted in some concave faces (Fig. 10, arrowheads).[9] Their morphology and distribution pattern are quite similar to those observed in *C. psittaci* EBs. The direct correlation between the craters and projections has been confirmed.[1]

When the inclusion membrane is exposed by the freeze-replica technique, several groups of fine particles, which are arranged hexagonally with a spacing of approximately 50 nm, are encountered (Fig. 11, arrowheads). These are quite similar to the ones observed on the surface of inclusions containing *C. psittaci* organisms at a stage of RB multiplication.[1] The *C. pneumoniae* RBs frequently make contact with the inclusion membrane, where a comb-like structure, perhaps consisting of projections, is evident (Fig. 11). It is, therefore, likely that the fine particles seen on the surface of the inclusion membrane are RB projections which pierce the inclusion membrane, resulting in a direct connection to the host cytoplasm.

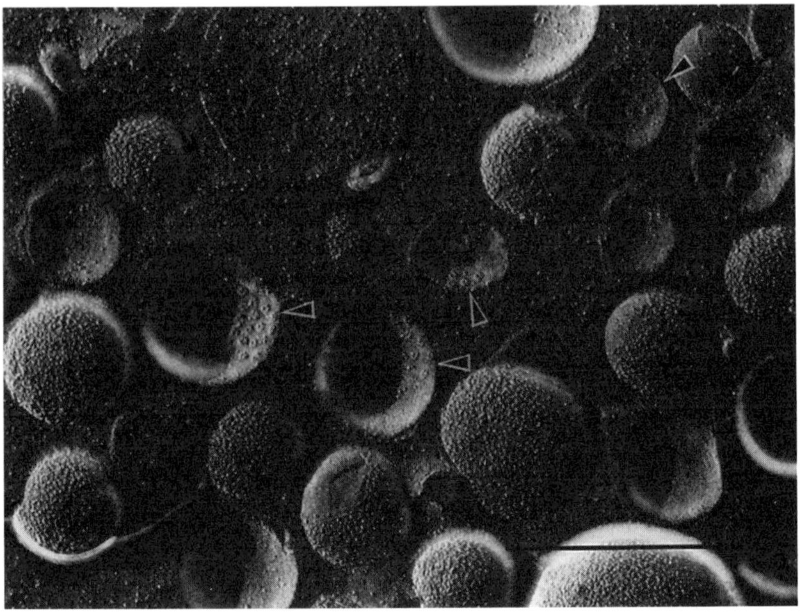

FIGURE 10. Freeze-replica images of *in situ* chlamydial bodies and inclusion members of the *Chlamydia pneumoniae* KKpn-15 strain at 60 h after infection. Chlamydial bodies are cleaved into convex or concave faces. Many B structures are seen (arrowheads). Bars indicate 1 μm.

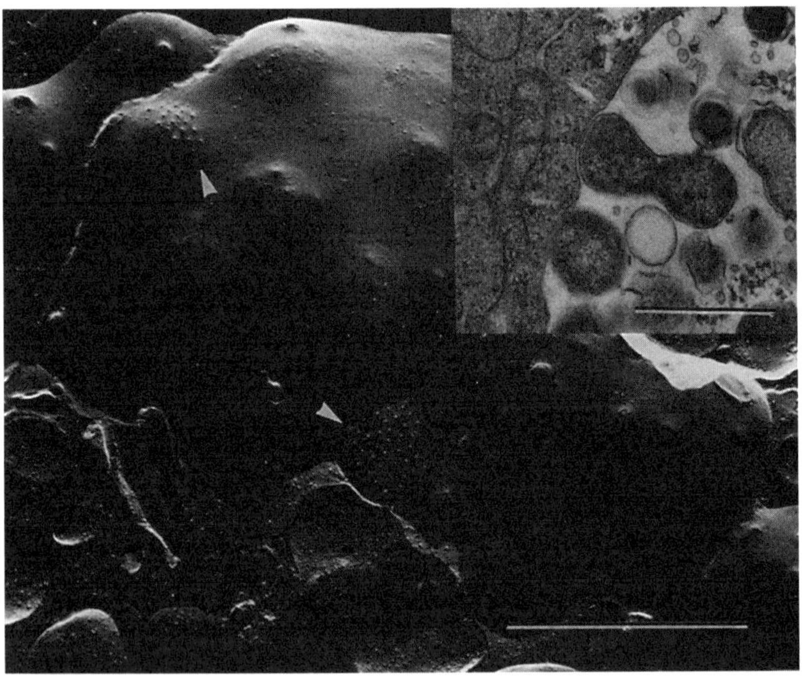

FIGURE 11. Freeze-replica images of an *in situ* chlamydial inclusion membrane containing the *Chlamydia pneumoniae* KKpn-15 strain at 60 h after infection. Fine particles in groups are indicated by arrowheads. The face is rugged over the RB outline (inset). Bars indicate 1 μm.

It was recently proposed that these projections may function in the secretion of proteins from developing RBs, possibly through a type III secretion mechanism.[27] Hsia *et al.*[28] demonstrated that *C. psittaci* and *C. trachomatis* encode proteins that likely participate in a type III secretion pathway, and that the structure of the surface projections resembles the base of a flagellar assembly, as well as the structures in Salmonella.[29] It is likely that the chlamydiae use an alternate secretory pathway, a type III secretion pathway. The surface projections are, therefore, possible candidates for that function. Although their actual function remains to be elucidated, it is reasonable to propose that the surface projections are involved in some aspect of the interaction between the intracellular environment and the infecting EBs and/or developing RBs.

6. PERSISTENT INFECTION AND ABERRANT FORMS

C. pneumoniae is regarded as a common cause of respiratory tract infections and it can cause prolonged or chronic infections which may be due to persistence for months or years.[14,30] These persistent infections have been implicated in the development of a number of chronic diseases including atherosclerosis, asthma, and obstructive pulmonary diseases (see other chapter). These persistent chlamydial infections can be established *in vitro* using several methods involving cytokines,[31–34] antibiotics,[35,36] and deprivation of certain nutrients.[37] Despite differences in treatment, chlamydiae respond to form inclusions containing atypical RBs, which occasionally have been shown to be pleomorphic forms, termed aberrant bodies (ABs). The ABs are generally larger in diameter than typical RBs, and display a sparse densinometric appearance (Fig. 12). No evidence of redifferentiation into EBs has been documented. However, when the growth inhibitory factors are removed, the ABs can be restored to normal RBs, which reorganize infectious EBs.

Beatty *et al.*[31,32] demonstrated that treatment with gamma interferon (IFN-γ) inhibited the intracellular growth of *C. trachomatis*. IFN-γ restricted the division of RBs and interrupted their differentiation into infectious EBs. In this cases, the development of large aberrant RB forms along with the absence of EBs is characteristic of persistent *C. trachomatis* infection. Aberrant chlamydial development is also concomitant with decreased levels of major outer membrane protein (MOMP), 60-kDa outer membrane protein and lipopolysaccharide. Ultrastructural examinations of persistent *C. pneumoniae* infection induced with IFN-γ have also been described. Pantoja *et al.*[38] demonstrated that *C. pneumoniae* inclusions that formed in the presence of IFN-γ did not contain any typical EB or RB in morphology, but instead contained only organisms with pleomorphic, RB-like structures of various sizes. Mathews *et al.*[39] also described the morphology of *C. pneumoniae* ABs in a persistent state induced IFN-γ. Interestingly, the morphology of chlamydial organisms varied depending on the concentration of IFN-γ. Recently, Mathews *et al.*[39] reported that *C. pneumoniae* upregulates the transcription of specific genes, such as *ompA*, *ompB*, *pyk*, *nlpD*, and *Cpn0585*, in response to IFN-γ treatment. This indicates the existence of an altered host

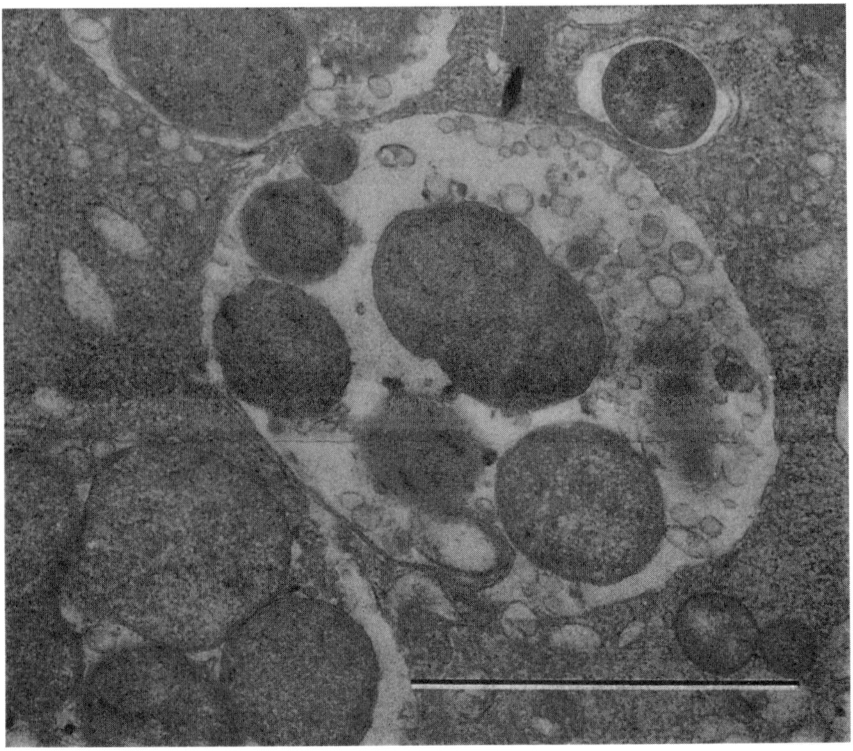

FIGURE 12. Aberrant RB forms produced during culture of the *Chlamydia pneumoniae* KKpn-15 strain in the presence of ampicillin. A thin section of ampicillin-treated, infected cells fixed 60 h after infection. Large and abnormal RBs are seen in inclusions. Bars indicate 1 μm.

cell environment in which a condition of nutritional stress has been created so that the *C. pneumoniae* organisms fall into a persistent state.

Treatment with antibiotics, such as penicillin and cycloserine, induces the formation of abnormally enlarged RBs. When cells having inclusions containing such large RBs are transferred into penicillin-free culture medium, normal RBs are restored and then infectious EBs are produced.[36] Other antibiotics have also been reported to induce abnormal chlamydial development. Dreses-Werringlore *et al.*[35] found that ciprofloxacin and ofloxacin at the minimal chlamydiacydal concentration induced persistent infection and that the expression of MOMP was significantly decreased while the production of Hsp 60 and chlamydial lipopolysaccharide was minimally affected. Removal of ciprofloxacin from the cell culture 10 or 14 days after infection reversed the effect, allowing differentiation of the persistent chlamydiae into infectious EBs. *C. pneumoniae* persistence is also induced by ampicillin. The RBs formed with ampicillin are abnormally large in size and appear to undergo little or no cell division.[40]

In general, it is likely that this aberrant developmental step leads to the persistence of viable but nonculturable chlamydiae within infected cells over long periods.[41] Removal of several stress factors described above results in the condensation of nuclei, the appearance of late proteins, and the production of viable, infectious EBs.[32] Most of the major sequelae of chlamydial disease are thought to arise from either repeated or persistent chlamydial infection of an individual. The persistence would allow constant presentation to the individual immune response of these potentially deleterious immune targets.[41] Since repeated infection can certainly be documented in many clinical settings, persistence is thought to also play a role.

7. SUMMARY

On the basis of our electron microscopic examinations, here we have made diagrams of the round- and pear-shaped EBs of *C. pneumoniae*, as shown in Fig. 13. These are quite similar to the diagram of *C. psittaci* EB previously proposed by Matsumoto.[1] Despite the difference between the round- and pear-shaped EBs in the sectioned profile, the basic morphology at the electron microscopic level in both EBs is obviously identical. The surface projections, the function of which has recently been identified, are arranged in a group and are distributed on the surface far from the nucleus, which is normally located in an eccentric region in the "cytoplasmic body." One end of each projection fits into the cytoplasmic membrane while the other end of the projection protrudes beyond the outer membrane. Although no distinct evidence has been obtained, DNA fibers may bind to the cytoplasmic membrane site where the projections are connected as seen in *C. psittaci* EBs. The connection between DNA fibers and the cytoplasmic

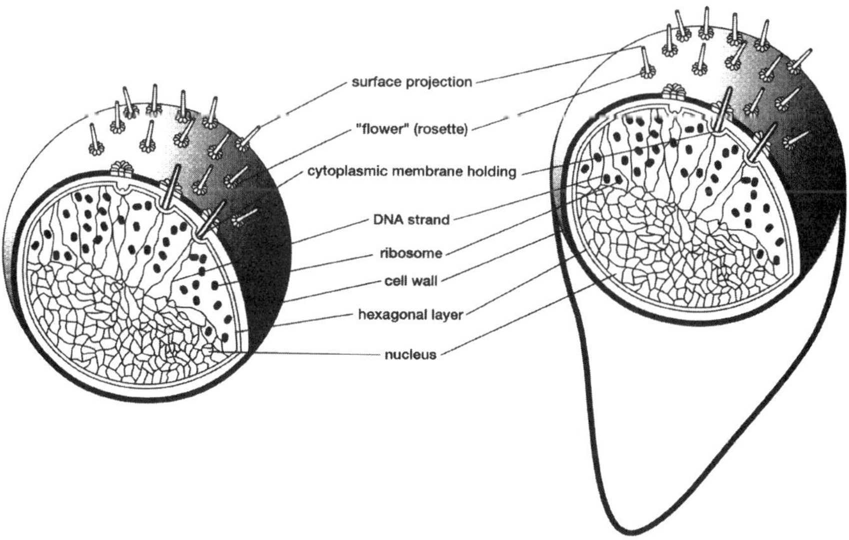

FIGURE 13. Diagrammatic representation of the morphology of the round (a) and pear-shaped (b) EBs of *Chlamydia pneumoniae.*

membrane disappeared with trypsin treatment in *C. psittaci* EBs, suggesting that the DNA molecule was not bound directly, but through a component sensitive to trypsin.

In isolated *C. psittaci* inclusions, the RBs close to the inclusion membrane are closely connected to the inside surface of the inclusion membrane by means of the projections, which appeared to pierce the inclusion membrane.[(42)] Interestingly, the arrangement of these projections in a hexagonal pattern was also observed on inclusion surfaces exposed by the freeze-replica technique, suggesting strongly that the *C. pneumoniae* RBs also connect directly with the host cytoplasm through these projections. This evidence evokes a subject of deep interest in the function of the projections, in special reference to the host–parasite relationship during chlamydial multiplication.

REFERENCES

1. Matsumoto, A., 1988, Structural characteristics of chlamydial bodies, in: *Microbiology of Chlamydia* (A. L. Barron, ed.), CRC Press, Florida, pp. 21–45.
2. Rockey, D. D., and Matsumoto, A., 1999, The chlamydial developmental cycle, in: *Prokaryotic Development* (Y. V. Brum, ed.), American Society for Microbiology Press, Washington, D.C., pp. 403–425.
3. Miyashita, N., Kubota, Y., Kimura, M., Nakajima, M., Niki, Y., Soejima, R., and Matsushima, T., 1994, Characterization of *Chlamydia pneumoniae* strain isolated from a 57-year-old man, *Microbiol. Immunol.* **38:**857–864.
4. Moulder, J. W., 1991, Interaction of chlamydiae and host cells *in vitro, Microbiol. Rev.* **55:**143–190.
5. Hatch, T. P., Al-Houssaing, E., and Silverman, J. A., 1982, Adenine nucleotide and lusin transport in *Chlamydia psittaci, J. Bacteriol.* **150:**662–670.
6. Read, T. D., Brunham, R. C., Shen, C., Gill, S. R., Heidelberg, J. E., White, O., Hickey, E. R., Peterson, J., Utterback, T., Berry, K., Bass, S., Linher, K., Weidman, J., Khouri, H., Craren, B., Bowman, C., Dadson, R., Gwinn, M., Nelson, E., DeBoy, R., Kolonay, J., McClartym, G., Salzberg, S. L., Eisen, J., and Fraser, C. M., 2000, Genome sequences of *Chlamydia trachomatis* MoPn and *Chlamydia pneumoniae* AR-39, *Nucleic Acid Res.* **28:**1399–1406.
7. Shirai, M., Hirakawa, H., Kimoto, M., Tabuchi, M., Kishi, F., Ouchi, K., Shiba, T., Ishii, K., Hattori, S., Kuhara, S., and Nakazawa, T., 2000, Comparison of whole genome sequences of *Chlamydia pneumoniae* J 138 from Japan and CWL 029 from USA, *Nucleic Acid Res.* **28:**2311–2314.
8. Matsumoto, A., Bessho, H., Uehira, K., and Suda, T., 1991, Morphological studies of the association of mitochondria with chlamydial inclusions and the fusion of chlamydial inclusions, *J. Electron Microsc.* **40:**356–363.
9. Miyashita, N., Kanamoto, Y., and Matsumoto, A., 1993, The morphology of *Chlamydia pneumoniae, J. Med. Microbiol.* **38:**418–425.
10. Rockey, D. D., Lenart, J., and Stephens, R. S., 2000, Genome sequencing and our understanding of chlamydiae, *Infect. Immun.* **68:**5473–5479.
11. Grayston, J. T., Kuo, C.-C., Campbell, L. A., and Wang, S.-P., 1989, *Chlamydia pneumoniae* sp. nov. for Chlamydia sp. strain TWAR, *Int. J. Syst. Bacteriol.* **39:**88–90.
12. Chi, E. Y., Kuo, C.-C., and Grayston, J. T., 1987, Unique ultrastructure in the elementary body of *Chlamydia* sp. strain TWAR, *J. Bacteriol.* **169:**3757–3763.
13. Miyashita, N., and Matsumoto, A., 1994, Microbiology of Chlamydiae—with emphasis on physicochemistry, antigenicity and drug susceptibility of *Chlamydia pneumoniae, Kawasaki Med. J.* **20:**1–17.

14. Miyashita, N., Matsumoto, A., Kubota, Y., Nakajima, M., Niki, Y., and Matsushima, T., 1996, Continuous isolation and characterization of *Chlamydia pneumoniae* from a patient with diffuse panbronchiolitis, *Microbiol. Immunol.* **40:**547–552.
15. Miyashita, N., Matsumoto, A., Soejima, R., Kishimoto, T., Nakajima, M., Niki, Y., and Matsushima, T., 1997, Morphological analysis of *Chlamydia pneumoniae, Jpn. J. Chemother.* **45:**255–264.
16. Miyashita, N., and Matsumoto, A., 1992, Establishment of a particle-counting method for purified elementary bodies of Chlamydiae and evaluation of sensitivities of the IDEIA Chlamydia Kit and DNA probe by using the purified elementary bodies, *J. Clin. Microbiol.* **30:**2911–2916.
17. Carter, M. W., Al-Mahdawi, S. A. H., Giles, I. G., Treharne, J. D., Ward, M. E., and Clarke, I. N., 1991, Nucleotide sequence and taxonomic value of the major outer membrane protein gene of *Chlamydia pneumoniae* IOL-207, *J. Gen. Microbiol.* **137:**465–475.
18. Popov, V. L., Shatkin, A. A., Pankratova, V. N., Smirnova, N. S., von Bonsdorff, C.-H., Ekman, M.-R., Morttinen, A., and Saikku, P., 1991, Ultrastructure of *Chlamydia pneumoniae* in cell culture, *FEMS Microbiol. Lett.* **84:**129–134.
19. Kuo, C.-C., Chi, E. Y., and Grayston, J. T., 1988, Ultrastructural study of entry of Chlamydia strain TWAR into HeLa cells, *Infect. Immun.* **56:**1668–1672.
20. Tamura, A., and Higashi, N., 1963, Purification and chemical composition of meningopneumonitis virus, *Virology* **20:**596–604.
21. Fox, A., Rogers, J. C., Gilbart, J., Morgan, S., Davis, C. H., Knight, S., and Wyrick, P. B., 1990, Muramic acid is not detectable in *Chlamydia psittaci* or *Chlamydia trachomatis* by gas chromatography mass spectrometry, *Infect. Immun.* **58:**835–837.
22. Evertt, K. D., and Hatch, T. P., 1995, Architecture of the cell envelope of *Chlamydia psittaci* 6BC, *J. Bacteriol.* **177:**877–882.
23. Manire, G. P., 1966, Structure of purified cell walls of dense forms of meningopneumonitis organisms, *J. Bacteriol.* **91:**409–413.
24. Tamura, A., and Manire, G. P., 1967, Preparation and chemical composition of the cell membranes of developmental reticulate forms of meningopneumonitis organisms, *J. Bacteriol.* **94:**1184–1188.
25. Matsumoto, A., and Manire, G. P., 1970, Electron microscopic observations on the fine structure of cell walls of *Chlamydia psittaci, J. Bacteriol.* **104:**1332–1337.
26. Matsumoto, A., Izutsu, H., Miyashita, N., and Ohuchi, M., 1998, Plaque formation by and plaque cloning of *Chlamydia trachomatis* biovar trachoma, *J. Clin. Microbiol.* **36:**3013–3019.
27. Bavoil, P., and Hsia, R., 1998, Type III secretion in Chlamydia: A case of *deja vu*? *Mol. Microbiol.* **28:**860–862.
28. Hsia, R., Pannekoek, Y., Ingerowski, E., and Bavoil, P., 1997, Type III secretion genes identify a putative virulence locus of Chlamydia, *Mol. Microbiol.* **25:**351–359.
29. Kubori, T., Matsushima, Y., Nakamusra, D., Uralil, J., Lara-Tejero, M., Sukhan, A., Galan, J. E., and Aizawa, S., 1998, Supramolecular structure of the *Salmonella typhimurium* type III protein secretion system, *Science* **280:**602–605.
30. Hammerschlag, M. R., Chirgwin, K., Roblin, P. M., Gelling, M., Dumornay, W., Mandel, L., Smith, P., and Schachter, J., 1992, Persistent infection with *Chlamydia pneumoniae* following acute respiratory illness, *Clin. Infect. Dis.* **14:**178–182.
31. Beatty, W. L., Byrne, G. I., and Morrison, R. P., 1993, Morphologic and antigenic characterization of interferon gamma mediated persistent *Chlamydia trachomatis* infection *in vitro, Proc. Natl. Acad. Sci. U.S.A.* **90:**3998–4002.
32. Beatty, W. L., Morrison, R. P., and Byrne, G. I., 1995, Reactivation of persistent *Chlamydia trachomatis* infection in cell culture, *Infect. Immun.* **63:**199–205.
33. Mehta, S. J., Miller, D., Ramirez, J. A., and Summersgill, J. T., 1998, Inhibition of *Chlamydia pneumoniae* replication in HEp-2 cells by interferon-γ role of tryptophan catabolism, *J. Infect. Dis.* **177:**1326–1331.

34. Summergill, J. T., Sahney, N. N., Gaydos, C. A., Quinn, T. C., and Ramirez, J. A., 1995, Inhibition of *Chlamydia pneumoniae* growth in HEp-2 cells pretreated with gamma interferon and tumor necrosis factor alpha, *Infect. Immun.* **63:**2801–2803.
35. Dreses-Werringloer, U., Padubrin, I., Jürgens-Saathoff, B., Hudson, A. P., Zeidler, H., and Köhler, L., 2000, Persistent of *Chlamydia trachomatis* is induced by ciprofloxacin *in vitro, Antimicrob. Agents Chemother.* **44:**3288–3297.
36. Matsumoto, A., and Manire, G. P., 1970, Electron microscopic observations on the effects of penicillin on the morphology of *Chlamydia psittaci, J. Bacteriol.* **101:**278–285.
37. Harper, A., Pogson, C. I., Jones, M. L., and Pearce, J. H., 2000, Chlamydial development is adversely affected by minor changes in amino acid supply, blood plasma amino acid levels, and glucose deprivation, *Infect. Immun.* **68:**1457–1464.
38. Pantoja, L. G., Miller, R. D., Ramirez, J. A., Molestina, R. E., and Summersgill, J. T., 2001, Characterization of *Chlamydia pneumoniae* persistence in HEp-2 cells treated with gamma interferon, *Infect. Immun.* **69:**7927–7932.
39. Mathews, S., George, C., Flegg, C., Stenzel, D., and Timms, P., 2001, Differential expression of ompA, ompB, pyk, nlpD and Cpn0585 genes between normal and interferon-gamma-treated cultures of *Chlamydia pneumoniae, Microb. Pathog.* **30:**337–345.
40. Wolf, K., Fischer, E., and Hackstadt, T., 2000, Ultrastructural analysis of developmental events in *Chlamydia pneumoniae*–infected cells, *Infect. Immun.* **68:**2379–2385.
41. Beatty, W. L., Byrne, G. I., and Morrison, R. P., 1994, Repeated and persistent infection with Chlamydia and the development of chronic inflammation and disease, *Trends Microbiol.* **2:**94–98.
42. Matsumoto, A., 1981, Isolation and electron microscopic observations of intracytoplasmic inclusions containing *Chlamydia psittaci, J. Bacteriol.* **145:**605–612.

3

Cell and Molecular Biology of *Chlamydia pneumoniae*

GUNNA CHRISTIANSEN, BRIAN VANDAHL, and SVEND BIRKELUND

1. INTRODUCTION

Chlamydiae are obligate intercellular bacteria with a biphasic developmental cycle in which they alternate between the infectious extracellular and metabolically inactive form, the elementary body (EB), and the intracellular, metabolically active replicating form, the reticulate body (RB). EB are small tight bodies with a diameter of 300 nm, which can attach to susceptible host cells and induce their uptake by endocytosis. Within the cytoplasm of the host cell, chlamydiae are localized within a phagosome, the *Chlamydia* inclusion, surrounded by the inclusion membrane, which does not fuse with lysosomes. Chlamydiae stay within the phagosome during their entire intracellular life. Shortly after uptake, EBs transform to RBs, which are larger than EBs (1 μm in diameter). They divide by binary fission and after a specific amount of time RBs transform to EBs. This is accompanied by the synthesis of the late phase proteins, the DNA-binding protein Hc1 and the two cysteine-rich outer membrane proteins Omp2 and Omp3. After approximately 72 h RB transform to EB. At the end of this process the inclusion bursts, releasing the infectious EB.[1]

Attachment of *Chlamydia pneumoniae* EB is primarily to the epithelial cells of the human respiratory tract;[2] however, the mechanism is not known. At the surface of *C. pneumoniae* EB the lipopolysaccharide molecule carrying the common, genus-specific LPS epitope, is present. Proteins covering the EB surface are folded in a complicated manner, leaving no linear epitopes at the surface.[3,4] The chlamydial major outer membrane protein MOMP, which is the major

GUNNA CHRISTIANSEN and SVEND BIRKELUND • Department of Medical Microbiology and Immunology, University of Aarhus, Denmark. BRIAN VANDAHL • Loke Diagnostics ApS, Science Park, Aarhus, Denmark.

constituent of the *Chlamydia trachomatis* surface, has a conformational epitope on MOMP trimers present at the surface,[4] but a family of polymorphic proteins (Pmps) are the major constituents of the *C. pneumoniae* outer membrane complex (COMC) and cover the *C. pneumoniae* surface.[3,5]

The primary site of human infection with *C. pneumoniae* is the upper respiratory tract, but severe cases of pneumonia with long recovery are also seen.[2] In experimentally infected mice *C. pneumoniae* inclusions are seen in the bronchial epithelial cells, and probably these cells can also be infected during human pneumonia. *C. pneumoniae* can also multiply in alveolar macrophages and *C. pneumoniae* DNA has been found by PCR in human blood monocytes.[6] Infection of blood monocytes induces differentiation of monocytes into macrophages.[7] When rabbits are infected, *C. pneumoniae* is capable of multiplying in endothelial cells, resulting in activation of growth factors and intimal thickening,[8] indicating that *C. pneumoniae* may play a role in the development of atherosclerosis.[9]

2. ULTRASTRUCTURE OF *Chlamydia* EB AND RB

Electron microscopy of infected cell cultures reveals both the localization and structure of the inclusion and the size and shape of RB and EB. Thin section of the developing inclusion shows that RB are spherical bodies surrounded by a double-layered membrane characteristic of bacteria. EB seen late in the developmental cycle are small, with a condensed nucleoid. EB of the *C. pneumoniae* TWAR strain can be pear-shaped with a wide periplasmic space[10] but other *C. pneumoniae* isolates are round. The inner layer of the EB cell wall is composed of hexagonal arrayed structures estimated to have a periodicity of 10–20 nm, and a granular outer layer. From the surface of both RB and EB envelopes cylindrical spike-like surface projections protrude that on RB connect the RB with the inclusion membrane. Such projections are present both on RB and EB but are more numerous on the RB.[10] Freeze-replica preparations of infected cells examined by electron microscopy reveal fine particles grouped together at several areas on the convex side of the inclusion membrane, suggesting direct connection between the RB and the inclusion membrane and the possibility that RB through the projections are in direct contact with the host cytoplasm.

3. COMPONENTS RESPONSIBLE FOR THE STRUCTURAL APPEARANCE

Because of the lack of tools for genetic manipulation of chlamydiae it is important to be able to study subfractions of the microorganism. The sarcosyl-insoluble chlamydial outer membrane complex (COMC)[11] can be used to identify components responsible for the structural appearance, and electron microscopy of purified COMC shows that the shape of EB is maintained (Fig. 1). SDS-PAGE and immunoblotting have shown that *C. pneumoniae* COMC contains

LPS, MOMP, Omp2, and Omp3 cysteine-rich proteins of 60 and 12.5 kDa, respectively. In addition, proteins with the size of 90–100 kDa were found.[12] These proteins are identified to belong to a family of proteins named polymorphic outer membrane proteins (Pmps).[3]

4. PHYSICAL LOCALIZATION OF THE COMPONENTS

Physical localization of the components can be studied by immunoelectron microscopy. Such studies show that LPS is surface localized both at the EB and RB of *C. pneumoniae.*[13] MOMP, which in *C. trachomatis* is surface localized in close association with LPS,[14] was found not to be surface localized in *C. pneumoniae,*[15] but recently Wolf *et al.*[4] identified conformational surface exposed epitopes on *C. pneumoniae* MOMP. It is suggested that the late-synthesized Omp2 is the structural element for the hexagonally arrayed structures seen at the inner surface of COMC. As suggested by Hatch,[16] immunoelectron microscopy of COMC, in which disulfide bonds were reduced by DTT, showed that Omp2 was present at the inner surface, in agreement with its localization in the periplasmic space.[17] Whether this is the case in *C. pneumoniae* remains to be determined. Omp3, which also is synthesized late in the developmental cycle, is a lipoprotein, which may be linked to Omp2 by disulfide cross-linking and to the outer membrane by its lipid anchor.[16] In *C. pneumoniae* also Pmps are shown to be surface localized.[3] A model of the membrane structure is shown in Fig. 2. The inner membrane is seen as a lipid bilayer. In the

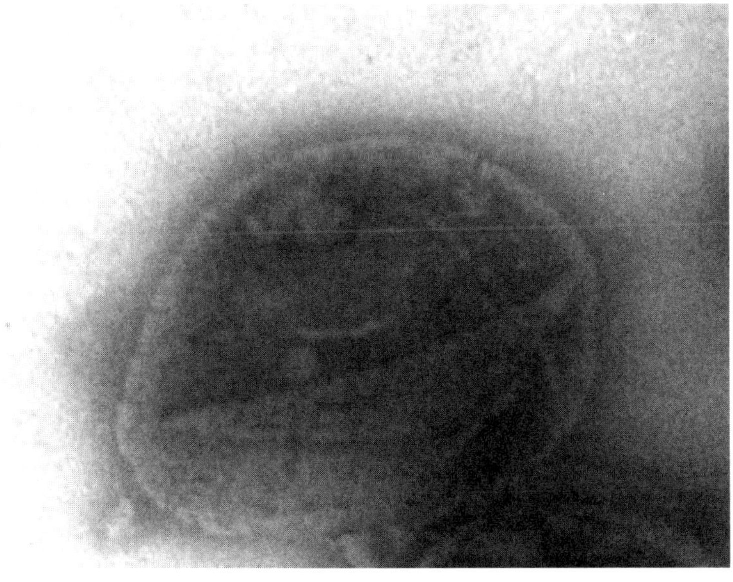

FIGURE 1. Electron microscopy of *C. pneumoniae* COMC stained with 1% ammonium molybdate. The diameter of COMC is 300 nm.

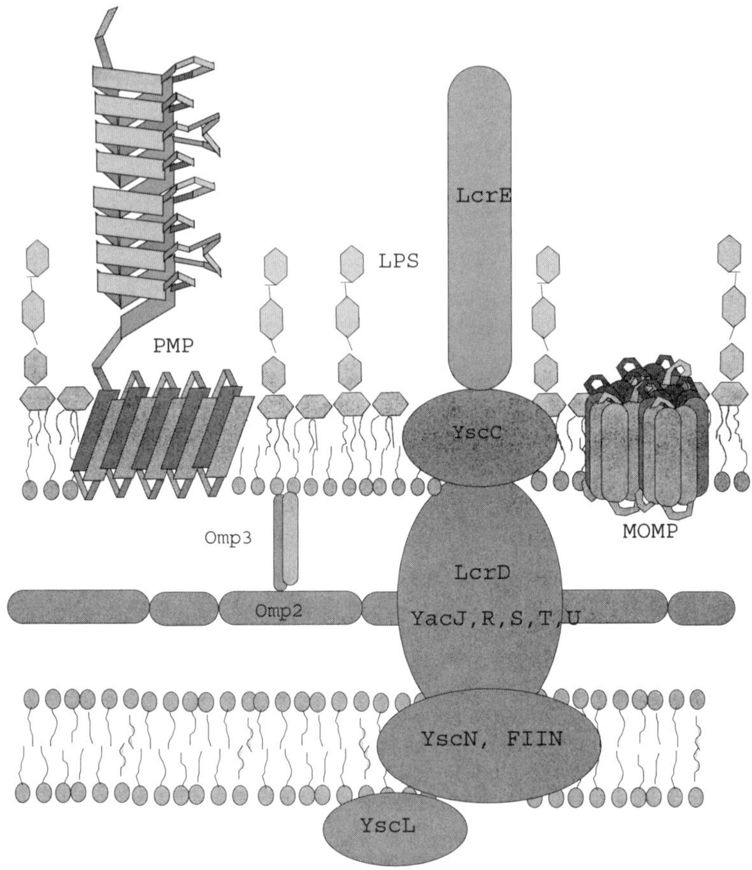

FIGURE 2. Schematic drawing of the hypothetical structure of the *C. pneumoniae* membrane modified from refs. 16 and 26.

periplasmic space Omp2 is seen connected to Omp3. The structure of COMC is complex, with LPS, Ppm, spike-like structures, and MOMP trimers being surface exposed.

5. CHLAMYDIA GENOMES

Sequencing of Chlamydia genomes has provided new means to analyze the molecular and structural biology of *Chlamydia.* There are now published five complete genome sequences: for *C. trachomatis* D,[18] *C. trachomatis* MoPn,[19] and *C. pneumoniae* CWL029,[20] J138,[21] and AR39.[19] *Chlamydia* is the microbe with most sequenced genomes. This has greatly expanded the information concerning the biology of these obligate intracellular pathogens. In addition to provide information on all potential protein products, the genome sequences unexpectedly identified several classes of putative proteins relevant for understanding the structure of chlamydiae. *Chlamydia* genomes are small and show striking similarity (1.230.230 nucleotides in *C. pneumoniae* CWL029, and 1.042.519 in *C. trachomatis* D). In *C. trachomatis*, 214 of the *C. pneumoniae* CWL029 protein-coding sequences are not found, most with no homology to known sequences. The *C. pneumoniae* genome has the capacity to encode 1052 proteins, of which genes are found that encode enzymes for essential functions such as DNA replication, repair, transcription, and translation, enzymes that function in aerobic respiration and enzymes central in the glycolytic pathway. Chlamydiae are described as "energy parasites" because they import ATP from the host cell. It was therefore surprising that the genome contained genes for phosphoglycerate kinase, pyruvate kinase, and succinate thiokinase. There are also genes encoding enzymes of fatty acid and phospholipid biosynthesis but only few genes involved in amino acid biosynthesis, and no genes for purine and pyrimidine *de novo* synthesis except for the synthesis of cytosine triphosphate. It is unknown how chlamydiae obtain nutrients from the inclusion, but genes for membrane transport systems, including ABC transporters for amino acids, oligopeptides, and ions are found. Special attention should be given to the large family of genes encoding the Pmp proteins, the Inc proteins, and genes encoding type III secretion systems, which will be described later.

Chlamydia genomes are very stable. Divergence between the genomes of *C. trachomatis* D and *C. pneumoniae* CWL029 was analyzed by Dalevi *et al.*[22] They found no examples of horizontal gene transfer, no evidence for an excessive number of tandem gene duplications, and approximately 60 events (mostly inversions and transpositions) were required for converting the gene order structure of one genome to the other. *C. pneumoniae* CWL029 and *C. pneumoniae* J138 are also very similar, *C. pneumoniae* J138 being 3665 nucleotides shorter than *C. pneumoniae* CWL029[21] with five DNA segments being absent from and three being present in *C. pneumoniae* J138. The *C. pneumoniae* AR39 chromosome is nearly identical to that of *C. pneumoniae* CWL029, with only a few small deletions and approximately 300 single nucleotide polymorphisms separating the two genomes. But the *C. pneumoniae* AR39 includes a sequence encoding a novel infecting bacteriophage[19]; that seems to be uncommon among isolates, but phagecontaining isolates may commonly infect individuals with vascular disease.[23]

6. THE *C. pneumoniae* PROTEOME

A genome provides important understanding of the biological potential of an organism, but the genome is static and gives no information on regulation of particular genes or on posttranslational modification of gene products. The actual protein content of any compartment of the organism is dynamic, and proteins may be processed in various ways. By separation of proteins by two-dimensional (2D) gel electrophoresis and identification of protein spots by mass spectrometry (MS) protein composition of RB, EB, and COMC can be studied. The protein spots are visualized by silver staining or by radioactive labeling of the proteins. To study intracellular growth of chlamydiae, cycloheximide, which efficiently blocks eukaryotic protein synthesis, is added to the labeling medium to restrict labeling to chlamydial proteins. Such labeling allows visualization of the labeled spots on the eukaryotic background. By pulse labeling at different time points it is possible to study protein synthesis over time during the developmental cycle, and by pulse-chase analysis the turnover of proteins can be determined.

To generate a proteome reference map of *C. pneumoniae* EB, Vandahl *et al.*[(24)] identified 263 proteins encoded from 167 chlamydial genes in the pH range from 3 to 11. The identified proteins were mostly products of housekeeping genes with homology to genes in other organisms, but 31 were so-called hypothetical proteins encoded by reading frames showing no homology to known proteins in other organisms. Among the products were key components of a type III secretion system, and members of the Pmp family (Fig. 3).

A new aspect of proteome analysis of *C. pneumoniae* is the development of a tool for identification of secreted proteins. It is not possible to isolate the cytoplasm of *Chlamydia*-infected cells without contamination by chlamydial proteins

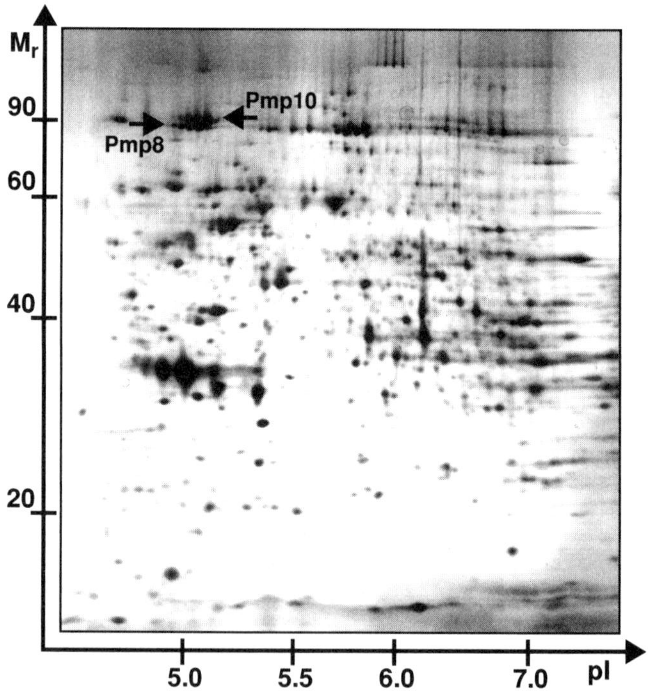

FIGURE 3. Silverstained two-dimensional gel of proteins from purified elementary body of Chlamydia pneumoniae cultivated in HEp-2 cells and harvested at 72 h postinfection. The position of horizontal rows of different isoforms of Pmp8 and Pmp10 are marked. An immobilized nonlinear pH gradient (pH 3–10) was used for isoelectrical focusing in the first dimension. The gel for second dimension was a 9–16% polyacrylamide SDS gel. Gels of radio-labeled chlamydiae in various pG intervals can be found at http://www.gram.au.dk.

form disrupted inclusions and RB. A way to circumvent such contamination is to subtract the 2D protein profile of purified EB and RB from that of chlamydial proteins present in total lysates of infected cells. The remaining proteins are likely to be secreted from chlamydiae into the host cell, or into the inclusion membrane. The suitability of this technique was demonstrated by the identification of the N- and C-terminal fragments of CPAF from *C. pneumoniae* and *C. trachomatis* A, D, and L2 in whole lysates of infected cells but not in purified EB or RB.[(25)] In this study it was also demonstrated how the turnover of proteins can be followed over time by proteome analysis by pulse-chase studies. Proteome reference maps of *C. pneumoniae* and *C. trachomatis* A, D, and L2 are available at http://www.gram.au.dk.

7. THE Pmp GENE FAMILY

All *Chlamydia* species harbor a family of distantly related *pomp/pmp* genes. The genes encode polymorphic membrane proteins whose function is unknown. Genome sequencing[(20)] revealed the presence of 21 such genes in *C. pneumoniae* CWL029 (most of the genes are located in two genomic clusters whereas the remaining genes are found as a pair and a single gene at different genomic localizations). Sixteen of the 21 genes were full length genes whereas one was truncated and 4 had small insertions or deletions that caused a premature stop. Most of the genes had a leader sequence with a cleavage site for signal peptidase I, but two of the genes, *Pmp*10 and *Pmp*11 differed from the others by having a cleavage site for signal peptidase II, indicating that these proteins probably were lipid modified.[(3)] The first identification of expressed *C. pneumoniae* Pmp proteins was by Knudsen *et al.*,[(3)] who identified *pmp*1, 2, 10 and 11 in an expression library with antibodies raised to denatured *C. pneumoniae* COMC. Antibodies to recombinant Pmp10 and 11 were used in immunofluorescence microscopy of *C. pneumoniae*–infected Hep2 cells. Antibodies to Pmp11 stained all inclusions but antibodies to Pmp10 stained only some of the inclusions,[(5)] indicating that Pmp10 is differentially expressed. The gene encoding Pmp10 is transcribed in the opposite reading direction of the other *pmp* genes present in the two genomic clusters, and the *pmp*10 gene is the only *pmp* gene with a poly-G tract within the open reading frame. Pedersen *et al.*[(5)] sequenced a number of clones containing the poly-G tract of the *Pmp*10 gene. The number of G-residues differed, and only 14 guanosine residues represented a correct open reading frame. From the sequenced genomes 14 guanosine residues were found in *C. pneumoniae* AR39 whereas 13 were found in CWL029. In TW183 Pmp10 was also expressed but here the number of guanosine residues was 11. Switching may be caused by a frequent change of the number of guanosine residues within the poly-G tract.

The genome sequence identified the Pmp gene family, but evidence for expression of *C. pneumoniae* Pmps came from the proteome analysis by Vandahl *et al.*.[(24,26)] By separation of the *C. pneumoniae* EB proteins by 2D-PAGE, 10 of the 21 Pmps were identified (Pmp2, 6, 7, 8, 10, 11, 13, 14, 20, and 21). Pmp2, 6, 7, 8, 10, 11, 13, and 14 were investigated with respect to time-dependent expression by pulsed labelling and all were found to be heavily upregulated between 36 and 48 h postinfection at the time of conversion from RB to EB. Pmp8 and Pmp10 are the Pmps expressed in the greatest amount when analyzed for spot intensity in autoradiographs. Three of the Pmps (Pmp6, 20, and 21) were cleaved, in agreement with the similarity to autotransporter proteins.[27] Antibodies against Pmp6, 8, 10, and 11 reacted in immunofluorescence microscopy after formalin fixation of infected cell cultures, and it is therefore likely that they are surface exposed.[26] A partwise characterization of the *C. pneumoniae* COMC by proteome analysis showed the Pmps to be the major constituent of this complex.[28]

8. Pmp STRUCTURE

The *C. pneumoniae pmp* gene family is a heterogeneous group of genes with low identity but common characteristics. The majority of the genes have the capacity to encode proteins with sizes 90–100 kDa, and they resemble members of the autotransporter family.[27] Analysis of the C-terminal part of the proteins showed that full length proteins ended with an amphipatic beta-sheet and a terminal phenylalanine residue. In addition, the program AMPHI[29] predicted the C-terminal part of the proteins to form a transmembranic beta-barrel,[27] supposed to be used for translocation of the N-terminal part of the protein to the chlamydial surface (Fig. 2). This is in agreement with the finding that it is the N-terminal part of the *Chlamydia psittaci* Pomp molecules that are surface exposed.[30] Another characteristic is the presence of repeats of the amino acid motif GGAI and FXXN in the N-terminal part of the proteins. Use of common prediction programs did not reveal any characteristic patterns for the N-terminal part of the Pmp proteins. A newly described program, BETAWRAP,[31] uses beta-strand interactions to detect proteins that may form a structure of parallel right-handed beta-helices. Such proteins fold into an elongated triangular prism shape where amino acid residues that are close in space in the folded protein are far apart in the extended form. BETAWRAP identified a high number of protein sequences of virulence factors from bacterial pathogens such as toxins, adhesins, and surface proteins including *Chlamydia* Pmps. Birkelund *et al.*[32] used bioinformatics to further analyze the *C. pneumoniae* Pmps and found support for these to elicit the structure of parallel beta-helices (Fig. 2). This is in agreement with the lack of linear epitopes at the *C. pneumoniae* surface.

9. TYPE III SECRETION SYSTEM

Surface projections are seen on all *Chlamydia* species. The projections extend through holes in the outer membrane and protrude from the surface of the chlamydiae. They have a physical resemblance to the Type III secretion apparatus found in other bacteria. The genome sequence identified a complete set of genes with homology to the type III secretion system both in *C. trachomatis* and in *C. pneumoniae.*[18,20] The genes were first identified in *C. psittaci,*[33] and Bavoil and Hsia[34] speculated that the projections represent a type III secretion system (Fig. 2). Analysis of the *C. pneumoniae* EB proteome demonstrated that members of the Type III secretion system were present.[24] A hypothetical structure of the chlamydial Type III secretion apparatus was presented by Rockey *et al.*,[35] who suggested YcsC to be anchored in the chlamydial outer membrane (Fig. 2). In agreement with this model Vandahl *et al.*[28] identified YscC in *C. pneumoniae* COMC proteome, but not YscL, YscN, and LcrE, the other Type III secretion system proteins identified in EB. These proteins are supposed to be localized in the chlamydial cytosol, the inner membrane, and in contact with the host cells (Fig. 2). There is no direct evidence for the protrusions to be constituted by the type III gene products but it is likely that they are. Since these are found to penetrate through the inclusion membrane it is likely that they are providing contact with the outer of the inclusion. Such contact may serve as a channel by which secreted proteins get access to the environment outside the chlamydial inclusion, where they may serve as regulators of host cell transcription, of vesicle transport, and as inhibitors of fusion of the chlamydial inclusion with host cell lysosomes. The projections may also serve as a communication system within the inclusion, leading to differentiation of RB to EB late in the developmental cycle.

10. SECRETED PROTEINS

The first demonstration of a secreted protein came from Rockey *et al.*[36] To identify proteins specific for the intracellular phase of the developmental cycle they screened an expression library of *C. psittaci* DNA with a convalescent serum from infected animals and a hyperimmune antiserum against formalin-killed purified EB. They identified and sequenced a clone that was recognized only by the convalescent serum and produced an antiserum to the recombinant protein. By immunofluorescence microscopy they found that the protein was present in the inclusion membrane and therefore named IncA for inclusion membrane protein. IncA is also present in the *C. pneumoniae* inclusion membrane (Fig. 4).

IncA has a characteristic secondary structure with a long bilobed hydrophobic region present in amino acids 65–116.[36] This motif was found in 46 potential members of inclusion membrane proteins in the *C. trachomatis* genome.[37] Searching the *C. pneumoniae* genome revealed an even higher number of hypothetical Inc proteins.[35] Several of these had homology to genes found in *C. trachomatis* but over half of the genes in each species had no homology to the genes in the other species. The potential for chlamydiae to export such a high number of Incs to the inclusion membrane generates the possibility that the inclusion membrane can serve functions as inclusion development, avoidance of lysosomal fusion, nutrient acquisition, signaling associated with EB–RB–EB reorganization and vesicle trafficking. How the Incs are secreted and transported to the inclusion membrane and whether this is done by the Type III secretion system remains to be determined. Subtil *et al.*[38] used a heterologous expression

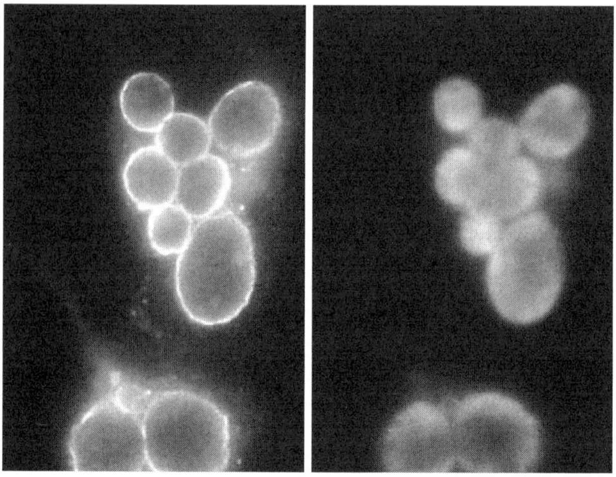

FIGURE 4. Immunofluorescence microscopy of Hep2 cells infected for 48 h with *C. pneumoniae* cultivated on coverslips. After methanol fixation the sample was incubated with a monoclonal antibody recognizing the *C. pneumoniae* surface[5,13] and a polyclonal antibody raised against recombinant *C. pneumoniae* IncA. Secondary antibodies were FITC-conjugated goat anti-rabbit and rhodamine-conjugated goat anti-mouse antibodies. The first part shows IncA lining the inclusion membrane (FITC) and the second part the *C. pneumoniae* inclusions (rhodamine).

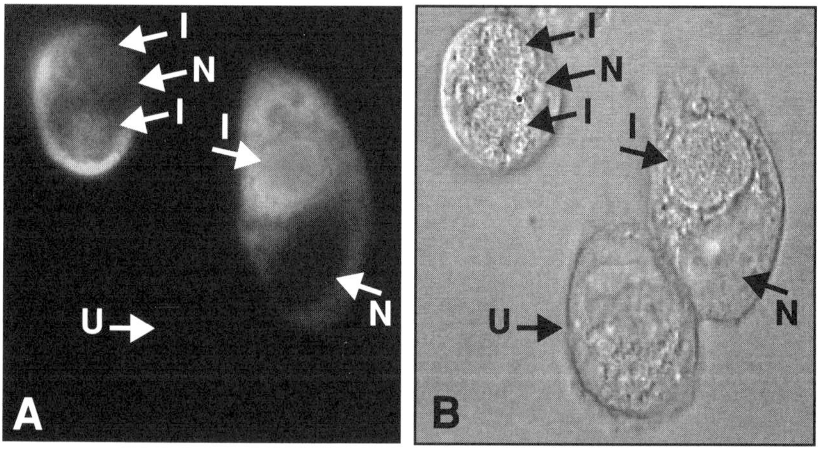

FIGURE 5. (A) Immunofluorescence mincroscopy showing the subcellular localization of CPAF in HEp-2 cells infected with *C. pneumoniae* CWL029 (fixed in formaldehyde at 72 h post infection). PAb253 against *C. pneumoniae* CPAF (visualized by FITC) and mAb26.4.23 against Pmp10 (visualized by rhodamine) were used for double staining.[25] I: inclusion; N: nucelus; U: uninfected cell. (B) Nomarsky microscopy image of the cells in (A). Note the uninfected cell in the lower part of the picture. CPAF is clearly detected in the host cell cytoplasm of *Chlamydia* infected cells, but not in uninfected cells.

system to export the N-terminal part of IncA, B, and C fused with a reporter, indicating that they are exported through the type III system.

An exciting observation came from Zhong *et al.*,[39] who for the first time described the secretion of a *Chlamydia*-encoded protein into the host cell cytoplasm. The protein is a protease-like activity factor, CPAF, which is present both in *C. trachomatis* and in *C. pneumoniae.*[40] The protein is cleaved and secreted into the host cell cytoplasm, and recombinant CPAF can degrade host cell transcription factors required for MHC expression, and thus provide the chlamydiae with protection from host immune recognition. The results by Zhong *et al.*[39] were verified and extended in studies by Shaw *et al.*[25] It was shown that both *C. trachomatis* A, D, and L2 and *C. pneumoniae* secrete CPAF from the middle of the developmental cycle (Fig. 5). CPAF has a very slow turnover that may reflect limited degradation of the secreted protein by the host cell proteasome. Thus, *Chlamydia* may secrete a protein, which inhibits MHC class I and II antigen presentation without itself being a target for proteasome degradation and MHC class I antigen presentation.

11. VESICLE TRAFFICKING

Vesicle transport has only been studied for *C. trachomatis.* When *C. trachomatis* L2 attaches to susceptible host cells a host cell protein of 70 kDa is phosphorylated[41] and the phosphorylation can be followed in the inclusion membrane during inclusion development. After uptake the phagosome migrates to the perinuclear space transported along microtubules where the Chlamydia-containing vesicles are positioned at the centre of the microtubular network, indicating a microtubule-dependent mode of chlamydial redistribution. Dynein, a microtubule-dependent motor protein known to be involved in directed vesicle transport along microtubules, was found to colocate with perinuclear aggregates of *C. trachomatis.*[42] Chlamydiae-containing vesicles do not fuse with lysosomes, but fuse with sphingomyelin-containing exocytic vesicles from the Golgi apparatus.[43,44] Analysis of sphingomyelin incorporation showed that no incorporation occurred the first 2 h after infection, but from 4 h after infection incorporation of sphingomyelin was seen.[43] Fusion of the vesicles exposes sphingomyelin on the interior of the inclusion, and here RB can adsorb and incorporate it into the cell wall.[45] The way the intersection of the exocytotic pathway is established by the chlamydiae is unknown. It could be speculated that either early-exported Inc proteins or the phosphorylated host cell protein may play a role in diverging the vesicle transport and thus allow the chlamydiae to establish themselves in an intracellular compartment protected from the normal destruction mechanisms of the cell.

12. HOST ATTACK ON *Chlamydia*

During human infections the immune system is activated, and both the cellular and humoral immune systems are important for development of immunity.

Antibodies to *C. pneumoniae* are developed during infection and a high number of the adult population has antibodies to *C. pneumoniae*. Whether this is caused by repeated or persistent infections, or both, remains to be determined. Antibodies can be detected with microimmunofluorescence microscopy, and the major immunogens are surface-exposed Pmps, LPS, and MOMP trimers. The cellular immune response is also activated, and influence of IFN-γ on the *C. trachomatis* trachoma serovars has a dramatic effect on the developmental cycle of these serovars. It is assumed that lack of a functional TrpBA operon at least in part is responsible for this.[46] *C. pneumoniae* also lacks this operon, and developmental changes were observed when cell cultures infected with *C. pneumoniae* were treated with IFN-γ.[47] Deviation from the normal developmental cycle may provide an intracellular environment supporting persistent infections, which may explain the high number of seropositive persons.

13. PERSPECTIVES

Genomics and proteomics have highly influenced our understanding of the molecular biology of *C. pneumoniae*, and in combination with cell biology a more complete picture is emerging. There are, however, many unknown factors still to be considered. Even though there are indications that persistent infections can be established because of the influence of IFN-γ, there are still many open questions as how other cytokines and host factors influence the developmental cycle. In case a persistent infection is established it is of interest to learn in what cells such an infection can be found. The discovery of secreted proteins opens new ways to study how chlamydiae communicates with the host both intracellularly and in the infected organism.

REFERENCES

1. Pearce, J. H., and Gaston, J. S. H., 2001, *Chlamydia*, in: *Molecular Medical Microbiology*, (M. Susmann, ed.), AP Academic Press London, UK, pp. 1825–1864.
2. Kuo, C. C., Jackson, L. A., Campbell, L. A., and Grayston, J. T., 1995, *Chlamydia pneumoniae* (TWAR), *Clin. Microbiol. Rev.* **8**:451–461.
3. Knudsen, K., Madsen, A. S., Mygind, P., Christiansen, G., and Birkelund, S., 1999, Identification of two novel genes: encoding 97- to 99-kilodalton outer membrane proteins of *Chlamydia pneumoniae*, *Infect. Immun.* **67**:375–383.
4. Wolf, K., Fischer, E., Mead, D., Zhong, G., Peeling, R., Whitmire, B., and Caldwell, H. D., 2001, *Chlamydia pneumoniae* major outer membrane protein is a surface-exposed antigen that elicits antibodies primarily directed against conformation-dependent determinants, *Infect. Immun.* **69**:3082–3091.
5. Pedersen, A. S., Christiansen, G., and Birkelund, S., 2001, Differential expression of Pmp10 in cell culture infected with *Chlamydia pneumoniae* CWL029, *FEMS Microbiol. Lett.* **203**:153–159.
6. Boman, J., and Hammerschlag, M. R., 2002, *Chlamydia pneumoniae* and atherosclerosis: critical assessment of diagnostic methods and relevance to treatment studies, *Clin Microbiol Rev.* 2002;**15**:1–20.

7. Yamaguchi, H., Haranaga, S., Widen, R., Friedman, H., and Yamamoto, Y., 2002, *Chlamydia pneumoniae* infection induces differentiation of monocytes into macrophages, *Infect. Immun.* **70:**2392–2398.
8. Coombes, B. K., Chiu, B., Fong, I. W., and Mahony, J. B., 2002, *Chlamydia pneumoniae* infection of endothelial cells induces transcriptional activation of platelet-derived growth factor-B: A potential link to intimal thickening in a rabbit model of atherosclerosis, *J. Infect. Dis.* **185:**1621–1630.
9. Grayston, J. T., 2000, Background and current knowledge of *Chlamydia pneumoniae* and atherosclerosis, *J. Infect. Dis.* **181** (Suppl. 3)**:**S402–S410.
10. Miyashita, N., Kanamoto, Y., and Matsumoto, A., 1993, The morphology of Chlamydia pneumoniae, *J. Med. Microbiol.* **38:**418–425.
11. Caldwell, H. D., Kromhout, J., and Schachter, J., 1981, Purification and partial characterization of the major outer membrane protein of, *Chlamydia trachomatis.*, *Infect. Immun.* **31:**1161–1176.
12. Melgosa, M. P., Kuo, C. C., and Campbell, L. A., 1993, Outer membrane complex proteins of *Chlamydia pneumoniae, FEMS Microbiol. Lett.* **112:**199–204.
13. Christiansen, G., Østergaard, L., and Birkelund, S., 1994, Analysis of the *Chlamydia pneumoniae* surface, in: *Chlamydial Infections*, Proceedings of the Eighth International Symposium on Human Chlamydial Infections (Orfila *et al.*, eds.), Societa Editrice Esculapio, pp. 173–176.
14. Birkelund, S., Lundemose, A. G., Christiansen, G., 1988, Chemical cross-linking of Chlamydia trachomatis, *Infect. Immun.* **56:**654–659.
15. Christiansen, G., Madsen, A. S., Knudsen, K., Mygind, P., and Birkelund, S., 1998, Stability of the outer membrane proteins of, *Chlamydia pneumoniae*, in: *Chlamydial Infections*, Proceedings of the ninth Symposium on Human Chlamydial Infections (Stephens *et al.*, eds.) International Chlamydia Symposium, San Francisco, CA 94110, pp. 271–274.
16. Hatch, T. P., 1996, Disulfide cross-linked envelope proteins: The functional equivalent of peptidoglycan in chlamydiae? *J. Bacteriol.* **178:**1–5.
17. Mygind, P., Christiansen, G., Birkelund, S., 1998, Topological analysis of *Chlamydia trachomatis* L2 outer membrane protein 2, *J. Bacteriol.* **180:**5784–5787.
18. Stephens, R. S., Kalman, S., Lammel, C., Fan, J., Marathe, R., Aravind, L., Mitchell, W., Olinger, L., Tatusov, R. L., Zhao, Q., Koonin, E. V., and Davis, R. W., 1998, Genome sequence of an obligate intracellular pathogen of humans: *Chlamydia trachomatis, Science* 282(5389), 754–759.
19. Read, T. D., Brunham, R. C., Shen, C., Gill, S. R., Heidelberg, J. F., White, O., Hickey, E. K., Peterson, J., Utterback, T., Berry, K., Bass, S., Linher, K., Weidman, J., Khouri, H., Craven, B., Bowman, C., Dodson, R., Gwinn, M., Nelson, W., DeBoy, R., Kolonay, J., McClarty, G., Salzberg, S. L., Eisen, J., and Fraser, C. M., 2000, Genome sequences of, *Chlamydia trachomatis* MoPn and *Chlamydia pneumoniae* AR39, *Nucleic Acids Res.* **28:**1397–1406
20. Kalman, S., Mitchell, W., Marathe, R., Lammel, C., Fan, J., Hyman, R. W., Olinger, L., Grimwood, J., Davis, R. W., and Stephens, R. S., 1999, Comparative genomes of *Chlamydia pneumoniae* and *C. trachomatis, Nat. Genet.* **21:**385–389.
21. Shirai, M., Hirakawa, H., Kimoto, M., Tabuchi, M., Kishi, F., Ouchi, K., Shiba, T., Ishii, K., Hattori, M., Kuhara, S., and Nakazawa, T., 2000, Comparison of whole genome sequences of *Chlamydia pneumoniae* J138 from Japan and CWL029 from USA, *Nucleic Acids Res.* **28:**2311–2314.
22. Dalevi, D. A., Eriksen, N., Eriksson, K., and Andersson, S. G., 2002, Measuring genome divergence in bacteria: a case study using chlamydian data, *J. Mol. Evol.* **55:**24–36.
23. Karunakaran, K. P., Blanchard, J. F., Raudonikiene, A., Shen, C., Murdin, A. D., and Brunham, R. C., 2002, Molecular detection and seroepidemiology of the *Chlamydia pneumoniae* bacteriophage (PhiCpn1), *J. Clin. Microbiol.* **40:**4010–4014.
24. Vandahl, B. B., Birkelund, S., Demol, H., Hoorelbeke, B., Christiansen, G., Vandekerckhove, J., and Gevaert, K., 2001, Proteome analysis of the *Chlamydia pneumoniae* elementary body. *Electrophoresis.* **22:**1204–1223.

25. Shaw, A., Vandahl, B. B., Larsen, M. R., Roepstorff, P., Gevaert, K., Vanderkerckhove, J., Christiansen, G., and Birkelund, S., 2002, Characterization of a secreted, *Chlamydia protease, Cell. Microbiol.* **4:**411–424.
26. Vandahl, B. B., Pedersen, A. S., Gevaert, K., Holm, A., Vandekerckhove, J., Christiansen, G., and Birkelund, S., 2002, The expression, processing and localization of polymorphic membrane proteins in *Chlamydia pneumoniae* strain CWL029, *BMC Microbiol.* **2:**36
27. s Henderson, I. R., Lam, A. C., 2001, Polymorphic proteins of *Chlamydia* spp.—autotransporters beyond the Proteobacteria, *Trends Microbiol.* **9:**573–578.
28. Vandahl, BB, Christiansen, G., and Birkelund, S., 2002, 2D-PAGE analysis of the *Chlamydia pneumoniae* outer membrane complex, in: *Chlamydial Infections,* Procedings of the 10th Symposium on Human Chlamydial Infections (Schachter *et al.*, eds.), International Chlamydia Symposium, San Francisco, CA, USA, pp. 547–550.
29. Jahnig, F., 1990, Structure predictions of membrane proteins are not that bad, *Trends Biochem. Sci.* **15:**93–95.
30. Longbottom, D., Findlay, J., Vretou, E., and Dunbar, S. M., 1998, Immunoelectron microscopic localisation of the OMP90 family on the outer membrane surface of *Chlamydia psittaci, FEMS Microbiol. Lett.* **164:**111–117.
31. Bradley, P., Cowen, L., Menke, M., King, J., and Berger, B., 2001, BETAWRAP: successful prediction of parallel beta-helices from primary sequence reveals an association with many microbial pathogens. *Proc. Natl. Acad. Sci. U.S.A.* **98:**14819–14824.
32. Birkelund, S., Christiansen, G., Vandahl, B., and Pedersen, A. S., 2002, Are the Pmp proteins parallel beta-helices? in: *Chlamydial Infections,* Proceedings of the 10th Symposium on Human Chlamydial Infections, (Schachter *et al.*, eds.) International Chlamydia Symposium, San Francisco, CA, pp. 551–554.
33. Hsia, R. C., Pannekoek, Y., Ingerowski, E., and Bavoil, P. M., 1997, Type III secretion genes identify a putative virulence locus of Chlamydia, *Mol. Microbiol.* **25:**351–359.
34. Bavoil, P. M., and Hsia, R. C., 1998, Type III secretion in Chlamydia: A case of deja vu? *Mol. Microbiol.* **28:**860–862.
35. Rockey, D. D., Lenart, J., and Stephens, R. S., 2000, Genome sequencing and our understanding of chlamydiae, *Infect. Immun.* **68:**5473–5479.
36. Rockey, D. D., Heinzen, R. A., and Hackstadt, T., 1995, Cloning and characterization of a *Chlamydia psittaci* gene coding for a protein localized in the inclusion membrane of infected cells, *Mol. Microbiol.* **15:**617–626.
37. Bannantine, J. P., Rockey, D. D., Hackstadt, T., 1998, Tandem genes of *Chlamydia psittaci* that encode proteins localized to the inclusion membrane, *Mol. Microbiol.* **28:**1017–1026.
38. Subtil, A., Parsot, C., and Dautry-Varsat, A., 2001, Secretion of predicted Inc proteins of *Chlamydia pneumoniae* by a heterologous type III machinery, *Mol. Microbiol.* **39:**792–800.
39. Zhong, G., Fan, P., Ji, H., Dong, F., and Huang, Y., 2001, Identification of a chlamydial protease-like activity factor responsible for the degradation of host transcription factors, *J. Exp. Med.* **193:**935–942.
40. Fan, P., Dong, F., Huang, Y., and Zhong, G., 2002, *Chlamydia pneumoniae* secretion of a protease-like activity factor for degrading host cell transcription factors required for major histocompatibility complex antigen expression, *Infect. Immun.* **70:**345–349.
41. Birkelund, S., Johnsen, H., and Christiansen, G., 1994, *Chlamydia trachomatis* serovar L2 induces protein tyrosine phosphorylation during uptake by HeLa cells, *Infect. Immun.* **62:**4900–4908.
42. Clausen, J. D., Christiansen, G., Holst, H. U., and Birkelund, S., 1997, *Chlamydia trachomatis* utilizes the host cell microtubule network during early events of infection, *Mol. Microbiol.* **25:**441–449.
43. Wolf, K.. Hackstadt, T., 2001, Sphingomyelin trafficking in *Chlamydia pneumoniae*–infected cells, *Cell. Microbiol.* **3:**145–152.
44. Hackstadt, T., 2000, Redirection of host vesicle trafficking pathways by intracellular parasites, *Traffic* **1:**93–99.

45. Hackstadt, T., Fischer, E. R., Scidmore, M. A., Rockey, D. D., and Heinzen, R. A., 1997, Origins and functions of the chlamydial inclusion, *Trends Microbiol.* **5:**288–293.
46. Shaw, A. C., Christiansen, G., Roepstorff, P., and Birkelund, S., 2000, Genetic differences in the *Chlamydia trachomatis* tryptophan synthase alpha-subunit can explain variations in serovar pathogenesis, *Microbes Infect.* **2:**581–592.
47. Pantoja, L. G., Miller, R. D., Ramirez, J. A., Molestina, R. E., and Summersgill, J. T., 2001, Characterization of *Chlamydia pneumoniae* persistence in HEp-2 cells treated with gamma interferon, *Infect. Immun.* **69:**7927–7932.

4

Antibiotic Susceptibility and Treatment of *Chlamydia pneumoniae* Infections

MARGARET R. HAMMERSCHLAG

1. INTRODUCTION

One of the major characteristics of *Chlamydia* spp. is the ability to cause prolonged, often subclinical infections. Chronic, persistent infection with *Chlamydia pneumoniae* has been implicated in the pathogenesis of several chronic diseases, initially not thought to be infectious, including asthma, arthritis, and atherosclerosis. However, studies of the association of *C. pneumoniae* and these disorders have been hampered by difficulty in diagnosing chronic, persistent infection with the organism which, in turn, makes it very difficult to determine the efficacy of antibiotic therapy.

2. *IN VITRO* SUSCEPTIBILITY OF *C. pneumoniae*

Although *in vitro* susceptibility testing of *C. pneumoniae* is not standardized, results from different laboratories have been remarkably similar despite use of different cell lines.[1] *C. pneumoniae* is susceptible to a wide range of antimicrobial agents, specifically those that interfere with protein and DNA synthesis such as tetracyclines, macrolides and quinolones (Table I). *C. pneumoniae*, like *C. psittaci* and *C. pecorum*, but unlike *C. trachomatis*, is constitutively resistant to sulfonamides. Although *Chlamydia* lacks peptidoglycan and muramic acid, β-lactam antibiotics, specifically penicillin, amoxicillin, and mecillinam have activity *in vitro*.[1,2] Amoxicillin has been found to be as effective as erythromycin for treatment of genital *C. trachomatis* infection in pregnancy.[3] This is often

MARGARET R. HAMMERSCHLAG • Department of Pediatrics, SUNY Downstate Medical Center, Brooklyn, New York 11203-2098.

TABLE I
Comparative *In Vitro* Activities of Various Antimicrobials Against *C. pneumoniae*[a]

Antimicrobial agent	MIC range (μg/l)
Doxycycline	0.015–0.25
Minocycline	0.03–0.06
Tigecycline (GAR-936)	0.06–0.125
Erythromycin	0.015–0.25
Roxithromycin	0.06–2
Azithromycin	0.05–0.25
Clarithromycin	0.004–0.03
Telithromycin	0.03–2
Cethromycin (ABT-773)	0.008–0.015
Ciprofloxacin	1–4
Ofloxacin	0.5–2
Levofloxacin	0.25–1
Moxifloxacin	0.125–1
Gemifloxacin	0.06–0.25
Gatifloxacin	0.125–0.25
Garenoxacin (BMS-284756)	0.015–0.03
Rifampin	0.0075–0.03
Rifalazil (ABI-1648)	0.00125–0.0025
NVP-PDF386 (VRC4887)	0.008–0.015

[a]From refs. 1, 7, 9–15, 17, 20, and 23.

termed the chlamydial anomaly.[(4)] *Chlamydiae* have three penicillin-binding proteins (PBPs).[(4,5)] Recent genomic studies have revealed that *Chlamydiae* have the genetic apparatus for synthesis of both peptidogylcan and muramic acid;[(6)] thus PBPs may be remnants of a time when the organism was able to synthesize these cell wall components. However, cephalosporins do not have any appreciable antichlamydial activity *in vitro*.[(7)]

Tetracyclines and macrolides are the antibiotics used most frequently to treat *C. pneumoniae* infections in humans.[(1)] Tetracyclines, including doxycycline and minocycline, which are very active *in vitro* with MICs ranging from 0.015 to 0.25 μg/ml, are now being used. Tigecycline (GAR-936), a novel glycylcycline tetracycline derivative, is currently in Phase 3 clinical trials. This drug was found to have activity *in vitro* against *C. pneumoniae* similar to doxycycline.[(7)]

Macrolides are thought to act against *C. pneumoniae* as other bacteria, by binding to 23S rRNA.[(8)] Macrolides and their derivatives, azalides and ketolides, have a wide range of activity against *C. pneumoniae*. Clarithromycin is the most active macrolide *in vitro* with MICs 0.004–0.03 μg/ml.[(1,9–11)] Azithromycin, which has probably become the most frequently used drug in the class for *C. pneumoniae* infections because of its tissue penetration and pharmacokinetics, has *in vitro* activity very similar to erythromycin.[(1,10–12)] Ketolides are a new class of macrolide derivatives. They have a number of unique structural features that differentiate it from the 14-membered ring macrolides, including a 3-keto function in place of the L-cladinose and an 11,12-carbamate-substituted side chain. Telithromycin has activity similar to clarithromycin, but there was wide interisolate variation in susceptibility, with recent clinical isolates of *C. pneumoniae* being more susceptible that older isolates that had been passaged more intensively *in vitro*.[13,14] Cethromycin is the most active drug in this class with an MIC_{90} of 0.015 μg/ml.[15]

Extrapolating from studies of *C. trachomatis*, DNA gyrase is probably the primary target for quinolones in *C. pneumoniae*.[16] The older quinolones, including ciprofloxacin and levofloxacin, are less potent *in vitro* against *C. pneumoniae* than are macrolides and tetracyclines.[1,17] Newer compounds, specifically the desfluoro-quinolone garenoxacin (BMS-284756), have activity that is comparable to macrolides with an MIC_{90} and MBC_{90} of 0.015 μg/ml.[18]

The most active antibiotics *in vitro* against *C. pneumoniae* are rifamycin derivatives, including rifalazil and its derivative compounds.[19,20] Rifampin and other rifamycins have been known for over 30 years to be very active *in vitro* against *C. trachomatis* and *C. pneumoniae*, with MICs ranging from 0.0075 to 0.03 μg/ml.[19–22] However, as early as 1973, Keshisyan *et al.*[21] demonstrated rapid single-step emergence of resistance to rifampin in *C. trachomatis* in eggs. Similar results were also obtained in tissue culture.[22] The risk of emergence of resistance does not occur with every rifamycin compound. Treharne *et al.*[22] found that although *C. trachomatis* rapidly developed resistance to rifampin after serial passage in subinhibitory concentrations, the organism remained susceptible to rifabutin even after 10 serial passages. There are no data on *in vitro* emergence of resistance in either *C. trachomatis* or *C. pneumoniae* with rifalazil or its derivative compounds, ABI-1657 and ABI-1131.

Peptide deformylase (PDF), a necessary enzyme in bacterial protein synthesis, is a novel target for antibacterial chemotherapy.[23] The gene encoding PDF (*def*) is essential for protein synthesis in a variety of pathogenic bacteria, including *Chlamydiae*, but is not required for mammalian protein synthesis. Preliminary *in vitro* data have demonstrated that one candidate compound, NVP-PDF386 (VRC4887), was very active against *C. pneumoniae*, with an MIC_{90} of 0.008 μg/ml.[23] Clinical trials of several PDF inhibitors have been undertaken.

2.1. Is There Antimicrobial Resistance in *C. pneumoniae*?

Although, antimicrobial resistance of *C. trachomatis* has been described, it has not as yet been described for *C. pneumoniae* in vivo. We previously reported three clinical isolates of *C. pneumoniae*, from two patients with pneumonia, who developed a 4-fold increase in MICs after treatment with azithromycin.[12] Phenotypic resistance to macrolides has been found in *C. trachomatis* from patients who failed therapy.[24] Tetracycline resistance has also been described in *C. trachomatis* and *C. suis*.[24,25] Dessus-Babus *et al.*[16] were able to induce high-level resistance to ofloxacin and sparfloxacin in *C. trachomatis* after four passages in subinhibitory concentrations (½ MIC) of the drugs. In a subsequent study, Morrisey *et al.*[26] were able to induce resistance to ofloxacin and ciprofloxacin in *C. trachomatis*, but not in *C. pneumoniae* after more than 30 passages in subinhibitory concentrations of the drugs. Repeated passage in subinhibitory concentrations of various antibiotics, including macrolides, quinolones and beta-lactams, will reproducibly induce resistance in other bacteria, especially *Streptococcus pneumoniae.*[27] Macrolide resistance often arises by (i) mutation of the 23S rRNA at the macrolide contact site, or (ii) methylation of this site by adenine *N*-methyltransferase enzymes encoded by the *erm* gene family, or (iii) an efflux pump mechanism.[8,28] We were unable to demonstrate any mutations in two candidate resistance-determining domains of the 23S rRNA gene of three clinical isolates of *C. pneumoniae* obtained after treatment with azithromycin, even though the MICs to azithromycin and erythromycin increased 4-fold from that of the isolates obtained before treatment.[28a] A systematic comparison of other proteins involved in macrolide resistance revealed significant homologies of 3 *Staphylococcus aureus* proteins to predicted gene products in the *C. pneumoniae* genome. ErmA (BAB42746, 243 amino acids (aa)), a 23S rRNA ribosomal methylase, has 24% indentity and 42% similarity over its first 215aa to the *C. pneumoniae* dimethyadenosine transferase gene, known as the kasugamycin resistance gene (ksgA, 565 aa). Second, the macrolide efflux gene product (MefA, AAL58635, 405 aa) has homology to a *C. pneumoniae* putative efflux protein (YgeD, NP_224932.1, 565aa), within its N-terminal one-third. Remarkably, the *S. aureus* NorA multi-drug efflux protein (BAB56857, 388 aa), though not homologous to MefA, also has 23% identity and 43% similarity to YgeD, in a central region of the protein spanning 195aa. Finally, the virginiamycin resistance gene product (VgaB,U82085, 552 aa), an ATP-binding efflux pump, shares 32% identity and 53% homology over its C-terminal two-thirds with the chlamydial yjjK gene product, a putative ATP-binding transporter. Of note, the macrolide-streptogramin resistance gene products MsrC of *E. fecalis* (NP_815134) and MsrA of *S. epidermidis* (P23212) are homologous to both VgaB and YjjK in a similar region. By contrast, no significant homologies could be found in the chlamydia genomes to macrolide hydrolases (vgbA, vgbB), esterases (ereA,ereB), phosphorylases(mphA), transferases (lnuA) or other multi-drug efflux proteins (AcrAB,TolC, LmrA, NorM, Mmr, EmrB) or their regulators (AcrR, BltR, BmrR, Mta). It is still possible that these genes have been acquired via transfer of extra-chromosomal elements. We were not able to detect the presence of an *erm*-like gene, using degenerate primers, which work across a broad range of species. The mechanism underlying the increase in MIC needs to be further defined, as it may be the first step toward genuine high-level resistance in *C. pneumoniae.*

We recently investigated the effect of serial passage of two *C. pneumoniae* isolates and two serotypes of *C. trachomatis* in subinhibitory concentrations of two rifamycins, rifalazil and rifampin, on the development of phenotypic and genotypic resistance. *C. trachomatis* developed resistance to both antimicrobials within 6 passages, with higher level resistance to rifampin (128-256 μg/ml) and lower level resistance to rifalazil (0.5-1 μg/ml). In contrast, *C. pneumoniae* TW-183 developed only low level resistance to rifampin (0.25 μg/ml) and rifalazil (0.016 μg/ml) after 12 passages. Another isolate, *C pneumoniae* CWL-029, failed to develop resistance to either drug. Two unique mutations emerged in the *rpoB* gene of rifampin- (L456I) and rifalazil- (D461E) resistant *C. pneumoniae* TW-183. A single mutation (H471Y) was detected in both rifampin- and rifalazil-resistant *C. trachomatis* UW-3/Cx/D, and a unique mutation (V136F) was found in rifalazil-resistant BU-434/L_2. No mutations were detected in the entire *rpoB* gene of rifampin-resistant BU-434/L_2. This is the first description of antibiotic resistance-associated mutations in *C. pneumoniae.* Given the favorable pharmacokinetic profile of rifalazil and its derivatives, this drug may maintain physiological levels well above the increased MIC in *C. pneumoniae* for the duration of therapy, making resistance and subsequent clinical failure less likely to occur.

2.2. Treatment of Persistent *C. pneumoniae* Infection *In Vitro*

Despite the finding that many antimicrobial agents have excellent activity *in vitro,* one may not be able to readily extrapolate from these results to microbiologic efficacy *in vivo.* Methods currently used for *in vitro* susceptibility testing are not analogous to the infection as it occurs *in vivo.* Kutlin *et al.*[29,30] established an *in vitro* model of continuous *C. pneumoniae* infection in Hep-2 cells. These long-term continuously infected cells have been maintained for over 7 years without addition of new cells or chlamydia, or centrifugation or addition of cycloheximide to the media, which is required for standard culture and *in vitro* susceptibility testing of *C. pneumoniae.* Ultrastructural studies demonstrated that these cultures contain a subpopulation of aberrant inclusions (10–20%) where the Chlamydia do not appear to be actively replicating.[30] These aberrant inclusions were similar in appearance to those seen in γ-interferon-induced persistent *C. pneumoniae* infection.[31] Kutlin *et al.*[30] reported that 6 days' treatment with 4 μg/ml ofloxacin or 0.5 μg/ml azithromycin, which exceeds achievable serum levels, reduced the concentration of *C. pneumoniae* from 10^6 to 10^3 inclusion-forming units/ml, but failed to completely eradicate the organism from long-term continuously infected HEp-2 cells. In a subsequent study using the same model, 30 days' treatment with levofloxacin, azithromycin or clarithromycin, at concentrations achievable in the pulmonary epithelial lining fluid (32–500 × MIC), reduced the concentrations of *C. pneumoniae* by 4 logs but, again, failed to eliminate the organism.[32] Similar results were also obtained with 30 days' treatment with gemifloxacin.[33] Thus antibiotic treatment may not be completely effective if a proportion of chlamydia inclusions are in a persistent state, which has been hypothesized to occur in chronic *C. pneumoniae* infection.

3. TREATMENT OF *C. pneumoniae* RESPIRATORY INFECTIONS

Data are limited on the treatment of respiratory infections due to *C. pneumoniae.* Despite the fact that many antimicrobial agents have excellent activity *in vitro,* as described above, *in vitro* activity may not always predict *in vivo* efficacy. Early anecdotal data suggested that prolonged therapy (i.e., at least 2 weeks') was necessary since recrudescent symptoms have been described following 2-week courses of erythromycin and even after 30 days' tetracycline or doxycycline.[34,35] However, practically all treatment studies of *C. pneumoniae* respiratory infection, including pneumonia, bronchitis, and asthma, presented or published to date have used serology alone for diagnosis, essentially limiting themselves to a clinical endpoint. Most studies have followed this premise: if the patient has serologic evidence of infection and clinically improves, the organism is presumed to have been eradicated. In 1990, Lipsky *et al.*[36]

described four patients with bronchitis and pneumonia, treated with a 10-day course of ofloxacin, who were retrospectively identified as having serologic evidence of *C. pneumoniae* infection [4-fold rise in IgG/IgM, single IgM $\geq$ 16 or IgG $\geq$ 512 by microimmunofluorescence (MIF)]. All reportedly demonstrated marked clinical improvement. On the basis of the MICs of three laboratory isolates to ofloxacin (1–2 μg/ml), the authors concluded that ofloxacin was effective in these patients as the MICs were less than achievable serum levels. In a subsequent prospective pneumonia treatment study, Plouffe *et al.*[37] found a clinical response rate of 83% in those patients with serologic evidence of *C. pneumoniae* infection who were treated with ofloxacin compared with 75% of those who received standard therapy, which was a β-lactam antibiotic plus erythromycin or a tetracycline. Similarly, File *et al.*[38] reported a clinical cure rate of 98% among patients with serologic evidence of *C. pneumoniae* infection who were treated with levofloxacin compared with 93% of those treated with ceftriaxone and/or cefuroxime axitiil, plus erythromycin or doxycycline, which was added to the treatment regimen at the investigator's discretion. In the latter group, the response rate did not differ between those patients who had erythromycin or doxycycline added to their treatment regimen and those who were treated with a cephalosporin alone. There was also no difference in the response rate among those patients who had "definite" serologic evidence of infection, i.e. a 4-fold rise in MIF IgG or IgM, compared with those who had "probable" infection, i.e. a single IgG $\geq$ 512 or IgM $\geq$ 32. The success of the cephalosporin regimens results raise some questions about the specificity of the serologic criteria as these antibiotics have no or poor activity against *Chlamydia* spp. *in vitro.*[6] A number of published treatment studies have claimed microbiologic eradication, despite the fact that cultures were not done, including a major, recently published study comparing sequential intravenous and oral moxifloxacin compared with sequential intravenous and oral Co-amoxiclav with or without clarithromycin.[39] Conversely, the results of studies that have assessed microbiologic efficacy have frequently shown that patients improve clinically despite persistence of the organism.[12,40–43]

There are only 6 published pneumonia treatment studies that have utilized *C. pneumoniae* culture and assessed microbiologic efficacy. Block *et al.*[40] found that 10 days' treatment with erythromycin or clarithromycin suspension eradicated *C. pneumoniae* from the nasopharynx of 86 and 79% of culture-positive children, 3 through 12 years of age, with community-acquired pneumonia, respectively. All these children improved clinically despite persistence of the organism. Persistence did not appear to be related to the development of antibiotic resistance as all the isolates remained susceptible to erythromycin and clarithromycin during and after treatment.[9] The experience with azithromycin has been similar, with eradication rates of 70–83%.[12,41] Harris *et al.*[41] reported that *C. pneumoniae* was eradicated after treatment from the nasopharynx of 19 of 23 (83%) culture-positive children with community-acquired pneumonia, 6 months through 16 years of age, who received azithromycin; four of four, and seven of seven who received amoxicillin–clavulanate and erythromycin, respectively ($p = 0.9$, chi square). The MICs and minimum bactericidal concentrations (MBCs) of three of nine isolates obtained after treatment from two of seven persistently infected patients in both studies who were treated with azithromycin increased 4-fold after treatment, although they were still within the range considered susceptible to the antibiotic.[12] It is not clear if this was an isolated event or suggestive of possible development of resistance. All patients improved clinically despite persistence of the organism. In both pediatric studies, over 70% of the *C. pneumoniae* culture-positive children had no detectable antibody by MIF, and less than 5% met the serologic criteria for acute infection.[40,41]

The results of two pneumonia treatment studies in adults, which evaluated levofloxacin and moxifloxacin, found eradication rates of 70–80%.[42,43] As seen with the children, all the patients who were microbiologic failures improved or were clinical cures. The MICs and MBCs of isolates of *C. pneumoniae* from the patients who were microbiologic failures to both drugs remained the same before and after treatment. Only 4 of 18 (22.2%) of the *C. pneumoniae* culture-positive patients enrolled in the moxifloxacin study met the serologic criteria for acute infection; 6 (33.3%) were seronegative and 10 (55.6%) had stable IgG titers.[42]

Although telithromycin is now approved for treatment of respiratory infections, studies of this drug for treatment of community-acquired pneumonia have only used serology for diagnosis of infections due to *C. pneumoniae*. There is one preliminary dose-ranging study of cethromycin (ABT-773) for this indication.[44] Patients with community-acquired pneumonia were enrolled in a randomized study comparing cethromycin at a dose of 150 mg q.d. with 150 mg bid, p.o. for 10 days. *C. pneumoniae* was eradicated from the nasopharynx of 10 of 10 (100%) microbiologically evaluable patients. *In vitro* susceptibility testing of 13 isolates of *C. pneumoniae* from 12 patients obtained before and after therapy was performed. The MIC_{90}s and MBC_{90}s for cethromycin were 0.015 μg/ml, which was identical to prior *in vitro* studies.[15]

Data on efficacy of antibiotic treatment of other respiratory infections due to *C. pneumoniae*, specifically asthma, are also limited. Only one study assessed microbial efficacy by culture. Emre *et al.*[45] isolated *C. pneumoniae* from the nasopharynx of 13 (11%) of 118 children presenting with acute exacerbations of reactive airway disease. Six of these children had positive cultures on multiple occasions, ranging from 1 to 6 months. Twelve of the culture-positive children were treated with either 14 days of erythromycin or 10 days of clarithromycin; all eventually became culture negative, although six required two to three courses of therapy. Nine (75%) of the treated children demonstrated significant improvement of their asthma, as demonstrated by improved pulmonary function tests and decreased requirements for medication, after eradication of their *C. pneumoniae* infection. As described above for the pneumonia studies, there was a very poor correlation between MIF serology and isolation of *C. pneumoniae* in these children, almost 60% had no detectable MIF IgG or IgM, including one child who was culture-positive five times over a 6-month period. Only three children met the serologic criteria for acute infection.

Two recent studies in adults have also attempted to assess the efficacy of antibiotics for the treatment of reactive airway disease.[46,47] Kraft *et al.*[46] obtained bronchoalveloar lavage specimens for culture and PCR of *C. pneumoniae* and *Mycoplasma pneumoniae* from 55 adults with stable asthma. The patients were then randomized to receive either clarithromycin, 500 mg bid or placebo for 6 weeks. *C. pneumoniae* was not isolated from any of the patients by culture, although seven (12.7%) were PCR positive. Treatment with clarithromycin was found to improve lung function, but only in those subjects who were PCR positive for *C. pneumoniae* or *M. pneumoniae* (>55% of patients). It was difficult to assess the microbiologic efficacy of therapy as data on the *C. pneumoniae* and *M. pneumoniae* were not presented separately. Of the 26 patients in the treatment arm, 15 were PCR positive for either organism at baseline, 9 became PCR negative after treatment, 5 remained PCR positive, and 2 patients who were PCR negative at baseline became PCR positive. Aside from the fact that we do not know if PCR can be used to assess the efficacy of antibiotic treatment, *C. pneumoniae* PCR is not standardized and there can be substantial problems with inter and intralaboratory reproducibility.(48) Another explanation for the observed effect of clarithromycin on pulmonary function in this study could be the antiinflammatory effect of macrolides, which is independent of their antimicrobial activity.(49)

In a larger multinational study, Black *et al.*(47) compared the effect of 6 weeks' treatment with roxithromycin, 150 mg, bid compared with placebo on pulmonary function and anti–*C. pneumoniae* IgG and IgA titers by MIF. A total of 232 subjects, 18 to 60 years of age who had anti–*C. pneumoniae* IgG $\geq$ 64 and/or IgA $\geq$ 64, were enrolled in four countries. Cultures for *C. pneumoniae* were not performed. The investigators found an initial beneficial effect of roxithromycin on pulmonary function tests that was not sustained beyond 3 months after treatment. The investigators hypothesized that the lack of sustained benefit was due to failure to eradicate *C. pneumoniae* from the subjects; however, this is really impossible to assess as cultures were not done, thus we really do not know how many patients were actually infected. As in the previous study, a more likely explanation is that any benefit seen was probably secondary to the antiinflammatory action of roxithromycin.(49)

4. RELEVANCE FOR CARDIAC CLINICAL TREATMENT INTERVENTION STUDIES

There are currently at least 11 prospective, randomized antibiotic intervention studies of coronary artery disease that have been either published or are still ongoing.(50) All have based the diagnosis of *C. pneumoniae* on a variety of serologic criteria. Preliminary results of four published studies have been conflicting, but most have not demonstrated an effect.(50) Azithromycin achieved levels ranging from 0.2 to 4 μg/ml in atherosclerotic plaque after dosages of either 500 or 1000 mg/day for 3 days in patients undergoing endarterectomy.(51) The third dose was given 4 h before the procedure. Although these concentrations exceed the MIC, it would be practically impossible to document the eradication of *C. pneumoniae* from a vascular focus. Melissano *et al.*(52) treated 32 patients with either roxithromycin, 150 mg twice a day, or placebo for 17–35 days before undergoing carotid endarterectomy. Using a seminested PCR, *C. pneumoniae* DNA was detected in plaques of 5 of 16 (31.3%) treated patients compared with 12 of 16 (75%) who received placebo ($p = .034$). The authors claimed that these results demonstrated that treatment with roxithromycin was effective in eradicating *C. pneumoniae* from the plaques. However, as the pretreatment status was unknown and given the problems observed with performance of

PCR, one cannot really assume that the lower prevalence of *C. pneumoniae* DNA in the treatment group was due to successful eradication. However, Geiffers *et al.*(53) reported that treatment with azithromycin, 500 mg/day for 3 days, followed by an additional 500-mg dose on either Day 14, 15, or –16, was ineffective in eliminating *C. pneumoniae*, as determined by culture, from peripheral blood monocytes in two patients with unstable angina.

Reasons for failure to find an effect of antibiotic treatment on adverse cardiac outcome in the initial intervention studies include low power due to small numbers, inadequate dosing and duration of treatment, or poor positive predictive value of *C. pneumoniae* serology.(50) In terms of dosage and duration of treatment, none of the antibiotic regimens used or currently in use has been demonstrated to be effective in the eradication of *C. pneumoniae* from the respiratory tract or any other site. Ironically, the only study that found a significant reduction in adverse cardiac outcome used an azithromycin regimen, 500 mg/day for 3 or 6 days,(27) that was the least likely to be effective, as 1.5 g azithromycin over 5 days had only 70% efficacy in eradicating *C. pneumoniae* from the nasopharynx of adults with pneumonia.(54) Furthermore, the weekly dosing regimens of azithromycin used in the published and ongoing intervention studies may result in long periods of subinhibitory levels of drug, which may lead to development of resistance, not only in *C. pneumoniae* but in other respiratory bacteria, as described earlier.(27) None of these studies are monitoring the effect on respiratory flora. Results of studies of long-term continuously infected cells found that 30 days' treatment with 4 μg/ml azithromycin, the concentration achievable in the pulmonary epithelial lining fluid, failed to eradicate *C. pneumoniae*.(32) Extrapolating from these data, one would also expect that these weekly azithromycin regimens would probably not be effective in eradicating a chronic vascular *C. pneumoniae* infection, if present. Finally, there are no reliable serologic markers for chronic or persistent *C. pneumoniae* infection.(55) In reality, this means that we really cannot determine who is infected and who is not. It also means that we cannot assume that any effect seen is due to successful treatment or eradication of *C. pneumoniae*.

REFERENCES

1. Hammerschlag, M. R., 1994, Antimicrobial susceptibility and therapy of infections due to *Chlamydia pneumoniae*, *Antimicrob. Agents. Chemother.* **38:**1873–1878.
2. Storey, C., and Chopra, I., 2001, Affinities of β-lactams for penicillin binding proteins of *Chlamydia trachomatis* and their antichlamydial activities, *Antimicrob. Agents. Chemother.* **45:**303–305.
3. Ghuysen, J.-M., and Goffin, C., 1999, Lack of cell wall peptidoglycan versus penicillin sensitivity: New insights into the chlamydial anomaly, *Antimicrob. Agents. Chemother.* **43:**2339–2341.
4. Turrentine, M. A., and Newton, E. R., 1995, Amoxicillin or erythromycin for the treatment of antenatal chlamydial infection: A meta-analysis, *Obstet. Gynecol.* **86:**1021–1025.
5. Rockey, D. D., Lenart, J., and Stephens, R. S., 2000, Genome sequencing and our understanding of chlamydiae, *Infect. Immun.* **68:**5473–5479.
6. Hammerschlag, M. R., and Gleyzer, A., 1983, The *in vitro* activity of a group of broad spectrum cephalosporins and other beta-lactam antibiotics against *Chlamydia trachomatis*, *Antimicrob. Agents Chemother.* **23:**492–493.

7. Roblin, P. M., and Hammerschlag, M. R., 2000, *In vitro* activity of GAR-936 against *Chlamydia pneumoniae* and *Chlamydia trachomatis, Int. J. Antimicrob. Agents* **16:**61–63.
8. Vester, B., and Douthwaite, S., 2001, Macrolide resistance conferred by base substitutions in 23S rRNA, *Antimicrob. Agents Chemother.* **45:**1–12.
9. Roblin, P. M., Montalban, G., and Hammerschlag, M. R., 1994, Susceptibility to clarithromycin and erythromycin of isolates of *Chlamydia pneumoniae* from children with pneumonia, *Antimicrob. Agents Chemother.* **38:**1588–1589.
10. Welsh, L., Gaydos, C., and Quinn, T. C., 1996, *In vitro* activities of azithromycin, clarithromycin and tetracycline against 13 strains of *Chlamydia pneumoniae, Antimicrob. Agents Chemother.* **40:**212–214.
11. Kuo, C. C., Jackson, L. A., Lee, A., and Grayston, J. T., 1996, *In vitro* activities of azithromycin, clarithromycin and other antibiotics against *Chlamydia pneumoniae, Antimicrob. Agents Chemother.* **40:**2669–2670.
12. Roblin, P. M., and Hammerschlag, M. R., 1998, Microbiologic efficacy of azithromycin and susceptibility to azithromycin of isolates of *Chlamydia pneumoniae* from adults and children with community acquired pneumonia, *Antimicrob. Agents Chemother.* **42:**194–196.
13. Roblin, P. M., and Hammerschlag, M. R., 1998, *In vitro* activity of a new ketolide antibiotic, HMR 3647, against *Chlamydia pneumoniae, Antimicrob. Agents Chemother.* **42:**1515–1516.
14. Miyashita, N., Fukano, H., Niki, Y., and Matshishima, T., 2001, *In vitro* activity of telithromycin, a new ketolide, against *Chlamydia pneumoniae, J. Antimicrob. Chemother.* **48:**403–405.
15. Strigl, S., Roblin, P. M., Reznik, T., and Hammerschlag, M. R., 2000, *In vitro* activity of ABT 773, a new ketolide antibiotic, against *Chlamydia pneumoniae, Antimicrob. Agents Chemother.* **44:**1112–1113.
16. Dessus-Babus, S., Bebear, C. M., Charron, A., Bebear, C., and de Barbeyrac, B., 1998, Sequencing of gyrase and topoisomerase IV quinolone-resistance-determining regions of *Chlamydia trachomatis* and characterization of quinolone-resistant mutants obtained *in vitro, Antimicrob. Agents Chemother.* **42:**2447–2481.
17. Hammerschlag, M. R., 2000, Activity of gemifloxacin and other new quinolones against *Chlamydia pneumoniae*: A review, *J. Antimicrob. Chemother.* **45**(Suppl. S1)**:**35–39.
18. Malay, S., Roblin, P. M., Reznik T., Kutlin, A., and Hammerschlag, M. R., 2002, *In vitro* activity of BMS-28476 against *Chlamydia trachomatis* and recent clinical isolates of *Chlamydia pneumoniae, Antimicrob. Agents Chemother.* **46:**517–518.
19. Jones, R. B., Ridgway, G. L., Boulding, S., and Hunley, K. L., 1983, *In vitro* activity of rifamycins alone and in combination with other antibiotics against *Chlamydia trachomatis, Rev. Infect. Dis.* **5:**S556–S561.
20. Roblin, P. M., Reznik, T., Kutlin, A., Hammerschlag, M. R., 2003, *In vitro* activity of rifamycin derivatives ABI-1648 (Rifalazil, KRM-1648), ABI-1657 and ABI-1131 against *Chlamydia trachomatis* and recent clinical isolates of *Chlamydia pneumoniae, Antimicrob. Agents Chemother.*, **47:** 1135–1136.
21. Keshishyan, H., Hanna, L., and Jawetz, E., 1973, Emergence of rifampin-resistance in *Chlamydia trachomatis, Nature* **244:**173–174.
22. Treharne, J. D., Yearsley, P. J., and Ballard, R. C., 1989, *In vitro* studies of *Chlamydia* trachomatis susceptibility and resistance to rifampin and rifabutin, *Antimicrob. Agents Chemother.* **33:**1393–1394.
23. Roblin, P. M., and Hammerschlag, M. R., 2003, *In vitro* activity of a novel new antibiotic, NVP-PDF386 (VRC4887), against *Chlamydia pneumoniae, Antimicrob. Agents Chemother.*, **47:** 1447–1448.
24. Somani, J., Bhullar, V. B., Workowski, K. A., Farshy, C. E., and Black, C. M., 2000, Multiple drug-resistant *Chlamydia trachomatis* associated with clinical treatment failure, *J. Infect. Dis.* **181:**1421–1427.
25. Lenart, J., Anderson, A. A., and Rockey, D. D., 2001, Growth and development of tetracycline-resistant *Chlamydia suis, Antimicrob. Agents Chemother.* **45:**2198–2203.

26. Morrisey, I., Salman, H., Bakker, S., Farrell, D., Bebear, C. M., and Ridgeway, G., 2002, Serial passage of *Chlamydia* spp. in sub-inhibitory fluoroquinolone concentrations, *J. Antimicrob. Chemother.* **49:**757–761.
27. Pankuch, G. A., Jueneman, S. A., Jacobs, M. R., and Appelbaum, P. C., 1998, *In vitro* selection of resistance to four β-lactams and azithromycin in *Streptococcus pneumoniae*, *Antimicrob. Agents Chemother.* **42:**2914–2918.
28. Roberts, M. C., Sutcliffe, J., Courvalin, P., Jensen, L. B., Rood, J., and Seppala, H., 1999, Nomenclature for macrolide and macrolide-lincosamide-streptogramin B resistance determinants, *Antimicrob. Agents Chemother.* **43:**2823–2830.
28a. Riska, P.F., Kutlin, A., Ajiboye, P., Cua, A., Roblin, P.M., and Hammerschlag, M. R., 2004. Genetic and culture-based approached for detecting macrolide resistance in *Chlamydia pneumoniae*. *Antimicrob. Agents Chemother*, in Press.
29. Kutlin, A., Roblin, P. M., and Hammerschlag, M. R., 1999, *In vitro* activity of azithromycin and ofloxacin against *Chlamydia pneumoniae* in a continuous infection model, *Antimicrob. Agents Chemother.* **43:**2268–2272.
30. Kutlin, A., Flegg, C., Stenzel, D., Reznik, T., Roblin, P. M., Mathews, S., Timms, P., and Hammerschlag, M. R., 2001, Ultrastructural study of Chlamydia pneumoniae in a continuous infection model, *J. Clin. Microbiol.* 39:3721–3723.
31. Summersgill, J. T., Sahney, N. N., Gaydos, C. A., Quinn, T. C., and Ramirez, J. A., 1995, Inhibition of *Chlamydia pneumoniae* growth in HEp-2 cells pretreated with gamma-interferon and tumor necrosis factor alpha, *Infect. Immun.* **63:**2801–2303.
32. Kutlin, A., Roblin, P. M., and Hammerschlag, M. R., 2002, Effect of prolonged treatment with azithromycin, clarithromycin and levofloxacin on *Chlamydia pneumoniae* in a continuous infection model, *Antimicrob. Agents Chemother.* **46:**409–412.
33. Kutlin, A., Roblin, P. M., and Hammerschlag, M. R., 2002, Effect of gemifloxacin on viability of *Chlamydia pneumoniae* (*Chlamydophila pneumoniae*) in an *in vitro* continuous infection model, *J. Antimicrob. Chemother.* **49:**763–767.
34. Grayston, J. T., Campbell, L. A., Kuo, C. C., Mordhorst, C. H., Saikku, P., Thom, D. H., and Wang, S. P., 1990, A new respiratory tract pathogen: *Chlamydia pneumoniae* strain, TWAR, *J. Infect. Dis.* **161:**618–625.
35. Hammerschlag, M. R., Chirgwin, K., Roblin, P. M., Gelling, M., Dumornay, W., Mandel, L., Smith, P., and Schachter, J., 1992, Persistent infection with *Chlamydia pneumoniae* following acute respiratory illness, *Clin. Infect. Dis.* **14:**178–182.
36. Lipsky, B. A., Tack, K. J., Kuo, C. C., Wang, S. P., and Grayston, J. T., 1990, Ofloxacin treatment of *Chlamydia pneumoniae* (strain TWAR) lower respiratory tract infections, *Am. J. Med.* **89:**722-724.
37. Plouffe, J. F., Herbert, M. T., File, T. M., Baird, I., Parsons, J. N., Kahn, J. B., and Reilly-Gauvin, K. T., 1996, Ofloxacin versus standard therapy in treatment of community-acquired pneumonia requiring hospitalization, *Antimicrob. Agents Chemother.* **40:**1175–1179.
38. File, T. M., Segreti, J., Dunbar, L., Player, R., Kohler, R., Williams, R. R., Kojak, C., and Rubin, A., 1997, A multicenter, randomized study comparing the efficacy and safety of intravenous and/or oral levofloxacin versus ceftriaxone and/or cefuroxime axetil in the treatment of adults with community-acquired pneumonia, *Antimicrob. Agents Chemother.* **41:**1965–1972.
39. Finch, R., Schurmann, D., Collins, O., Kubin, R., McGivern, J., Bonnaers, H., Izquierdo, J. L., Nikolaides, P., Ogundare, F., Raz, R., Zuck, P., and Hoeffken, G., 2002, Randomized controlled trial of sequential intravenous (i.v.) and oral moxifloxacin compared with sequential i.v. and oral Co-amoxiclav with or without clarithromycin in patients with community-acquired pneumonia requiring initial parenteral treatment, *Antimicrob. Agents Chemother.* **46:**1746–1754.
40. Block, S. J., Hedrick, J., Hammerschlag, M. R., Cassell, G. H., and Craft, C., 1995, *Mycoplasma pneumoniae* and *Chlamydia pneumoniae* in pediatric community-acquired

pneumonia: Comparative efficacy and safety of clarithromycin vs. erythromycin ethylsuccinate, *Pediatr. Infect. Dis. J.* **14:**471–477.

41. Harris, J.-A., Kolokathis, A., Campbell, M., Cassell, G. H., and Hammerschlag, M. R., 1998, Safety and efficacy of azithromycin in the treatment of community acquired pneumonia in children, *Pediatr. Infect. Dis. J.* **17:**865–871.
42. Hammerschlag, M. R., and Roblin, P. M., 2000, Microbiologic efficacy of moxifloxacin for the treatment of community-acquired pneumonia due to *Chlamydia pneumoniae, Int. J. Antimicrob. Agents* **15:**149–152.
43. Hammerschlag, M. R., and Roblin, P. M., 2000, Microbiological efficacy of levofloxacin for treatment of community-acquired pneumonia due to *Chlamydia pneumoniae, Antimicrob. Agents Chemother.* **44:**1409.
44. Hammerschlag, M. R., Reznik, T., Roblin, P. M., Ramirez, J., Summersgill, J., and Bukofzer, S., 2003, Microbiologic efficacy of ABT-773 (Cethromycin) for the treatment of community-acquired pneumonia due to *Chlamydia pneumoniae, J. Antimicrob. Chemother.*, **51**: 1025–1028.
45. Emre, U., Roblin, P. M., Gelling, M., Dumornay, W., Rao, M., Hammerschlag, M. R., and Schachter, J., 1994, The association of *Chlamydia pneumoniae* infection and reactive airway disease in children, *Arch. Pediatr. Adolesc. Med.* **148:**727–731.
46. Kraft, M., Cassell, G. H., Pak, J., and Martin, R. J., 2002, *Mycoplasma pneumoniae* and *Chlamydia pneumoniae* in asthma: Effect of clarithromycin, *Chest* **121:**1782–1788.
47. Black, P. N., Blasi, F., Jenkins, C. R., Scicchitano, R., Mills, G. D., Rubenfeld, A. R., Ruffin, R. E., Mullins, P. R., Dangain, J., Cooper, B. C., Bem David, D., and Allegra, L., 2002, Trial of roxithromycin in subjects with asthma and serological evidence of infection with *Chlamydia pneumoniae, Am. J. Respir. Crit. Care Med.* **164:**536–541.
48. Apfalter, P., Blasi, F., Boman, J., Gaydos, C. A., Kundi, M., Maass, M., Makristathis, A., Meijer, A., Nadrchal, R., Persson, K., Rotter, M. L., Tong, C. Y. W., Stanek, G., and Hirschl, A. M., 2001,Multicenter comparison trial of DNA extraction methods and PCR assays for detection of *Chlamydia pneumoniae* in endarterectomy specimens, *J. Clin. Microbiol.* **39:**519–524.
49. Scaglione, F., and Rossoni, G., 1998, Comparative anti-inflammatory effects of roxithromycin, azithromycin and clarithromycin, *J. Antimicrob. Chemother.* **41** (Suppl. B)**:**47–50.
50. Boman, J., and Hammerschlag, M. R., 2002, *Chlamydia pneumoniae* and atherosclerosis—A critical assessment of diagnostic methods and the relevance to treatment studies, *Clin. Microbiol. Rev.* **15:**1–20.
51. Schneider, C. A., Diedrichs, H., Reidel, K.-D., Zimmerman, T., and Hopp, H.-S., 2000, *In vivo* uptake of azithroycin in human coronary plaques, *Am. J. Cardiol.* **86:**89–791.
52. Melissano, G., Blasi, F., Esposito, G., Tarsia, P., Dordoni, L., Arosio, C., Tshomba, Y., Fagetti, L., Allegra, L., and Chiesa, R., 1999, *Chlamydia pneumoniae* eradication from carotid plaques. Results of an open, randomised treatment study, *Eur. J. Vasc. Endovasc. Surg.* **18:**355–335.
53. Geiffers, J., Fullgraf, H., Jahn, J., Klinger, M., Dalhoff, K., Katus, H. A., Solbach, W., and Maass, M., 2001, *Chlamydia pneumoniae* infection in circulating human monocytes is refractory to antibiotic treatment, *Circulation* **103:**351–356.
54. Gupta, S., Leatham, E. W., Carrington, D., Mendall, M. A., Kaski, J. C., and Camm, A. J., 1997, Elevated *Chlamydia pneumoniae* antibodies, cardiovascular events, and azithromycin in male survivors of myocardial infarction, *Circulation* **96:**404–407.
55. Dowell, S. F., Boman, J., Carlone, G. M., Fields, B. S., Guarner, J., Hammerschlag, M. R., Jackson, L. A., Kuo, C. C., Maass, M., Messmer, T. O., Peeling, R. W., Talkington, D., Tondella, M. L., Zaki, S. R., and the *C. pneumoniae* workshop (2000) participants, 2001, Standardizing *Chlamydia pneumoniae* assays: Recommendations from the Centers for Disease Control and Prevention (USA), and the Laboratory Centre for Disease Control (Canada), *Clin. Infect. Dis.* **33:**492–503.

5

Pneumonia Caused by *Chlamydia pneumoniae*

FRANCESCO BLASI, ROBERTO COSENTINI, PAOLO TARSIA, and LUIGI ALLEGRA

1. INTRODUCTION

In 1989, the previously labelled *Chlamydia* strain TWAR was recognised as a third species of the *Chlamydia* genus on the basis of ultrastructural and DNA homology analysis and named *Chlamydia pneumoniae.*[1]

Like other *Chlamydia,* this agent is an obligate intracellular, Gram-negative bacterium present in two developmental forms: infective elementary bodies (EB) and reproductive reticulate bodies (RB). *Chlamydia* possess a specific replication cycle that differs from conventional bacteria. They multiply within membrane-bound vacuoles in eucaryotic host cells but are unable to generate adenosine triphosphate (ATP) and are therefore dependant on the host cell ATP deposits for all energy requirements. Moreover, they are incapable of *de novo* nucleotide biosynthesis and are dependent on host nucleotide pools.

Chlamydia pneumoniae has been recognised as a cause of respiratory tract infections and is considered the most common nonviral intracellular human respiratory pathogen. *C. pneumoniae* is involved in a wide spectrum of respiratory infections of the upper respiratory tract (pharyngitis, sinusitis, and otitis) and lower respiratory tract [acute bronchitis, exacerbations of chronic bronchitis and asthma, and community-acquired pneumonia (CAP)] in both immunocompetent and immunocompromised hosts.

FRANCESCO BLASI, PAOLO TARSIA and LUIGI ALLEGRA • Institute of Respiratory Medicine, Università degli Studi di Milano, IRCCS Ospedale Maggiore di Milano, Milan, Italy. ROBERTO COSENTINI • Department of Emergency Medicine, IRCCS Ospedale Maggiore di Milano, Università degli Studi di Milano, Milan, Italy.

TABLE I
Common Signs and Symptoms of *C. pneumoniae* Pneumonia and Frequency of Presentation[1,3–9]

Signs and symptoms	Frequency of presentation (%)
Dry cough	75–90
Sore throat	70–80
Hoarseness	65–75
Headache	25–60
Fever (>37.8°C)	25–45
Abnormal breath sounds	65–78
Leukocytes >10,000/mm^3	15–25
Sedimentation rate >15 mm/h	50–70

2. EPIDEMIOLOGY

C. pneumoniae accounts for 6–20% of CAP in adults, depending on several factors such as setting of the studied population, age group examined, and diagnostic methods used.[2–5] The clinical course may vary from a mild, self-limiting illness to a severe form of pneumonia, particularly in elderly patients, and in patients with coexisting cardiopulmonary diseases.[6–8]

This agent is present as part of a coinfection involving other bacterial agents in approximately 30% of cases.[4] Presenting symptoms most frequently reported by patients with *C. pneumoniae* pneumonia are sore throat and hoarseness (Table I).[5] After a period of up to a week, dry persistent cough often sets in.[9] Body temperature is generally slightly increased, seldom going higher than 38–39°C. Fever may be often missed if the patient is not seen early in the course of infection.

Physical examination does not often show abnormalities and, if present, physical findings are generally not specific. Pulmonary rales, ronchi, or signs of pulmonary consolidation are sometimes found.

Chest X-ray generally reveals small pulmonary infiltrates, sublobar or segmental at presentation (Table II). Multiple infiltrates may sometimes be seen and are often bilateral. Extensive lobar involvement is uncommon, whereas pleural effusion may be present in up to 20% of cases.

TABLE II
Most Commonly Occurring Radiographic Presentations of *C. pneumoniae* Pneumonia[1,5–7]

Radiographic involvement	Frequency of presentation (%)
Normal	0–6
Single sublobar lesion	70–85
Lobar	4–7
Bilateral involvement	7–14

TABLE III
Incidence of *C. pneumoniae* Infection in Community-Acquired Pneumonia in Children

Reference	Number of patients	Type of patients	Methods	Incidence of *C. pneumoniae* (%)
Block *et al.*, 1995[20]	260	Outpatients	Culture and serology	28 (13% by culture)
Harris *et al.*, 1998[21]	456	Outpatients	Serology and culture	7
Heiskanen-Kosma *et al.*, 1998[10]	201	Inpatients	Serology	10
Wubbel *et al.*, 1999[11]	168	Outpatients	Serology and PCR	6
Esposito *et al.*, 2001[12]	203	Inpatients	Serology and PCR	9.5

3. CAP IN PEDIATRIC PATIENTS

Table III shows the main studies on CAP in children.

In a study in eastern Finland involving 201 children with pneumonia, bacterial infection was demonstrated in 102 cases (51%) and *C. pneumoniae* was involved in 20/102 cases.[10] The incidence of this pathogen was more than 2-fold higher in children ≥5 years than in those <5 years of age. Wubbel *et al.* recently obtained similar results in 168 outpatient children with CAP.[11] *C. pneumoniae* acute infection was found in 6% of cases, ranking fourth after *Streptococcus pneumoniae*, viruses, and *Mycoplasma pneumoniae*. There was no difference in the effectiveness of the antibiotics used in the study (azithromycin, amoxicillin–clavulanate, or erythromycin estolate), even among those infections attributed to *M. pneumoniae*, *C. pneumoniae*, or *S. pneumoniae*. However, beta-lactam treatment was used only in children under 5 years of age, whereas children older than 5 were randomized to either azithromycin or erythromycin.

Esposito *et al.* recently studied 203 children aged 2–14 years admitted for pneumonia.[12] Evidence for *C. pneumoniae* infection as single pathogen (by means of serology and PCR) was obtained in 8/203 patients, whereas mixed infection with *M. pneumoniae* was found in a further 11 cases. Table IV shows a comparison between clinical and radiological characteristics associated with specific etiology in these young patients with pneumonia. In another study the authors confirmed the limited role of clinical, laboratory, and radiological features in predicting the etiology of CAP in pediatric patients.[13] C-reactive protein levels and WBC counts were significantly associated with pneumococcal infection, but a high degree of overlapping of the individual values in the different groups was reported. Radiological findings had the same limitations in terms of sensitivity and specificity; segmental or lobar consolidation, usually associated with *S. pneumoniae* pneumonia, and interstitial involvement, which is considered an indicator of atypical bacterial infection, were also demonstrated in all of the other etiologic groups. This study also evaluated the response to different antibiotic treatment of the different etiologic agents. Macrolides either alone or in combination for the treatment of atypical bacterial infections lead to a better clinical outcome than beta-lactams alone (Table V).

TABLE IV
Clinical and Radiographic Characteristics of 87 Children with Pneumonia due to Atypical Pathogens (Modified from Ref. 12)

Characteristics	*C. pneumoniae* ($n = 8$)	*M. pneumoniae* ($n = 68$)	Mixed *M.* and *C. pneumoniae* ($n = 11$)
Age (years)	5.76 + 2.94	6.32 + 3.34	6.90 + 3.36
Gradual onset	6 (75)[a]	41 (60)	6 (55)
Fever	5 (63)	58 (85)	10 (91)
Cough	4 (50)	44 (65)	7 (64)
Tachypnea	0	8 (12)	1 (9)
Rales	7 (88)	60 (88)	10 (91)
Wheezing	1 (13)	10 (15)	1 (9)
Days of illness	10.75 + 4.27	13.17 + 6.67	13.00 + 5.50
Days of hospitalization	6.38 + 2.88	6.51 + 2.74	6.33 + 3.39
Chest x-ray: Linear opacities	4 (50)	41 (60)	8 (73)
Reticulo-nodular infiltrate	1 (13)	27 (40)	5 (46)
Segmental/lobar consolidation	3 (37)	19 (28)	2 (18)
Bilateral consolidation	0	5 (7)	0
Pleural effusion	0	4 (6)	0

Note: No significant differences were observed.
[a]Figures in parenthesis represent percentages.

TABLE V
Clinical Outcomes for Children with CAP, According to Treatment and Etiology (Modified from Ref. 13)

	S. pneumoniae (44 pts)	Atypicals (42 pts)	Mixed infections (15 pts)
Macrolide alone			
Cure	6	13	5
Failure	1	1	0
Beta-lactam alone			
Cure	27*	11**	2
Failure	1	10	2
Beta-lactam + macrolide			
Cure	9	7	6
Failure	0	0	0

*$p = 0.0003$ compared with atypicals infection treated with beta-lactam monotherapy; $p = 0.34$ compared with mixed infection treated with beta-lactam monotherapy.
**$p = 0.30$ compared with atypical infection treated with beta-lactam + macrolide; $p = 0.023$ compared with atypical infection treated with macrolide monotherapy.

4. CAP IN ADULTS

Table VI shows the main studies on CAP in adults.

In a prospective study on 109 adult patients with CAP, noninvasive and invasive (including transthoracic needle aspiration) methods were applied to identify the causal agents.[4] Conventional methods provided a microbial etiology in

TABLE VI
Incidence of *Chlamydia pneumoniae* Infection in Community-Acquired Pneumonia in Adults

Reference	Number of patients	Type of patients	Methods	Incidence of *C. pneumoniae* (%)
Marrie *et al.*, 1987[22]	301	Inpatients	Serology	6
Grayston *et al.*, 1989[5]	198	Inpatients	Serology	10
Fang *et al.*, 1990[23]	359	Inpatients (including HIV-infected)	Serology	6
Almirall *et al.*, 1993[24]	105	Outpatients	Serology	13
Mundy *et al.*, 1995[25]	385	Inpatients (including HIV-infected)	PCR	4
Cosentini *et al.*, 1996[7]	61	Inpatients	Serology + DFA	10
Martson *et al.*, 1997[2]	2775	Inpatients	Serology	9
Ishida *et al.*, 1998[26]	318	Inpatients	Serology	3
Ruiz-Gonzalez *et al.*, 1999[4]	109	Inpatients	Serology + PCR	13

54 patients (50%), *C. pneumoniae* ranking second (17%) after *M. pneumoniae* (35%). However, when samples obtained from transthoracic needle aspiration were analyzed, evidence of microbial etiology was found in 36 other patients. The use of conventional testing plus needle aspiration allowed the authors to determine specific microbiologic causes in more than 80% of CAP cases. *S. pneumoniae* was the most frequent pathogen (30%), followed by *M. pneumoniae* (22%) and *C. pneumoniae* (13%). In three cases needle aspiration provided new evidence of *C. pneumoniae* infection, in four cases it confirmed the results of conventional methods, and in three cases with microbial diagnosis of *C. pneumoniae* infection by conventional methods other pathogens were identified (two *S. pneumoniae* and one *S. viridans*). These data confirm the role of *C. pneumoniae* in the etiology of CAP and that this agent is often present as part of a coinfection involving other bacterial agents.

The possible role of coinfection is elucidated in two studies on adult patients and in one study on children.[13–15] Kauppinen *et al.*[14] showed that coinfection with *S. pneumoniae* was associated with more severe disease and a longer hospital stay. The authors confirmed that in their series, differential diagnosis between *C. pneumoniae, S. pneumoniae,* and mixed infection was not possible on the basis of clinical, laboratory, and radiological features. More recently, Miyashita *et al.* found a mixed infection, mainly with *S. pneumoniae,* in 22/62 (35%) cases of *C. pneumoniae* CAP.[15] These studies suggest that mixed infection should be taken into account when planning empirical antimicrobial treatment for CAP, as suggested for children by Esposito *et al.*[13]

C. pneumoniae has also been involved in nursing home outbreaks.[16] This retrospective cohort study was performed in three nursing homes in Canada involving 549 residents. The authors report a high incidence of acute respiratory tract infections (RTIs), with 16 cases of pneumonia and 6 deaths. Using serologic testing on paired sera and nasopharyngeal swab culture, *C. pneumoniae* was identified as the etiological agent of this outbreak. This study provides new insight on the epidemiology of nursing homes infective outbreaks and underlines that *C. pneumoniae* pneumonia is not always benign and that some patients may die.

5. INFECTION IN IMMUNOCOMPROMISED SUBJECTS

Recent papers have analyzed the potential role of *C. pneumoniae* as a respiratory pathogen in both pediatric and adult immunocompromised patients. Cosentini *et al.* report the rate of seroconversion to *C. pneumoniae* of 26 HIV-infected children and 14 seroreverter children (HIV-negative children born to HIV-positive mothers) over a 3-year study period.[17] The study showed a high incidence of *C. pneumoniae* infection in HIV-1-infected children. The incidence of *C. pneumoniae* infection appears to correlate with the degree of immunosuppression and with the viral burden. Because in most cases the infection was asymptomatic, and in symptomatic cases the outcome was often favorable irrespective of the treatment used, it is still unclear whether *C. pneumoniae* may cause severe forms of infection in this population. This high rate of infection was not confirmed by a different study on adult HIV-positive patients with pneumonia.[18] In this retrospective study, involving 103 episodes of pneumonia of 83 patients, no *Chlamydia* spp. could be detected by PCR methods in BAL fluid specimens. In a different prospective study on hospitalized patients (including HIV-positive subjects) with pneumonia, *C. pneumoniae* was frequently identified using culture and PCR on BAL fluid.[19] Given the presence of asymptomatic carriage and isolation of copathogens, the authors conclude that in HIV-positive patients the association between *C. pneumoniae* detection and acute pulmonary infection is less clear than in immunocompetent patients. These data indicate that further epidemiological, clinical, and laboratory studies are needed to address the possible role of *C. pneumoniae* in the HIV-infected population.

6. CONCLUSIONS

A lot of circumstantial evidence now suggests that *C. pneumoniae* infection is important in patients with CAP: it is commonly present when looked for, it often coexists with bacterial pathogens, it may lead to more complicated course when not treated in the presence of bacterial copathogens, and outcomes, including mortality, are improved if macrolides are incorporated in CAP therapy regimens. Future studies are needed to document the role of this organism more clearly. This would require routine serologic testing, along with cultures and molecular biology techniques, to confirm the high frequency of infection with this organism. In addition, it would be necessary to conduct outcome studies that correlate the use of therapy for *C. pneumoniae* with improved outcomes in patients who are actually documented to have CAP due to this organism. In the meantime, it is reasonable to suggest empirical treatment of patient with CAP with drugs active against *C. pneumoniae*, such as macrolides alone or in combination with beta-lactams, or new fluoroquinolones (levafloxacin, gatifloxacin, and moxifloxacin).

REFERENCES

1. Grayston, J. T., Kuo, C. C., Campbell, L. A., and Wang, S. P., 1989, *Chlamydia pneumoniae* sp. nov. for *Chlamydia* sp. strain TWAR, *Int. J. Sys. Bacteriol.* **39:**88–90.
2. Marston, B. J., Plouffe, J. F., File, T. M., Hackman, B. A., Salstrom, S. J., Lipman, H. B., Kolczac, M. S., and Breiman, R. F., 1997, Incidence of community-acquired pneumonia requiring hospitalisation, *Arch. Intern. Med.* **157:**1709–1718.
3. Marrie, T. J., Peeling, R. W., Fine, M. J., Singer, D. E., Coley, C. M., and Kapoor, W. N., 1996, Ambulatory patients with community-acquired pneumonia: The frequency of atypical agents and clinical course, *Am. J. Med.* **101:**509–515.
4. Ruiz-Gonzalez, A., Falguera, M., Nogues, A., and Rubio-Caballero, M., 1999, Is *Streptococcus pneumoniae* the leading cause of pneumonia of unknown etiology? A microbiologic study of lung aspirates in consecutive patients with community-acquired pneumonia, *Am. J. Med.* **106:**385–390.
5. Grayston, J. T., 1989, *Chlamydia pneumoniae* strain TWAR, *Chest* **95:**664–669.
6. Steinhoff, D., Lode, H., Ruckdeschel, G., Heidrich, B., Rolfs, A., Fehrembach, F. J., Mauch, H., Hoffken, G., and Wagner, J., 1996, *Chlamydia pneumoniae* as a cause of community-acquired pneumonia in hospitalised patients in Berlin, *Clin. Infect. Dis.* **22:**958–964.
7. Cosentini, R., Blasi, F., Raccanelli, R., Rossi, S., Arosio, C., Tarsia, P., Randazzo, A., and Allegra, L., 1996, Severe community-acquired pneumonia: A possibile role for *Chlamydia pneumoniae*, *Respiration* **63:**61–65.
8. Pacheco, A., Gonzales, S. J., Aroncena, C., Rebollar, M., Antela, A., Guerrero, A., 1991, Community-acquired pneumonia caused by *Chlamydia pneumoniae* strain TWAR in chronic cardiopulmonary disease in the elderly, *Respiration* **58:**316–320.
9. Grayston, J. T., Campbell, L. A., Kuo, C. C., Mordhorst, C. H., Saikku, P., Thom, D. H., Wang, S. P., 1990, A new respiratory tract pathogen: *Chlamydia pneumoniae* strain TWAR, *J. Infect. Dis.* **161:**618–625.
10. Heiskanen-Kosma, T., Korppi, M., Jokinen, C., Kurki, S., Heiskanen, L., Juvonen, H., Kallinen, S., Sten, M., Tarkiainen, A., Ronnberg, P. P., Kleemola, M., Makel, P. H., Leinonen, M., 1998, Etiology of childhood pneumonia: Serologic results of a prospective, population-based study, *Pediatr. Infect. Dis. J.* **17:**986–991.
11. Wubbel, L., Muniz, L., Ahmed, A., Trujillo, M., Carubelli, C., McCoig, C., Abramo, T., Leinonen, M., McCracken, G. H., Jr. 1999, Etiology and treatment of community-acquired pneumonia in ambulatory children, *Pedriatr. Infect. Dis. J.* **18:**98–104.
12. Esposito, S., Blasi, F., Bellini, F., Allegra, L., Principi, N., and the Mowgli Study Group, 2001, *Mycoplasma pneumoniae* and *Chlamydia pneumoniae* infections in children with pneumonia, *Eur. Respir. J.* **17:**241–245.
13. Esposito, S., Bosis, S., Lavagna, R., Faelli, N., Begliatti, E., Marchisio, P., Blasi, F., Bianchi, C., and Principi, N., 2002, Characteristics of *Streptococcus pneumoniae* and atypical bacterial infections in children 2–5 years of age with community-acquired pneumonia, *Clin. Infect. Dis.* **35:**1345–1352.
14. Kauppinen, M. T., Saikku, P., Kujala, P., Herva, E., and Syrjala, H., 1996, Clinical picture of community-acquired *Chlamydia pneumoniae* pneumonia requiring hospital treatment: A comparison between chlamydial and pneumococcal pneumonia, *Thorax* **51:**185–189.
15. Miyashita, N., Fukano, H., Okimoto, N., Hara, H., Yoshida, K., Niki, Y., and Matsushima, T., 2002, Clinical presentation of community-acquired *Chlamydia pneumoniae* pneumonia in adults, *Chest* **121:**1776–1781.

16. Troy, C. J., Peeling, R. W., Ellis, A. G., Hockin, J. C., Bennet, D. A., Murphy, M. R., and Spika, J. S., 1997, *Chlamydia pneumoniae* as a new source of infectious outbreaks in nursing homes, *JAMA* **277:**1214–1218.
17. Cosentini, R., Esposito, S., Blasi, F., Clerici Schoeller, M., Pinzani, R., Tarsia, P., Fagetti, L., Arosio, C., Principi, N., Allegra, L., 1998, Incidence of *Chlamydia pneumoniae* infection in vertically HIV-1 infected children, *Eur. J. Clin. Microbiol. Infect. Dis.* **17:**720–723.
18. Tarp, B., Jensen, J. S., Østergaard, L., and Andersen, P. L., 1999, Search for agents causing atypical pneumonia in HIV-positive patients by inhibitor-controlled PCR assays, *Eur. Respir. J.* **13:**175–179.
19. Dalhoff, K., and Maass, M., 1996, *Chlamydia pneumoniae* pneumonia in hospitalized patients, *Chest* **110:**351–356.
20. Block, S., Hedrick, J., Hammerschlag, M. R., and Cassel, G. H., 1995, *Mycoplasma pneumoniae* and *Chlamydia pneumoniae* in community-acquired pneumonia in children: Comparative safety and efficacy of clarithromycin and erythromycin suspensions, *Pediatr. Infect. Dis. J.* **14:**471–477.
21. Harris, J. A., Kolakathis, A., Campbell, M., Cassel, G. H., Hammerschlag, H. R., 1998, Safety and efficacy of azithromycin in the treatment of community-acquired pneumonia in children, *Pediatr. Infect. Dis. J.* **17:**865–871.
22. Marrie, T. J., Grayston, J. T., Wang, S.-P., Kuo, C. C., 1987, Pneumonia associated with TWAR strain of Chlamydia, *Ann. Intern. Med.* **106:**507–511.
23. Fang, G. D., Fine, M., Orloff, J., Arisumi, D., Yu, V. C., Kapoor, W., Grayston, J. T., Wang, S. P., Kohler, R., Muder, R. R., 1990, New and emerging etiologies for community-acquired pneumonia with implications for therapy. A prospective multicenter study of 359 cases, *Medicine* **69:**307–315.
24. Almirall, J., Moratò, I., Riera, F., Verdaguer, A., Priu, R., Coll, P., Vidal, J., Murgui, L., Vells, F., Catalan, F., 1993, Incidence of community-acquired pneumonia and *Chlamydia pneumoniae* infection: A prospective multicentre study, *Eur. Respir. J.* **6:**14–18.
25. Mundy, L. M., Auwaerter, P. G., Oldach, D., Warner, M. L., Burton, A., Vance, E., Gaydos, C. A., Joseph, J. M., Gopalan, R., Moore, R. D., 1995, Community-acquired pneumonia: Impact of immune status, *Am. J. Respir. Crit. Care Med.* **152:**1309–1315.
26. Ishida, T., Hashimoto, T., Arita, M., Ito, J., Osawa, M., 1998, Etiology of community-acquired pneumonia in hospitalized patients: A 3-year prospective study in Japan, *Chest* **114:**1588–1593.

6

Chlamydia Detection in Blood

YOSHIMASA YAMAMOTO

1. INTRODUCTION

Chlamydia pneumoniae has been established as a common and important pathogen causing upper and lower respiratory tract infections. As described in this book, current studies have suggested that *C. pneumoniae* may be implicated in not only respiratory infections but also some chronic inflammatory diseases, including atherosclerosis. Detection of the causative pathogens in clinical specimens obtained from patients with chronic inflammatory disease is essential for establishing the involvement of a certain pathogen in the disease. Peripheral blood specimens are widely utilized for diagnosis of diseases, since they can be easily collected from patients and give valuable information in many aspects, including microbiological. *C. pneumoniae* has been cultured from blood specimens of patients with chronic inflammatory diseases, including atherosclerosis. However any successful culture of *C. pneumoniae* has been reported, even the bacterial antigens, such as DNA, can be detected in blood samples.(13)

Properly performed serology is considered of value for diagnosing acute *C. pneumoniae* infection. However, in the case of persistent *C. pneumoniae* infection, which can be expected in chronic inflammatory diseases associated with this pathogen, serology is less useful because of the preexistence of anti–*C. pneumoniae* antibodies in the population(45) as well as the discrepancy between culture-documented chlamydia infections and detection of antibodies.(6) Therefore, there is an obvious need for methods that can identify individuals persistently infected by *C. pneumoniae.*

The polymerase chain reaction (PCR) is the most sensitive of the existing rapid methods to detect microbial pathogens in clinical specimens. In particular, when specific pathogens that are difficult to culture *in vitro* or require a

YOSHIMASA YAMAMOTO • Division of Molecular Microbiology, Department of Basic Laboratory Sciences, Graduate School of Medicine, Osaka University, Osaka, Japan.

prolonged cultivation period are expected to be present in specimens, the diagnostic value of PCR is known to be significant. However, the application of PCR to clinical specimens has many potential pitfalls because of the susceptibility of PCR to inhibitors, contamination, and experimental conditions. For instance, it is known that the sensitivity and specificity of a PCR assay is dependent on target genes, primer sequences, PCR techniques, DNA extraction procedures, and PCR product detection method. Because *C. pneumoniae* is difficult to culture *in vitro*, often low numbers of bacteria may be detected in the blood of patients with chronic inflammatory diseases such as atherosclerosis. In this chapter general PCR protocols for detection of bacteria in clinical specimens, as well as a specific example of using PCR for detection of *C. pneumoniae* in blood, is discussed.

The detection of bacterial DNA by PCR is not definitive proof of the actual presence of bacterial organisms. Because the persistent form of *C. pneumoniae* may not be detected by cultures, demonstration of bacterial antigen by staining with highly specific antibodies is another choice to demonstrate the presence of bacterial organisms besides demonstration of Chlamydia DNA by PCR. For this purpose, immunofluorescence staining of blood specimens with chlamydia antibody is utilized in several reports and will also be discussed in this chapter.

2. PCR—METHODOLOGICAL ASPECTS

The PCR assay for diagnosis involves several critical steps, such as DNA extraction from blood specimens, PCR amplification, and detection of amplicons. In particular, when blood with only a few bacteria present is tested by PCR, each procedure must be carefully designed and performed.

2.1. Contamination

Since PCR is based on DNA amplification, false-positive or -negative outcomes may easily occur. In particular, a single PCR cycle results in very large numbers of amplifiable molecules that can potentially contaminate subsequent amplifications of the same target sequence.[28] In fact, a primary source of false-positive reactions has been identified as a carryover of amplified product from previous reactions.[29] Carryover contamination of reagents, pipetting devices, laboratory surfaces, or even the skin of workers,[26] can yield false-positive results. To control such carryover contamination, one must prevent physical transfer of DNA between amplified samples and between positive and negative experimental controls. For this purpose, preparation of samples for PCR assay must be in a separate room or biosafety hood from that in which the reactions are performed. Using a pipette tip with an aerosol barrier is essential for avoiding cross-contamination as well as carryover contamination. UV exposure can also to destroy contaminating amplicons but is efficient only on surfaces and with amplicons greater than 300 bp in size.[14] Use of uracil *N*-glycosylase (UNG) to cleave the deoxyuridine triphosphate (dUTP) incorporated in PCR products is considered a powerful protocol to prevent carryover amplicon contamination enzymatically,[29] particularly in a clinical laboratory that is extensively performing PCR.

2.2. DNA Extraction

Because clinical specimens have PCR inhibitors, such as hemin, which binds to *Taq* polymerase and inhibits its activity,[8] DNA purification is important to avoid such effects. Extraction yield of target DNA is also a critical factor in the PCR detection assay for bacteria in clinical specimens, particularly when only a few bacteria are expected to be in specimens. Because bacteria have a rigid cell wall that may resist ordinary digestion protocol for DNA extraction, the extraction protocol for bacterial DNA in clinical specimens should be an additional consideration for sample preparation.

The classical DNA extraction protocol is based on purification with organic solvents like phenol/chloroform, followed by precipitation with ethanol. Such protocol is labor intensive and takes time, providing more chance for contamination during the extraction. In this regard, a new protocol for purification of DNA using solid-phase carriers has been developed. This protocol is based on the nature of nucleic acids, which can bind to silica or glass particles in the presence of chaotropic agents such as NaI or $NaClO_4$. There are several different types of such DNA extraction kit commercially available. Daugharty *et al.*[10] examined the efficacy of chlamydial DNA isolation from buffy coats spiked with *C. pneumoniae* elementary bodies (EB) using six different commercial DNA extraction kits. It was concluded in the report that QIAamp Blood Kit (QIAGEN, Valencia, CA) was the most sensitive among the kits tested, including the ELU-QUIK DNA Purification Kit (Schleicher & Schuell, Keene, NH), IsoQuick Nucleic Acid Extraction Kit (MicroProbe Corp., Bothell, WA), DNA/RNA Isolation Kit (United States Biochemical Corp., Cleveland, OH), Rapid Prep Micro Genomic (Pharmacia Biotech Inc., Piscataway, NJ), and ASAP Genomic DNA Isolation Kit (Boehringer Mannheim Corp., Indianapolis, IN). Similar evaluation for DNA extractions utilizing commercial kits was also performed by other groups[27,47] and the QIAamp kit was found to be more suitable than other commercial and noncommercial methods evaluated for the extraction of DNA for PCR.

2.3. Target Genes

Choice of target genes as well as desing of oligonucleotide primers are critical elements, in determining PCR sensitivity. Even though the same gene is selected as a target, the sensitivity of PCR with different primer sets shows a 100- to 1,000-fold sensitivity difference between primer sets.[20] Therefore, the sequence of primers is important in the sensitivity and specificity of PCR. The sensitivity of PCR is also dependent on the target gene selected, because copy numbers of genes or operons per bacterium vary. In this regard, if only the sensitivity of PCR is considered, reverse transcription (RT)-PCR would be another selection method because of the multiple copy numbers of mRNAs per bacterium. However, the practical value of RT-PCR in diagnosis is limited because of the short life span and the vulnerability of bacterial mRNAs.

Sequence polymorphism of a target gene is another concern in regard to PCR specificity. Some bacterial genes, such as the *C. trachomatis* outer membrane protein gene, have hypervariable regions within the gene.[17] Therefore, PCR products of different sizes as well as different sequences may occur between clinical isolates of the bacterium when such a gene is selected as a target for PCR.

2.4. PCR Protocols

There are several PCR protocols to enhance sensitivity, especially when dealing with small numbers of bacteria as the target. Nested PCR is one of these protocols for detection of only a few bacteria in clinical specimens. The process utilizes two consecutive PCRs. The first PCR contains an external pair of primers, whereas the second contains two nested primers that are internal to the first primer pair, or one of the first primers and a single nested primer. The larger fragment produced by the first reaction is used as a template for the second PCR. The sensitivity and specificity of DNA amplification can be considerably improved using such nested PCR, sometimes with 1,000 times more sensitivity than using a standard PCR. However, in the case of detection of *C. pneumoniae* by nested PCR in a standard solution spiked with bacteria, sensitivity was not always improved compared with a standard single PCR. For example, nested verses single PCR with primers specific for *C. pneumoniae omp-1* gene showed the same sensitivity (0.005 inclusions or 2.5 elementary bodies per PCR).[1]

A frequently encountered problem in PCR amplification of target gene sequences is the appearance of spurious smaller bands in the product spectrum.[12] This is usually interpreted to be due to mispriming by one or both of the oligonucleotide amplimers to the target template. Touchdown PCR is designed to avoid such problems and provides a clearly specific PCR band. The touchdown PCR utilizes a protocol with decreasing annealing temperatures at every cycle from above to below the expected annealing temperature. The application of this technique to detection of *C. pneumoniae* provided an improved analytical sensitivity (0.004 to 0.063 inclusion-forming unit per PCR).[31]

Fluorescent-probe-based PCR assays with labeled primers or specific probes labeled with a fluorescent dye have been developed with the advantages of a closed system that avoids carryover contamination during the PCR reaction and the increased detection sensitivity for amplicons. There are two types of assay, real-time and endpoint, readings. The real-time PCR in particular, which provides quick and accurate information regarding target genes, has been increasingly utilized. This approach has the advantage of quantitating the PCR in the exponential phase rather than the endpoint accumulation of PCR product or trying to capture the PCR in the exponential phase, as was done previously in many quantitative PCRs. This non-gel-based technique has several other advantages over ordinary agarose gel-based technique. For instance, this system allows

Primer and probe sequences (5′→3‘)

Sense: GGA CCT TAC CTG GAC TTG ACA TGT

Antisense: CCA TGC AGC ACC TGT GTA TCT G

Probe: FAM-CCT TGC GGA AAG CTG TAT TTC TAC AGT TGT CA-TAMRA

PCR amplification curves

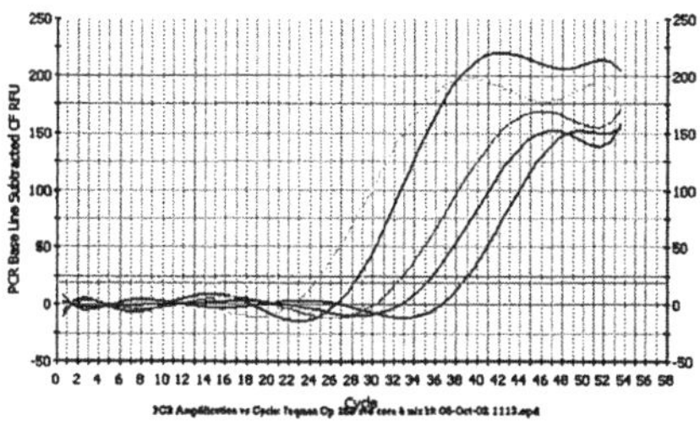

Standard curve for *C. pneumoniae*

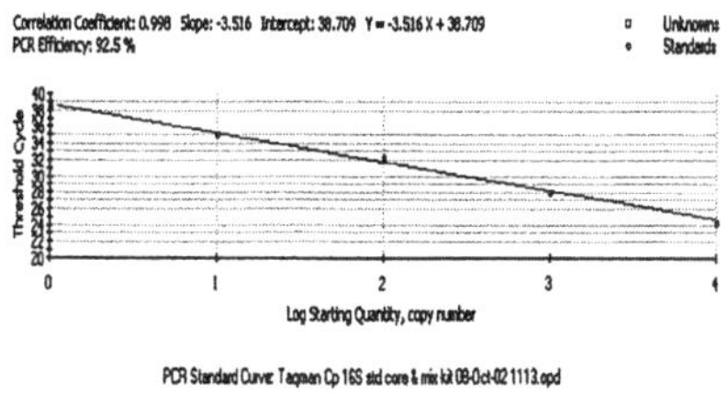

FIGURE 1. Real-time PCR specific for *C. pneumoniae* 16S rRNA using TaqMan technology. The PCR can quantify between 1 and 10,000 Chlamydia elementary bodies.

a large increase in throughput. The fluorescent probe assay is run in a 96-well, or even 384-well, format and many of the steps in the assay are automated. The assay is a closed system in which the reaction tube is never opened postamplification. In addition, it uses an automated detection system that quantitates and calculates the degree of fluorescence over that for the control at each cycle and, hence, accurately defines the cycle number and linear range for a positive result. Even though currently there are only few reports on the detection and quantitation of chlamydia in blood specimens by real-time PCR, this technique has excellent potential as a major protocol for PCR detection of bacteria in clinical specimens owing to these advantages. Figure 1 shows an example of real-time PCR specific for *C. pneumoniae* 16S rRNA originally developed by Berger *et al.*[3] The real-time PCR by using TaqMan technology enabled us to quantify from 1 to at least 10,000 elementary bodies of *C. pneumoniae* per PCR assay.

2.5. Detection of PCR Products

There are several different detection protocols reported for PCR products beside the traditional electrophoresis method on an ethidium bromide containing agarose gel. Southern hybridization with a specific probe labeled with a radioisotope or nonradioisotope marker to PCR amplicons has been widely utilized for the study of PCR specificity. This detection protocol also provides a higher sensitivity than that of the ethidium bromide detection method, but requires extra blotting and hybridization steps. The digoxigenin (DIG)-PCR enzyme linked immunosorbent assay (ELISA) is one of the PCR amplicon detection methods utilizing a microtiter plate and is now commercially available (Roche Molecular Biochemicals, Indianapolis, IN). This method involves capturing amplicons labeled with DIG by the probe immobilized onto the surface of a streptavidin-coated ELISA plate. The bound hybrid is detected with an anti-DIG peroxidase conjugate and the colorimetric substrate. This ELISA system has been shown to be 10 to 100 times more sensitive than the traditional electrophoresis method.[35] The PCR-immunoassay detection method utilizing a special small device (Clearview Immunoassay Detection Device; Oxoid Inc., Ogdensburg, NY), which holds a membrane and a sample application pad containing latex beads labeled with an anti-DNP antibody, is another type of detection method for amplicons designed to detect specific bacteria in clinical isolates.[9] The membrane utilized in this system is coated with lines of anti-biotin and anti-DIG antibodies. Therefore, an evaluation of PCR results in utilizing this detection kit in clinical laboratories, which do not have electrophoresis equipment, as positive or negative judgment is relatively easy. The application of this kit for detection of bacteria showed a detection limit of 1–3 organisms per PCR, which is 10 times more sensitive than detection of PCR products on traditional electrophoresis with agarose gels.[38]

3. DETECTION OF CHLAMYDIA IN BLOOD BY PCR

To date, there have been 18 reports concerning detection of *C. pneumoniae* in blood from patients with various diseases, including coronary artery disease, as well as control healthy subjects by PCR (Table I). The results of studies have shown a very wide variation in the positive rate, ranging from 0 to 72.2%. Even in the specimens obtained from healthy subjects the occurrence of *C. pneumoniae* DNA presence was found but the positive rate also varies between studies from 0 to 46%. Such variation of the *C. pneumoniae* positive rate in blood specimens may be dependent on the population of patients and/or the PCR protocol utilized. Because there is no standard PCR protocol for *C. pneumoniae* detection and no consistent pattern of positive results among various laboratories determined by a multicenter comparison trial of PCR assays for detection of *C. pneumoniae*,[1] each step of the PCR protocol utilized in each study should be carefully reviewed.

3.1. *C. pneumoniae* DNA Extraction

The amount of starting material as well as the final volume of DNA solution appears to affect the sensitivity of overall detection. Therefore, whether the protocol utilized is sensitive enough or not cannot be evaluated until the information of these factors is provided. Five of the 18 published papers mentioned in the table did not provide information as to the amount of blood tested. One paper used only 200 μl whole blood for study. The majority of other studies listed in the table utilized 5.0–10.0 ml of blood for isolation of mononuclear cells (PBMCs) followed by DNA extraction. For example, Tondella, *et al.*[43] utilized 4.0 ml of whole blood for preparation of PBMCs. DNAs were then extracted from the isolated PBMCs by using QIAamp DNA mini kit and eluted in 100 μl of elution buffer, which means a 40-fold concentration of the original blood. In contrast, Smieja *et al.*[39] utilized 8.0 ml blood and DNAs were extracted into 50–10 μl of elution buffer using the same DNA extraction kit as that of Tondella's protocol. The final outcome of this study was, therefore, an approximately 80- to 160-fold concentration of the original blood. In addition, the volume of extracted DNA applied to PCR also varies between studies, as shown in the table. The minimal blood volume converted from the data reported per PCR is as low as 18 μl whole blood per PCR. In contrast, the largest blood volume per PCR is as high as 4.0 ml. Thus, the difference of blood concentrations per PCR between studies is apparent and such difference should affect the final sensitivity of the PCR assay. If the level of *C. pneumoniae* in blood is low and close to the low detection limit of PCR utilized, even a few-fold difference only of blood concentration between studies may result in different positive rate.

Most studies utilized the DNA extraction protocol with solid-phase carriers, such as Qiagen columns that hold a silica gel membrane. In some studies, DNA extraction was performed by a standard extraction protocol such as phenol–chloroform and ethanol precipitation. *C. pneumoniae* is a Gram-negative bacterium and has lipopolysaccharide and other outer membrane components in its cell wall, which contribute to osmotic stabilities as well as to rigidity, particularly of elementary bodies, the infectious form that resists physical and chemical pressures in the extracellular environment. Therefore, the procedure for extraction of *C. pneumoniae* DNA from clinical specimens should be designed for bacterial DNA extraction. In fact, our study showed that when two extraction protocols were examined, one designed for extraction of mammalian DNA from blood samples and one designed for extraction of bacterial DNA, the protocol for bacterial DNA extracted the microbial DNA more efficiently.[22]

3.2. PCR Method

The major outer membrane protein (MOMP) genes, such as *omp-1 (ompA),*[24] of *C. pneumoniae* have been utilized widely in PCR as a target gene for detection of this bacterium. *C. pneumoniae* has many outer membrane proteins, including cysteine-rich proteins OmcA and OmcB[15,16] as well as the MOMP

TABLE I

Detection of *C. pneumoniae* in Blood by PCR

Author (ref. #)	Year	Subject #	Disease[a]	DNA positive	Blood volume	DNA extraction	Resulting DNA volume	Extracted DNA/PCR (original blood vol./PCR)	Target gene
1. Tsirpanlis *et al.*(44)	2003	130	CAD	9 (6.9%)	5.0 ml	Clonit kit	?[b]	?/nested-PCR (?/PCR)	RNA polymerase
2. Sessa *et al.*(36)	2003	18 33	CAD-symptomatic CAD-asymptomaotic	13 (72.2%) 10 (30.3%)	?	Phenol/chloroform	50 μl	10 μl/nested-PCR (?/PCR)	HL-1/HR-1
3. Vainio *et al.*(46)	2002	46	CAD	0 (0%)	200 μl	QIAamp Blood kit	50 μl	15 μl/touchdown-nested (60 μl/PCR)	ompA
4. Huhtinen *et al.*(21)	2002	64	Anterior uveitis	1 (1.5%)	?	QIAamp DNA mini kit	?	?/touchdown (?/PCR)	ompA
5. Tondella *et al.*(43)	2002	228		1 (0.4%)	4.0 ml	QIAamp DNA mini kit	100 μl	5 μl/real-time (TaqMan probe) (200 μl/PCR)	ompA
6. Smieja *et al.*(39)	2002	100	COPD	24 (24%)	8.0 ml	QIAamp DNA mini kit	50–100 μl	5-10 μl/nested (800 μl/PCR)	ompA
7. Muller *et al.*(34)	2002	196 114 342	Dialysis patients Renal transplant Healthy subjects	19 (16.3%) 7 (9.6%) 19 (8.5%)	5.0 ml	Phenol/chloroform	40 μl	10 μl/nested (1.25 ml/PCR)	53 kDa protein
8. Haranaga *et al.*(19)	2001	237	Healthy subjects	21(8.9%)	5.0 ml	QIAamp DNA mini kit	50 μl	0.2 ug DNA/touchdown (?/PCR)	16S rRNA
9. Sessa *et al.*(37)	2001	93 42	CAD Healthy subjects	24 (25.8%) 2 (4.8%)	5.0 ml	Phenol/chloroform	50 μl	10 μl/nested (1.0 ml/PCR)	HL-1/HR-1
10. Maraha *et al.*(32)	2001	88 88	AAA Control subjects	18 (20%) 8 (9%)	?	QIAamp DNA mini kit	200 μl	5 μl/regular (?/PCR)	16S rRNA

11. Apfalter *et al.*[82]	2001	15 57	Respiratory infections Healthy subjects	6 (40%) 0 (0%)	?	Phenol/chloroform	40 μl	1 ug DNA/nested (?/PCR)	16S rRNA
12. Iliescu *et al.*[23]	2000	55	Peritoneal dialysis	33 (60%)	10.0 ml	QIAamp Blood kit	?	?/nested (?/PCR)	53 kDa protein
13. Maass *et al.*[30]	2000	188 50	Unstable angina Myocardial infarction	52 (28%) 13(26%)	CD14/8.0 ml	Qiagen kit	100 μl	50 μl/nested (4.0 ml/PCR)	HL-1/HR-1
14. Kaul *et al.*[25]	2000	28 19	CAD Healthy subjects	13 (46%) 5 (26%)	20–30 ml	SDS-proteinase K	?	?/nested (?/PCR)	omp1/hsp60
15. Bodetti and Timms[5]	2000	60	Healthy subjects	10 (16.7%)	9.0 ml	Heating	1.0 ml	2 μl/nested (18 μl/PCR)	ompA
16. Blasi *et al.*[4]	1999	41	Aortic aneurysm	19 (46.3%)	8.0 ml	Boehringer DNA kit	50 μl	10 μl/touchdown-nested (1.6 ml/PCR)	ompA
17. Wong *et al.*[48]	1999	913 292	CAD Control subjects	79 (8.6%) 21 (7.2%)	?	Phenol/chloroform	50 μl	3 μl/nested (?/PCR)	ompA
18. Boman *et al.*[7]	1998	101 52	CAD Control subjects	60 (59%) 24 (46%)	10.0 ml	Phenol/chloroform	50 μl	5 μl/nested (1.0 ml/PCR)	ompA

[a]CAD, coronary artery disease; COPD, chronic obstructive pulmonary disease; AAA, abdominal aortic aneurysm.
[b]?, data not provided.

encoded by *omp-1*. It is known that *C. trachomatis* MOMP has genetic variation, including in amino acid sequences,[42] but not much information regarding *C. pneumoniae* MOMP is available. Therefore, the design of primers for MOMP genes must be undertaken with special care. The species-specific region of the 16S rRNA gene is also frequently utilized as a target gene in PCR for detection of *C. pneumoniae.*[11,13] In this regard, it is noteworthy that the detection sensitivity of the two PCRs with *omp-1* versus 16S rRNA gene primers under each set of optimal conditions was different.[22] The PCR for *omp-1* was at least 10 times more sensitive than that for the 16S rRNA gene.

It is generally accepted that nested PCR may be more sensitive than single PCR because of the utilization of two consecutive PCRs. However, in practice, nested PCR does not always give a higher sensitivity than single PCR.[1] In addition, nested PCR is much more prone to contamination. Therefore, detection of *C. pneumoniae* in blood that may not contain many bacteria by nested PCR must be carefully performed; otherwise, no other method presently can confirm positive PCR results.

Multicenter comparison trial of DNA extraction methods and PCR assay for detection of *C. pneumoniae* in tissue samples has been conducted by Apfalter *et al.*[1] Even though the study did not focus on the detection of *C. pneumoniae* in blood specimens, it is noteworthy that there was no consistent pattern of *C. pneumoniae*–positive results among the nine different laboratories by 16 different PCR testing methods, and there was no correlation between the detection rates and the sensitivity of the assay used, including nested, single, and real-time PCR protocols, as determined with a panel of spiked specimens. Since a random distribution of *C. pneumoniae* within the clinical tissues utilized may be expected, some of the discrepancies in the detection of this organism in tissue samples can be explained by the random distribution of bacteria.

3.3. Detection of *C. pneumoniae* in Blood

Controversy surrounds the detection rate of *C. pneumoniae* in blood obtained from patients with coronary artery disease, primarily because of the lack of a definitive test for detecting the small numbers of *C. pneumoniae* present. Culture is always considered the "gold standard" in microbiology but is difficult to perform for certain fastidious bacteria such as *C. pneumoniae* in specific clinical specimens. For instance, this bacterium has not been successfully cultured from blood samples, although its DNA can be detected in blood.[13] Even though PCR enables the detection of low concentrations of bacteria in clinical specimens, great variability of detection is usually found in blood, as shown in the table. In this regard, a recent study conducted by Smieja *et al.*[41] demonstrated the relationship between target concentrations and PCR detection rate; that is, lower concentrations of *C. pneumoniae* were only intermittently PCR-positive, and this relationship was predictable from a statistical viewpoint. From this point of view, theoretically a larger number of replicates of a PCR assay may result in a better chance for detecting low numbers of bacteria. In other words, the negative PCR results obtained from a single PCR test may not be a true negative because of the low validity of detection with a lower concentration of target. Because the majority of the papers reporting results of *C. pneumoniae* detection by PCR in blood provided any replicate number of PCR tests, the negative results reported may possibly not be true negatives but could indicate that there were few bacteria, if any, present.

The validation of the results, which showed that PCR for *C. pneumoniae* of blood is positive, is not yet solved in the majority of reports. The report conducted by Haranaga *et al.*[19] validated the PCR results by immunostaining of blood samples with anti-chlamydia antibody. Comparison of PCR results with at least another validated PCR assay that targets a different gene or a different sequence of the same gene is another possible way to validate the results.

3.4. Location of *C. pneumoniae* in Blood

Because *C. pneumoniae* is an obligate intracellular bacterium, the location of this bacterium in the blood should be in leukocytes, such as monocytes. Moazed *et al.*[33] reported that in experimental infection of mice with *C. pneumoniae* the bacteria were spread via peripheral-blood mononuclear cells and the authors speculated that the responsive cell vehicle may be the monocytes/macrophages. However, since that study did not fractionate the peripheral-blood mononuclear cells to determine the cell vehicle for *C. pneumoniae*, it was not clear which peripheral-blood mononuclear cells acted as the vehicle. In this regard, it is currently reported that *C. pneumoniae* DNA can be recovered from CD3-positive peripheral-blood leukocytes from patients,[25] suggesting that lymphocytes may serve as a host cell for *C. pneumoniae*. In fact, our study also demonstrated that *C. pneumoniae* can infect and proliferate in lymphocytes *in vitro*.[18] Because lymphocytes do not have any antimicrobial system within a cell, this cell can be a good harbor for *C. pneumoniae in vivo*. In addition, the lymphocytes infected with *C. pneumoniae* may be an important host cell for dissemination of the organisms as well as possibly alter lymphocyte functions and certain immune mechanisms in an infected individual.

3.5. Association of Blood *C. pneumoniae* with Cardiovascular Disease

Because a number of reports in terms of seroepidemiological, pathological, microbiological, experimental infection model, and clinical trial studies implicate that *C. pneumoniae* may be involved in the pathogenesis of atherosclerosis. If the prevalence of *C. pneumoniae* in peripheral blood specimens obtained from patients with cardiovascular disease is high compared with control subjects who do not have any atherosclerotic disease, the detection of *C. pneumoniae* in blood can be a superior epidemiologic tool for prospective studies assessing the contribution of infection to human cardiovascular disease.[40] As shown in the table, a number of groups have conducted studies to determine *C. pneumoniae* DNA in peripheral blood mononuclear cells from patients with or without cardiovascular disease. The largest size of patients tested for detection of *C. pneumoniae* in blood is that of the study by Wong *et al.*[48] More than 900 patients with coronary

FIGURE 2. Detection of *C. pneumoniae* DNA in PBMCs obtained from healthy donors by PCR. Arrow, the specific PCR product (197 bp); M, molecular weight marker; N, negative control; P, positive control; 1 through 14, donor number. (From ref. 19)

artery disease were assessed for the presence of *C. pneumoniae* in blood. They found an association between coronary atherosclerosis and *C. pneumoniae* detected in men. However, when the subjects were expanded to both genders, there was no significance concerning the prevalence of *C. pneumoniae* in blood. Smieja *et al.*[40] examined the nine published studies by meta-analysis and found a pooled prevalence of 252 of 1763 (14.3%) in cardiovascular disease patients versus 74 of 874 (8.5%) in controls.[40] The pooled odds ratio, using a random-effects model, was 2.03 (95% CI: 1.34, 3.08, $P<0.001$). However, the studies did not consider other cardiovascular risk factors. Without adjustment for such other risk factors, they concluded that cardiovascular patients had a higher overall prevalence of *C. pneumoniae* in blood than control subjects.

It is noteworthy that control healthy subjects also showed the presence of *C. pneumoniae* in their blood (Fig. 2) but the prevalence was low in some studies. The majority of studies, including ours,[19] showed a prevalence of around 9% in control healthy subjects. Thus, it seems likely that circulating *C. pneumoniae* DNA in blood is present at least in a significant fraction of the "healthy" population. The correlation of *C. pneumoniae* detection in blood and atherosclerosis is, therefore, difficult to be concluded because of the limited number of patients as well as variation of technical factors affected by the PCR utilized. Nevertheless, it is obvious that *C. pneumoniae* DNA is present in the blood of a significant fraction of the adult population, including both patients and healthy people.

4. CHLAMYDIA ANTIGEN DETECTION BY IMMUNOSTAINING

Demonstration of bacterial DNA by PCR is considered only indirect evidence because the PCR is based on the amplified results. Therefore, many technical pitfalls can affect the final results and false-positive or -negative results are easily introduced, as discussed above. In this regard, direct demonstration of *C. pneumoniae* by immunostaining with specific antibody is considered more specific in some studies. However, the chance to capture chlamydia antigen by staining should be low because the level of bacteria in blood is predicted to be minimal. Two studies by Bodetti & Timms[5] and ourselves[19] have been conducted to detect *C. pneumoniae* antigen in blood specimens by immunostaining with genus- and species-specific FITC-conjugated monoclonal antibodies. Both studies demonstrated the presence of chlamydia antigen by staining in blood samples obtained from healthy subjects. As shown in Fig. 3, even though they were few in number, the usually small chlamydia inclusion bodies in the PCR-positive sample preparations showed the typical apple-green stain.

(A) (B)

FIGURE 3. Representative micrographs of PBMC specimens obtained from healthy donors stained with FITC-conjugate chlamydia antibody. Arrow indicates chlamydia antigen. Magnification, ×1,000. (From ref. 19)

In contrast, no inclusion bodies stained with chlamydia antibody were observed in PCR-negative sample preparations.

5. CONCLUSIONS

It has been widely believed that the blood of healthy individuals should be germ-free. However, as discussed in this chapter at least a significant fraction of the healthy population carries *C. pneumoniae* in their blood without any clinical symptoms. Because the detection of bacterial antigen, including DNA, in blood may not prove the viability of the bacteria, validation of viability of the organisms determined by PCR is essential for study of Chlamydia infection in certain inflammatory diseases, including atherosclerosis. In addition, the level of *C. pneumoniae* in blood is also important in determining whether this bacterium is involved in certain diseases because healthy subjects also carry this pathogen in their blood. Thus, there are still many questions which are currently being investigated by several study groups, including ours.

REFERENCES

1. Apfalter, P., Blasi, F., Boman, J., Gaydos, C. A., Kundi, M., Maass, M., Markristathis, A., Meijer, A., Nadrchal, R., Persson, K., Rotter, M. L., Tong, C. Y., Stanek, G., and Hirschl, A. M., 2001, Multicenter comparison trial of DNA extraction methods and PCR assays for detection of *Chlamydia pneumoniae* in endarterectomy specimens, *J. Clin. Microbiol.* **39:**519–524.
2. Apfalter, P., Boman, J., Nehr, M., Hienerth, H., Makristathis, A., Pauer, J., Thalhammer, F., Willinger, B., Rotter, M. L., and Hirschl, A. M., 2001, Application of blood-based polymerase chain reaction for detection of *Chlamydia pneumoniae* in acute respiratory tract infections, *Eur. J. Clin. Microbiol. Infect. Dis.* **20:**584–546.
3. Berger, M., Schroder, B., Daeschlein, G., Schneider, W., Busjhn, A., Buchwalow, I., Luft, F. C., and Haller, H., 2000, *Chlamydia pneumoniae* DNA in non-coronary atherosclerotic plaques and circulating leukocytes, *J. Lab. Clin. Med.* **136:**194–200.
4. Blasi, F., Boman, J., Esposito, G., Melissano, G., Chiesa, R., Cosentini, R., Tarsia, P., Tshomba, Y., Betti, M., Alessi, M., Morelli, N., and Allegra, L., 1999, *Chlamydia pneumoniae*

DNA detection in peripheral blood mononuclear cells is predictive of vascular infection, *J. Infect. Dis.* **180:**2074–2076.

5. Bodetti, T. J., and Timms, P., 2000, Detection of *Chlamydia pneumoniae* DNA and antigen in the circulating mononuclear cell fractions of humans and koalas, *Infect. Immun.* **68:**2744–2747.
6. Boman, J., and Hammerschlag, M. R., 2002, *Chlamydia pneumoniae* and atherosclerosis: Critical assessment of diagnostic methods and relevance to treatment studies, *Clin. Microbiol. Rev.* **15:**1–20.
7. Boman, J., Soderberg, S., Forsberg, J., Birgander, L. S., Allard, A., Persson, K., Jidell, E., Kumlin, U., Juto, P., Waldenstrom, A., and Wadell, G., 1998, High prevalence of *Chlamydia pneumoniae* DNA in peripheral blood mononuclear cells in patients with cardiovascular diseae and in middle-aged blood donors, *J. Infect. Dis.* **178:**274–277.
8. Byrnes, J. J., Downey, K. M., Esserman, L., and So, A. G., 1975, Mechanism of hemin inhibition of erythroid cytoplasmic DNA polymerase. *Biochemistry* **14:**796–799.
9. Curran, R., Talbot, D. C., and Towner, K. J., 1996, A rapid immunoassay method for the direct detection of PCR products: Application to detection of TEM beta-lactmase genes, *J. Med. Microbiol.* **45:**76–78.
10. Daugharty, H., Skelton, S. K., and Messmer, T., 1998, Chlamydia DNA extraction for use in PCR: Stability and sensitivity in detection, *J. Clin. Lab. Anal.* **12:**47–53.
11. Derfuss, T., Gurkov, R., Then Bergh, F., Goebels, N., Hartmann, M., Barz, C., Wilske, B., Autenrieth, I., Wick, M., Hohlfeld, R., and Meinl, E., 2001, Intrathecal antibody production against *Chlamydia pneumoniae* in multiple sclerosis is part of a polyspecific immune response, *Brain* **124:**1325–1335.
12. Don, R. H., Cox, P. T., Wainwright, B. J., Baker, K., and Mattick, J. S., 1991, 'Touchdown' PCR to circumvent spurious priming during gene amplification, *Nacleic Acids Res.* **19:** 4008.
13. Dowell, S. F., Peeling, R. W., Boman, J., Carlone, G. M., Fields, B. S., Guarner, J., Hammerschlag, M. R., Jackson, L. A., Kuo, C. C., Maass, M., Messmer, T. O., Talkington, D. F., Tondella, M. L., and Zaki, S. R., 2001, Standardizing *Chlamydia pneumoniae* assays: Recommendations from the Centers for Disease Control and Prevention (USA) and the Laboratory Centre for Disease Control (Canada), *Clin. Infect. Dis.* **33:**492–503.
14. Espy, M. J., Smith, T. F., and Persing, D. H., 1993, Dependence of polymerase chain reaction product inactivation protocols on amplicon length and sequence composition, *J. Clin. Microbiol.* **31:**2361–2365.
15. Everett, K. D., Desiderio, D. M., and Hatch, T. P., 1994, Characterization of lipoprotein EnvA in *Chlamydia psittaci* 6BC, *J. Bacteriol.* **176:**6082–6087.
16. Everett, K. D., and Hatch, T. P., 1991, Sequence analysis and lipid modification of the cysteine-rich envelop proteins of *Chlamydia psittaci* 6BC, *J. Bacteriol.* **173:**3821–3830.
17. Gaydos, C. A., Bobo, L., Welsh, L., Hook, E. W., 3rd, Viscidi, R., and Quinn, T. C., 1992, Gene typing of *Chlamydis trachomatis* by polymerase chain reaction and restriction endonuclease digestion, *Sex. Transm. Dis.* **19:**303–308.
18. Haranaga, S., Yamaguchi, H., Friedman, H., Izumi, S., and Yamamoto, Y., 2001, *Chlamydia pneumoniae* infects and multiplies in lymphocytes *in vitro*. *Infect. Immun.* **69:**7753–7759.
19. Haranaga, S., Yamaguchi, H., Leparc, G. F., Friedman, H., and Yamamoto, Y., 2001, Detection of *Chlamydia pneumoniae* antigenin PBMNCs of healthy blood donors: *Transfusion* **41:**1114–1119.
20. He, Q., Marjamaki, M., Soini, H., Mertsola, J., and Viljanen, M. K., 1994, Primers are decisive for sensitivity of PCR, *Biotechniques* **17:**82, 84, 86–87.
21. Huhtinen, M., Laasila, K., Granfors, K., Puolakkainen, M., Seppala, I., Laasonen, L., Repo, H., Karma, A., and Leirisalo-Repo, M., 2002, Infectious background of patients with a history of acute anterior uveitis, *Ann. Rheum. Dis.* **61:**1012–1016.
22. Ikejima, H., Haranaga, S., Takemura, H., Kamo, T., Takahashi, Y., Friedman, H., and Yamamoto, Y., 2001, PCR-based method for isolation and detection of *Chlamydia pneumoniae* DNA in cerebrospinal fluids, *Clin. Diagn. Lab. Immunol.* **8:**499–502.

23. Illiescu, E. A., Fiebig, M. F., Morton, A. R., and Sankar-Mistry, P., 2000, *Chlamydia pneumoniae* DNA in peripheral blood mononuclear cells in peritoneal dialysis patients, *Perit. Dial. Int.* **20:**722–726.
24. Jantos, C. A., Roggendorf, R., Wuppermann, F. N., and Hegemann, J. H., 1998, Rapid detection of *Chlamydia pneumoniae* by PCR-enzyme immunoassay, *J. Clin. Microbiol.* **36:**1890–1894.
25. Kaul, R., Uphoff, J., Wiedeman, J., Yadlappalli, S., and Wenman, W. M., 2000, Detection of *Chlamydia pneumoniae* DNA in CD3+ lymphocytes from healthy blood donors and patients with coronary artery disease, *Circulation* **102:**2341–2346.
26. Kitchin, P. A., Szotyori, Z., Fromholc, C., and Almond, N., 1990, Avoidance of PCR false positives [corrected], *Nature* **344:**201.
27. Klein, A., Barsuk, R., Dagan, S., Nusbaum, O., Shouval, D., and Galun, E., 1997, Comparison of methods for extraction of nucleic acid from hemolytic serum for PCR amplification of hepatitis B virus DNA sequences, *J. Clin. Microbiol.* **35:**1897–1899.
28. Kwok, S., and Higuchi, R., 1989, Avoiding false positives with PCR, *Nature* **339:**237–238.
29. Longo, M. C., Berninger, M. S., and Hartley, J. L., 1990, Use of uracil DNA glycosylase to control carry-over contamination in polymerase chain reactions, *Gene* **93:**125–128.
30. Maass, M., Jahn, J., Gieffers, J., Dalhoff, K., Katus, H. A., and Solbach, W., 2000, Detection of *Chlamydia pneumoniae* within peripheral blood monocytes of patients with unstable angina or myocardial infarction, *J. Infect. Dis.* **181**(Suppl. 3)**:**S449–S451.
31. Madico, G., Quinn, T. C., Boman, J., and Gaydos, C. A., 2000, Touchdown enzyme time release-PCR for detection and identification of *Chlamydia trachomatis, C. pneumoniae,* and *C. psittaci* using the 16S and 16S–23S spacer rRNA genes, *J. Clin. Microbiol.* **38:**1085–1093.
32. Maraha, B., den Heijer, M., Wullink, M., van der Zee, A., Bergmans, A., Verbakel, H., Kerver, M., Graafsma, S., Kranendonk, S., and Peeters, M., 2001, Detection of *Chlamydia pneumoniae* DNA in buff-coat samples of patients with abdominal aortic aneurysm, *Eur. J. Clin. Microbiol. Infect. Dis.* **20:**111–116.
33. Moazed, T. C., Kuo, C. C., Grayston, J. T., and Campbell, L. A., 1998, Evidence of systemic dissemination of *Chlamydia pneumoniae* via macrophages in the mouse, *J. Infect. Dis.* **177:** 1322–1325.
34. Muller, J., Nyvad, O., Larsen, N. A., Lokkegaard, N., Pedersen, R. S., Soilling, J., and Pedersen, E. B., 2002, *Chlamydia pneumoniae* DNA in peripheral blood mononuclear cells in dialysis patients, renal transplant recipients and healthy controls, *Scand. J. Clin. Lab. Invest.* **62:**503–509.
35. Radstrom, P., Backman, A., Qian, N., Kragsbjerg, P., Pahlson, C., and Olcen, P., 1994, Detection of bacterial DNA in cerebrospinal fluid by an assay for simultaneous detection of *Neisseria meningitidis-Haemophilus influenzae*-and streptococci using a seminested PCR strategy, *J. Clin Microbiol.* **32:**2738–2744.
36. Sessa, R., Di Pietro, M., Schiavoni, G., Santino, I., Benedetti-Valentini, F., Perna, R., Romano, S., and Del Piano, M., 2003, *Chlamydia pneumoniae* DNA in patients with symptomatic carotid atherosclerotic disease, *J. Vasc. Surg.* **37:**1027–1031.
37. Sessa, R., Di Pietro, M., Schiavoni, G., Santino, I., Cipriani, P., Romano, S., Penco, M., and del Piano, M., 2001, Prevalence of *Chlamydia pneumoniae* in peripheral blood monounclear cells in Italian patients with acute ischaemic heart disease, *Atherosclerosis* **159:**521–525.
38. Seward, R. J., and Towner, K. J., 2000, Evaluation of a PCR-immunoassay technique for detection of *Neisseria meningitidis* in cerebrospinal fluid and peripheral blood, *J. Med. Microbiol.* **49:**451–456.
39. Smieja, M., Leigh, R., Petrich, A., Chong, S., Kamada, D., Hargreave, F. E., Goldsmith, C. H., Chernesky, M., and Mahony, J. B., 2002, Smoking, season, and detection of *Chlamydia pneumoniae* DNA in clinically stable COPD patients, *BMC. Infect. Dis.* **2:**12.
40. Smieja, M., Mahony, J., Petrich, A., Boman, J., and Chernesky, M., 2002, Association of circulating *Chlamydia pneumoniae* DNA with cardiovascular disease: A systematic review, *BMC. Infect. Dis.* **2:**21.

41. Smieja, M., Mahony, J. B., Goldsmith, C. H., Chong, S., Petrich, A., and Chernesky, M., 2001, Replicate PCR testing and probit analysis for detection and quantitation of *Chlamydia pneumoniae* in clinical specimens, *J. Clin. Microbiol.* **39:**1796–1801.
42. Stothard, D. R., Boguslawski, G., and Jones, R. B., 1998, Phylogenetic analysis of the *Chlamydia trachomatis* major outer membrane protein and examination of potential pathogenic determinants, *Infect. Immun.* **66:**3618–3625.
43. Tondella, M. L., Talkington, D. F., Holloway, B. P., Dowell, S. F., Cowley, K., Soriano-Gabarro, M., Elkind, M. S., and Fields, B. S., 2002, Development and evaluation of real-time PCR-based fluorescence assays for detection of *Chlamydia pneumoniae, J. Clin. Microbiol.* **40:**575–583.
44. Tsirpanlis, G., Chatzipanagiotou, S., Ioannidis, A., Moutafis, S., Poulopoulou, C., and Nicolaou, C., 2003, Detection of *Chlamydia pneumoniae* in peripheral blood mononuclear cells: Correlation with inflammation and atherosclerosis in haemodialysis patients, *Nephrol. Dial. Transplant,* **18:**918–923.
45. Tuuminen, T., Varjo, S., Ingman, H., Weber, T., Oksi, J., and Viljanen, M., 2000, Prevalence of *Chlamydia pneumoniae* and *Mycoplasma pneumoniae* immunoglobulin G and A antibodies in a healthy Finnish population as analyzed by quantitative enzyme immunoassays, *Clin. Diagn. Lab. Immunol.* **7:**734–738.
46. Vainio, K., Vengen, O., Hoel, T., Fremstad, H., and Anestad, G., 2002, Failure to detect *Chlamydia pneumoniae* in aortic valves and peripheral blood mononuclear cells from patients undergoing aortic valve replacement in Norway, *Scand. J. Infect. Dis.* **34:**660–663.
47. Vince, A., Poljak, M., and Seme, K., 1998, DNA extraction from archival Giemsa-stained bone-marrow slides: comparison of six rapid methods, *Br. J. Haematol.* **101:**349–351.
48. Wong, Y., Thomas, M., Tsang, V., Gallagher, P. J., and Ward, M. E., 1999, The prevalence of *Chlamydia pneumoniae* in atherosclerotic and nonatherosclerotic blood vessels of patients attending for redo and first time coronary artery bypass graft surgery, *J. Am. Coll. Cardiol.* **33:**152–156.

7

Chlamydia pneumoniae Infection and Diseases: Immunity to *Chlamydia pneumoniae*

HELJÄ-MARJA SURCEL

1. INTRODUCTION

Chlamydia pneumoniae is an obligate intracellular pathogen that has a unique biphasic life cycle including extracellular but metabolically quiescent elementary bodies (EBs) and intracellular metabolically active reticulate bodies (RBs). Although a detailed knowledge of the protective immunity for *C. pneumoniae* infection is limited, available literature indicates that the most important parts of the immune mechanisms are comparable to those in *C. trachomatis* infections. Cell-mediated immunity and especially participation of type 1 T cells and interferon-γ (IFN-γ) are crucial for eradication or limitation of the infection at a culture negative stage. Primary infection induces development of antigen-specific immunity, which mediates an improved recovery from reinfection in mice. In humans, immune protection is difficult to evaluate but the reduced incidence of serious pneumonia in the case of reinfection can be regarded as a sign of partial protection. This review summarizes the current understanding of the basic characteristics of *C. pneumoniae*–induced immune response.

HELJÄ-MARJA SURCEL • National Public Health Institute, Department of Microbiology, 90101 Oulu, Finland.

2. INNATE IMMUNE MECHANISMS

C. pneumoniae is generally transmitted from person to person via the respiratory route,[1,2] where mechanical barriers and innate immune mechanisms comprise the first defense systems of the host. Epithelial cells lining the trachea and nasopharynx are the first cellular barriers against inhaled pathogens. Infection of airway epithelial cells can trigger a preliminary cascade of pro- and antiinflammatory immune reactions (secretion of IL-8, prostaglandin-E_2 and expression of the epithelial adhesion molecule-1) that initiate drifting of polymorphonuclear neutrophils (PMN) and acute inflammation.[3–7] Accordingly, experimental *C. pneumoniae* pneumonia in mice begins with local infiltration of PMN in the early stages and mononuclear cells in the later stages of the infection.[8,9] The onset of acquired antigen-specific immune responses depends on the speed of microbial dissemination from the initial site to local lymph nodes.[10]

A preliminary pulmonary infection occurs in the alveoli, where the invading *Chlamydia* undergoes phagocytosis by dendritic cells[11] or alveolar macrophages.[12,13] Alveolar macrophages are important in regulating both antimicrobial immunity and nonspecific inflammation in the lungs. Prompt destruction of intracellular microbes is dependent on the macrophage activation and associated microbicidal processes by reactive oxygen and nitrogen intermediates. Many intracellular pathogens have, however, evolved specific strategies to resist the antimicrobial activities. *Chlamydia* can modify the metabolic pathways within eukaryotic cells by inhibiting the phagolysosomal fusion and replicating in a special nonacidic chlamydial inclusion.[14–17] The inhibition involves modification of vacuolar membrane by microbe-specific proteins that are expressed early in the infectious process.[16,17] Thus, *C. pneumoniae* can survive and be metabolically active and can even multiply within phagocytes.[18–21]

The innate immunity during *C. pneumoniae* infection is hardly sufficient to eliminate *Chlamydia*. As the number of chlamydial particles eventually increases, inclusion becomes maturate and chlamydial EBs that are released from infected host cells disseminate further into the lymph nodes and into other tissues.[22,23] The migration of some infected macrophages and dendritic cells to regional lymph nodes contributes significantly to the initiation of the antigen-specific immune response.[10]

It is not known how often *C. pneumoniae* infection proceeds to symptomatic illness, but experimental animal models suggest that this microorganism is quite invasive. *C. pneumoniae* can be isolated by culture from mice lungs within a few days after inoculation. Isolation is successful from all infected animals independent on the genetic type, but certain differences can be detected in the characteristics and kinetics of developing immune responses.[24–27] It is notable, though, that the inoculation doses used in the experimental models are often high, and thus it is difficult to draw clear-cut conclusions into the events following *in vivo* exposure in mice, not to mention in humans. *C. pneumoniae* disseminates rapidly from the lungs to the spleen and aorta[22,23,28] and can pass from monocytes to endothelial cells[29] and to smooth muscle cells.[30] Infection is self-limiting in mouse and induces protective immunity that is seen as faster microbial eradication and stronger inflammation response upon reinfection than in primary infection.[8,25,26,31] The acquired immune response, both humoral and cell-mediated, is detectable from 1 to 2 weeks after the primary infection, it occurs in parallel with detected inflammation mediated by mononuclear lymphocyte accumulation in the lungs,[24–26] and is comparable to the dynamics of immune response to most intracellular bacteria, including *C. trachomatis.*[66,67,70]

3. THE ROLE OF MACROPHAGES IN *C. pneumoniae* IMMUNITY

Monocytes/macrophages and dendritic cells participate in the development of antigen-specific cell-mediated immunity by (1) acting as antigen-presenting cells (APC) and by (2) secretion of pro- (TNF-α, IL-6, IL-1, IL-12) and antiinflammatory (IL-10, IL-13, TGF-β) cytokines,[7,13,32–34] which mediate the influence of APC into other cells of the immune system. In particular, the balance of IL-12 and IL-10 is crucial in regulating the development and functional characteristics of T cell responses.

According to the traditional concept of antigen processing and presentation, the APC present antigens to CD4+ and CD8+ T cells in the context of major histocompatibility complex (MHC) class II and class I molecule, respectively.[35] In general, MHC class II molecule binds and presents antigens that are processed in the endocytic pathway whereas class I molecule binds peptides that are processed in the cellular cytoplasm derived from endogenous antigens. *Chlamydia* grows in the endocytic vesicular pathway and generally avoids a fusion with MHC class II–containing endolysosome,[14] but chlamydial peptides processed by professional APC are more likely to be presented by MHC class II molecule to CD4+ cells.[34,36] Description of genes in the chlamydial genome that encode proteins for type III secretion system[37,38] and demonstration of *C. pneumoniae* proteins secreted into the host cells' cytoplasm[39,40] suggest that some chlamydial peptides are available for MHC class I presentation. This supports the observations of class I restricted CD8+ activation during *C. pneumoniae* infection in humans[34,41] or in mouse.[25,42,43]

Cytokine-mediated regulation of cell-mediated immunity occurs either by the supporting or inhibiting of T cell responses by the proinflammatory IL-12 and IL-18 or the antiinflammatory IL-10, respectively. IL-12 and IL-18 induce activation of type 1 T cell responses and secretion of IFN-γ, which is acknowledged as the most important cytokine controlling *Chlamydia* infections,[44–46] including *C. pneumoniae* infection.[25,47–49] IFN-γ controls *Chlamydia* infection, by not only regulating the cytotoxic T cells but also through direct induction of nitric oxide synthase and the nitric oxide production in the macrophages as well as tryptophan depletion that inhibit chlamydial growth.[48,50,51] On the other hand, dominant secretion of IL-10 from macrophages and subsequently IL-4 secretion from type 2 T cells, are associated with susceptibility to *C. trachomatis* infection, poor elimination of the microorganism and appearance of granulomatous and fibrotic reactions in mice[44,52] and scarring trachoma[53] or tubal factor infertility in humans.[54] Corresponding relationship of IL-10 or other type 2 cytokines with increased susceptibility to *C. pneumoniae* infection has been demonstrated in mice.[26,42]

As type 1 and type 2 cytokines are functionally antagonistic,[55,56] early presence of IL-12 and IFN-γ is crucial for bacterial clearance and T cell polarization. It can be postulated that the early stage of emerging *Chlamydia* immunity depends on the innate defense mechanisms. Secretion of IL-12 by NK cells[57] and IL-18 by infected epithelial cells[4] interacts synergistically and promotes a type 1 cytokine differentiation pattern of macrophages (the balance of IL-12 and IL-10) and other APC. The observation that IFN-γ gis produced also by macrophages as a response to stimulation with live bacteria[58,59] or with IL-12, IL-18, or IFN-γ itself[60] suggests that autocrine IFN-γ release from macrophages participates in the control of early stages of immune response providing that the macrophages have capacity for rapid recruitment to the site of infection.

The basic mechanisms that ultimately control the balance between secreted pro- and antiinflammatory cytokines are not well known. The role of primarily infected host cells is probably more important than presently acknowledged, and the activation stage of APC may well play a crucial role in the developing immune response although *in vivo* evidence for the association of different macrophages and fate of immune response are lacking. *Chlamydia*-infected dendritic cells show upregulated MHC class II expression *in vitro*[11] and are superior in terms of antigen processing and inducing type 1 T cell responses during *Chlamydia* infection.[61,62] Although monocytes are superior over macrophages in suppressing the development of infectious progeny of *C. pneumoniae* Ebs,[20,63] the cytokine secretion is comparable between the cell types upon *in vitro* infection.[64] Accordingly, alveolar macrophages are known as poor APC.[65] Presumably, when acting as the primary host cell for chlamydial EBs, alveolar macrophages may provide a reservoir for lengthened *C. pneumoniae* infection[13] and support inflammatory responses that are characteristic to *Chlamydia*-associated diseases.

4. ACQUIRED IMMUNITY

4.1. Effector Mechanisms of Protective Immunity

Since the middle of 1990, when cell-mediated immunity was acknowledged to be crucial for protection against *C. trachomatis* infection,[45,66–68] animal models have been developed to investigate the cell-mediated immunity and its effect mechanisms in *C. pneumoniae* infection. The experimental approaches have so far disclosed the role of CD4+ and CD8+ cells and that of different cytokines in protection against *C. pneumoniae* infection (Table I).

T cells are essential for protection against *C. pneumoniae* infection in a mouse model[49] and their number increases dramatically in mouse lungs during reinfection. The number of both CD4+ and CD8+ T cells is increased,[25] both

TABLE I
Summary of Studies to Define Immunological Parameters in Acquired Immunity to *C. pneumoniae* Infection

Method	Deficiency	Major outcome	Refs.
Adoptive transfer of			
IgG antibodies	None	No protection to nude mice	69
T cells		Not studied	
In vivo depletion of			
CD4+	CD4+ T cells	Infection kinetics comparable to wild type mice in primary and reinfection	69
CD8+	CD8+ T cells	Slightly impaired clearance at the early stage of primary infection but infection clearance comparable to wild type mice	69
		Impaired protection in reinfection; infection kinetics comparable to primary infection in wild type mice	
Neutralization of			
IFN-γ		Increased number of bacteria in the lungs	47
IL-12		Impaired kinetics at the early stage of primary infection but infection clearance comparable to wild type mice	129
Gene knockout			
Nude	T cells	No clearance	69
Scid	T- and B-cell immunity	No clearance up to 60 days of follow-up	26
$CD8^{-/-}$	CD8 T cells	Impaired kinetics at the early stage of primary infection; infection cleared in 60 days	26
$CD4^{-/-}$	CD4 T cells	Impaired kinetics but primary infection cleared in 60 days	
$CD4^{-/-}$; $CD8^{-/-}$	T cells	No clearance of primary infection up to 60 days of follow-up	26
IFN-$\gamma R^{-/-}$	IFN-γ receptor	No clearance	26, 48
TNFRp55	TNF receptor p55	Impaired clearance	26
IL-$12^{-/-}$	IL-12	Impaired clearance	48

the subpopulations are apparently operating in protective immunity and neither of them is absolutely essential for clearance of primary infection in BALB/c or C57Bl mouse strains.[26,69] However, genetically altered mice lacking CD8+ cells are slightly more susceptible to infection than immunocompetent wild-type mice or mice lacking CD4+ cells, and CD8+ T cells are especially crucial at the early phase of infection.[26] Furthermore, the acquired protection detected in reinfection is abolished if mice are treated with monoclonal antibodies to deplete CD8+ cells.[69] There are less data for the role of CD4+ T cells although they are important at a later stage in the primary infection and for protection.[26] Worth noticing is that the relative role of T cell subpopulations in resolution of and protection from the infection is dependent on host species, the site of infection, inoculation dose, and chlamydial strain used in the experimental studies.[70]

Protective impact of T cells involves promotion of type 1 responses by activating other inflammatory cells, monocytes and macrophages, and cytotoxic T cells and B cells via cytokine secretion. Absence of IFN-γ in knockout mice or in mice treated with IFN-γ-neutralizing antibody has demonstrated the central role of IFN-γ for the control *C. pneumoniae* infection in mice.[26] The conventional view is that IFN-γ is derived from CD4+ T cells but experimental evidence from *C. pneumoniae*–infected mice has revealed that the impact of CD8+ may also base on their IFN-γ-secreting capacity.[26] CD8+ T cells can differentiate and produce similar cytokine patterns as CD4+ cells.[71] However, the secreted levels are often lower compared to that from CD4+ cells. CD8+ cells may kill their antigen-presenting target cells before full cytokine activation occurs.[72] According to Rottenburg *et al.*[26] CD8-mediated protective immunity to *C. pneumoniae* infection was not related with perforin, suggesting that the cytotoxic role of CD8 cells was less significant than their regulatory one mediated by their IFN-γ-secreting capacity.

In terms of the role of CD4+ or CD8+ cells in *C. pneumoniae* immunity, the key question is in the route of antigen presentation and which host cells are eventually infected. MHC class II molecules are constitutionally expressed on professional phagocytes (macrophages and dendritic cells) and only occasionally on other cells such as activated endothelial cells, which means that MHC class II–restricted T cell defense is somewhat limited in terms of *C. pneumoniae*, which can invade and live in multiple different host cells. MHC class I molecules are expressed on all nucleated cells. Thus CD8+ cells have potential for more comprehensive antigen recognition than the MHC class II–restricted CD4+ cells, which may expound their observed role in the early defense during infection. In addition, CD8+ are particularly prominent in mucosal tissue, where they may participate in the first line defense against potential pathogens and fulfill an immunological gatekeeper function.[73,74]

4.2. Cell-Mediated Immunity in Humans

C. pneumoniae infection induces development of a definite antigen-specific cell-mediated immune response that can be detected in humans[34,75,76] and in mice[25] by a positive lymphocyte proliferation (LP) or IFN-γ[34] response against recalled chlamydial antigen *in vitro*. The LP responses appear positive at the same time as the antibody responses starting to increase at 3 weeks and peak at 8 to 9 weeks after the onset of pneumonia symptoms.[34,75] As expected, after the active phase of infection and the immunization, the specific LP response decays but it remains at a clearly higher level than before infection—lasting so for years without known reinfection (Surcel *et al.*, unpublished data). High prevalence of *C. pneumoniae* infection can be confirmed by analyzing the LP reactivity, which shows an even higher prevalence of positives than the analysis of IgG antibodies does in Finland (90 vs. 78%).[77]

CD4+ T cell activation generally dominates over CD8+ when circulating lymphocytes of *C. pneumoniae*–responding healthy people are infected *in vitro* with *C. pneumoniae* (Surcel *et al.*, unpublished data).[34] Studying the development of cell-mediated immunity to *C. pneumoniae*, Halme *et al.* have found that *C. pneumoniae*–induced T cell activation was linked with both CD4+ and CD8+ cells during the active stages of a primary infection, but at later stages the relative proportion of CD4+ cells is higher than those of CD8+ cells.[34]

LP response to both *C. pneumoniae* and *C. trachomatis* cell-mediated reactivity too is mainly species-specific during the primary infection.[34,75–79] The specificity of the LP response is supported by the lack of correlation between *C. pneumoniae*– and *C. trachomatis*–induced response.[75,76] In general, the *C. pneumoniae* responses are clearly higher in seemingly healthy, seropositive population (Fig. 1), whereas the *C. trachomatis*–induced LP responses are significantly higher in women with *Chlamydia*-associated tubal factor infertility (Surcel *et al.*, unpublished data).[54] In contrast, in primary infection in healthy people, circulating lymphocytes of angiographically confirmed coronary heart disease patients respond vigorously not only to *C. pneumoniae* but also to *C. trachomatis* antigens, suggesting that conservative chlamydial structures dominate over the species-specific ones as targets for the cell-mediated responses.[80]

In the general population, there is no correlation between the levels of cell-mediated responses and the *C. pneumoniae*–specific serum IgG and IgA antibodies,[77] which is in agreement with the corresponding data on the relationship between humoral and cellular responses in patients with oculogenital[81] or genital *C. trachomatis* infections.[78,79] In contrast, in a recent *C. pneumoniae* infection, elevated IgG antibody levels correlate with significant LP reactivity to *C. pneumoniae.*[77,82] *C. pneumoniae*–induced LP reactivity tends to appear, however, at a low level in individuals with elevated IgA antibodies compared with the individuals with elevated IgG antibodies. In general, the

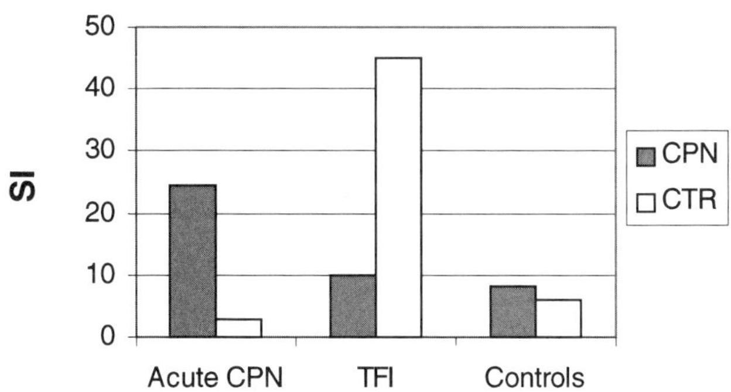

FIGURE 1. Cell-mediated responses [median stimulation index (SI) of antigen-stimulated lymphocytes *in vitro*] to *C. pneumoniae* (CPN) and to *C. trachomatis* (CTR) EB antigen in subjects with acute respiratory infection (acute CPN), with chlamydial infertility (TFI) or in healthy controls.

relationship of lowered LP response and elevated IgA levels is particularly related to smoking.[82] As smoking is clearly associated with chronic *C. pneumoniae* infection[83,84], the increased susceptibility may result from cigarette smoke–induced impairment of the immune system[85] or its switch to type 2 immune reactivity that is not optimal for the eradication of *Chlamydia.*

As expected, the LP responses in humans are not induced by only a single antigen but by multiple *C. pneumoniae* proteins. To search the antigens that are central in inducing the *C. pneumoniae*–specific T cell responses, Halme *et al.*[36] used size-fractionated (SDS-PAGE) chlamydial EB proteins as lymphocyte-stimulating antigens in *in vitro* cell cultures. Species-specific antigens were located at molecular weight (MW) ranges 92–98, 51–55, 43–46, and 31–33 kDa, and the genus-specific antigens were at MW ranges 65–70, 46–49, 26 and 12 kDa. The most frequently recognized antigens were found at MW ranges 100–125 kDa and at 50–55 kDa in one study with 30 subjects.[36,80] Characterization of the T cell–stimulating *C. pneumoniae* proteins has not yet been performed, and no data are available on the protective or immunopathogenic role of individual *C. pneumoniae* proteins in human immunity. Characterization of the protective T cell antigens in mice are in progress in connection with vaccine development and are described in more detail elsewhere in this book.

4.3. Immune Escape

The immune escape of *Chlamydia* within the host include mechanisms for evasion of T cell recognition by interfering with the expression of MHC molecules. *C. pneumoniae* can suppress MHC class II expression on human monocytes[64] or MHC class I expression on a monocytic cell line.[86] In this process, *C. pneumoniae* secretes a proteolytically active molecule,[39] a homologue of chlamydial protease-like activity factor of *C. trachomatis,* which can degrade constitutive transcription factors RFX5 and USF-1 molecules needed for MHC antigen transcription during microbial infection.[87–89] On the other hand, *Chlamydia*-induced IFN-β[90] or IL-10[86] secretion by infected macrophages has been linked with the inhibition of IFN-γ-dependent expression of MHC molecules. This kind of survival strategies may be augmented by an enhanced infectious dose. In a recent study, increasing infectious doses of *C. pneumonia* EBs in human monocytes was associated with an inverted relationship of IL-12 and IL-10, impaired T cell responses, and reduced MHC class II expression.[64] This suggests that locally accelerated growth of *C. pneumoniae* may interact with the developing immune responses and support conditions under which the infection can progress and remain chronic in susceptible individuals.

5. *C. pneumoniae*–SPECIFIC ANTIBODIES

Detection of *C. pneumoniae*–specific antibodies in human sera was the key to discovery of a new *Chlamydia* species in the 1980s[91] first associated with pneumonitis.[92,93] Since the identification of *C. pneumoniae* and the development of a microimmunofluorescence (MIF) test for the antibody analysis,[94,95] several seroepidemiological studies have been conducted to enlarge the epidemiological and immunological knowledge on *C. pneumoniae* infections. The MIF assay recognizes conformational epitopes on intact chlamydial EBs and is generally considered species-specific and the gold standard for *C. pneumoniae* serology.[96] Enhanced production of antibodies against genus-specific LPS, especially at an acute stage of *Chlamydia* infection, may, however, interfere with this specificity. Moreover, some of the surface exposure epitopes may be conservative (such as Hsp60 antigen), and primary induction of such response by other *Chlamydia* species or even other bacteria may add to the interference.

The *C. pneumoniae*–induced antibody response pattern comprise some diagnostically useful properties that are generally utilized to differentiate primary infection from reinfection. Primary *C. pneumoniae* infection is characterized by the IgM response that appears 2–3 weeks after the first symptoms of the illness. Delayed IgG response appears 6–8 weeks after the illness.[97–99] Reinfection is usually recognized by the absence of the IgM response and more rapidly appearing IgG response than in the primary infection.[1] Continuously detected high levels of IgG antibodies can be indicative for chronic or repeated infections, and differentiation of these two may often be impossible.

IgA antibodies are only occasionally found during the primary infection but appear frequently in the reinfection.[95,98,99] Activation of mucosal immunity against *C. pneumoniae* has been observed. Level of *C. pneumoniae*–specific secretory IgA detected in sputum correlate with the levels of circulating IgA.[82] Some of the antibodies are bound to immuno complexes. The significance of immuno complexes in the course of immune mechanisms is not clear.[1,100]

A positive antibody response is not necessarily linked with the symptoms of illness but it can be detected as well in asymptomatic subjects without clinical evidence of respiratory infection.[100] Although detection of chlamydial organisms in clinical specimens correlate with the presence of antibodies,[1] absence of seroconversion has been documented in children who were diagnosed with an acute, culture-positive *C. pneumoniae* infection.[101–103] This suggests that *C. pneumoniae* carriage may occur even without immune responses. *C. pneumoniae* carriage that can be demonstrated using PCR technique appears to be uncommon in subjectively healthy seropositive adults, the prevalence being less than 5% of that in the general population.[104–107] Asymptomatic carriers may, however, spread the infection as has been suggested on the basis of military outbreaks in Finland[99] or in Seattle.[108]

C. pneumoniae–specific antibodies are principally acquired in developed countries at the early school age. In the United States, first *C. pneumoniae* infection occurs between ages 5 and 14 years[109] and positive antibody levels are usually detectable for 3–5 years.[110] Reinfections are common, and the seroprevalence increases by age in the population.[109–111] The prevalence and the geometric mean titers of *C. pneumoniae*–specific antibodies are often higher in men than in women.[1] This suggests that the exposure rate and the immune responses or the general susceptibility to *C. pneumoniae* infection are different between men and women. The reasons for this gender dependency are not known but cannot be explained merely by smoking habits.[82,112]

The contribution of antibodies in protection against *C. pneumoniae* infection is not known. Passive transfer of *C. pneumoniae*-specific IgG antibodies in nude mice did not help to clear the infection from the respiratory tract.[69] Antibodies that are targeted to the antigen determinants of the *C. pneumoniae* EB surface are able to neutralize the infectivity of *C. pneumoniae in vitro*[113–118] and *in vivo.*[119] The actual contribution of neutralizing antibodies to protective immunity against *C. pneumoniae* infection is undefined. The effect of neutralizing antibodies to *C. trachomatis*[120–123] is mediated by preventing chlamydial EBs to attach to the host cells[124,125] or replication.[126,127] A recent report of Moore *et al.*[128] emphasizes the role of antibodies as supportive to T cell activation by mechanisms that involve enhanced antigen uptake, processing, and presentation.

6. CONCLUSION

C. pneumoniae has excited considerable attention during the last decade, not only as a respiratory pathogen but because of its association with a number of acute and chronic diseases, including atherosclerosis. The true linkage and causality of *C. pneumoniae* infection in the development of chronic manifestations remain puzzling. Efficient activation of type 1 T cell responses and secretion of IFN-γ mediates immune defense and provides improved recovery from reinfection. However, the factors that expose some individuals to impaired clearance of *Chlamydia* or drive immune responses to pathogenic ones are not known. The role of host genetic background, HLA molecules and cytokine gene polymorphism, environmental and epidemiological factors, mixed infections, and species or dose of the infecting agent probably all interact in a final balance of the immune defense mechanisms. A better understanding of the immunoregulatory processes during chlamydial infections in humans is necessary to furnish a novel immunotherapeutic strategy for treatment of acute *C. pneumoniae* infection and thereby to prevent development of disease manifestations that are associated with chronic or repeated infections. The understanding is facilitated by answering to the question whether the susceptibility for repeated infections is due to impaired type 1 response or activation of type 2 responses. The impact of primary host cells into developing immunity is a question of major importance and should be studied in humans.

REFERENCES

1. Kuo, C.-C., Jackson, L. A., Campbell, L. A., and Grayston, J. T., 1995, *Chlamydia pneumoniae* (TWAR), *Clin. Microbiol. Rev.* **8**:451.
2. Grayston, J. T., 2002, Background and current knowledge *of Chlamydia pneumoniae* and atherosclerosis, *J. Infect. Dis.* **181**:S402.

3. Coombes, B. K., and Mahony, J. B., 2001, cDNA array analysis of altered gene expression in human endothelial cells in response to *Chlamydia pneumoniae* infection, *Infect. Immun.* **69:**1420.
4. Rasmussen, S. J., Eckmann, L., Quayle, A. J., Shen, L., Zhang, Y. X., Anderson, D. J., Fierer, J., Stephens, R. S., and Kagnoff, M. F., 1997, Secretion of proinflammatory cytokines by epithelial cells in response to *Chlamydia* infection suggests a central role for epithelial cells in chlamydial pathogenesis, *J. Clin. Invest.* **99:**77.
5. Jahn, H. U., Krull, M., Wuppermann, F. N., Klucken, A. C., Rosseau, S., Seybold, J., Hegemann, J. H., Jantos, C. A., and Suttorp, N., 2000, Infection and activation of airway epithelial cells by *Chlamydia pneumoniae*, *J. Infect. Dis.* **182:**1678.
6. Molestina, R. E., Miller, R. D., Ramirez, J. A., and Summersgill, J. T., 1999, Infection of human endothelial cells with *Chlamydia pneumoniae* stimulates transendothelial migration of neutrophils and monocytes, *Infect. Immun.* **67:**1323.
7. Yang, J., Hooper, W. C., Phillips, D. J., Tondella, M. L., and Talkington, D. F., 2003, Induction of proinflammatory cytokines in human lung epithelial cells during *Chlamydia pneumoniae* infection, *Infect. Immun.* **71:**614.
8. Yang, Z. P., Kuo, C. C., and Grayston, J. T., 1993, A mouse model of *Chlamydia pneumoniae* strain TWAR pneumonitis, *Infect. Immun.* **61:**2037.
9. Yang, Z. P., Cummings, P. K., Patton, D. L., and Kuo, C. C., 1994, Ultrastructural lung pathology of experimental *Chlamydia pneumoniae* pneumonitis in mice, *J. Infect. Dis.* **170:**464.
10. Lipscomb, M. F., Bice, D. E., Lyons, C. R., Schuyler, M. R., and Wilkes, D., 1995, The regulation of pulmonary immunity, *Adv. Immunol.* **59:**369.
11. Prebeck, S., Kirschning, C., Durr, S., da Costa, C., Donath, B., Brand, K., Redecke, V., Wagner, H., and Miethke, T., 2001, Predominant role of toll-like receptor 2 versus 4 in *Chlamydia pneumoniae*–induced activation of dendritic cells, *J. Immunol.* **167:**3316.
12. Nakajo, M. N., Roblin, P. M., Hammerschlag, M. R., Smith, P., and Nowakowski, M., 1990, Chlamydicidal activity of human alveolar macrophages, *Infect. Immun.* **58:**3640.
13. Redecke, V., Dalhoff, K., Bohnet, S., Braun, J., and Maass, M., 1998, Interaction of *Chlamydia pneumoniae* and human alveolar macrophages: Infection and inflammatory response, *Am. J. Resp. Cell. Mol. Biol.* **19:**721.
14. Hackstadt, T., Fischer, E. R., Scidmore, M. A., Rockey, D. D., and Heinzen, R. A., 1997, Origins and functions of the chlamydial inclusion [Review] [46 refs], *Trends Microbiol.* **5:**288.
15. Al-Younes, H. M., Rudel, T., and Meyer, T. F., 1999, Characterization and intracellular trafficking pattern of vacuoles containing *Chlamydia pneumoniae* in human epithelial cells, *Cell Microbiol.* **1:**237.
16. Hackstadt, T., 2000, Redirection of host vesicle trafficking pathways by intracellular parasites, *Traffic* **1:**93.
17. Scidmore, M. A., Fischer, E. R., and Hackstadt, T., 2003, Restricted fusion of *Chlamydia trachomatis* vesicles with endocytic compartments during the initial stages of infection, *Infect. Immun.* **71:**973.
18. Gaydos, C. A., Summersgill, J. T., Sahney, N. N., Ramirez, J. A., and Quinn, T. C., 1996, Replication of *Chlamydia pneumoniae in vitro* in human macrophages, endothelial cells, and aortic artery smooth muscle cells, *Infect. Immun.* **64:**1614.
19. Kaukoranta-Tolvanen, S.-S. E., Teppo, A.-M., Laitinen, K., Saikku, P., Linnavuori, K., and Leinonen, M., 1996, Growth of *Chlamydia pneumoniae* in cultured human peripheral blood mononuclear cells and induction of a cytokine response, *Microb. Pathog.* **21:**215.
20. Airenne, S., Surcel, H. M., Alakärppä, H. Laitinen, K., Paavonen, J., Saikku, P., and Laurila, A., 1999, *Chlamydia pneumoniae* infection in human monocytes, *Infect. Immun.* **67:** 1445.
21. Kuo Cc, C. C., Puolakkainen, M., Lin, T. M., Witte, M., and Campbell, L. A., 2002, Mannose-receptor positive and negative mouse macrophages differ in their susceptibility to infection by *Chlamydia* species, *Microb. Pathog.* **32:**43.

22. Moazed, T. C., Kuo, C. C., Grayston, J. T., and Campbell, L. A., 1998, Evidence of systemic dissemination of *Chlamydia pneumoniae* via macrophages in the mouse, *J. Infect. Dis.* **177:**1322.
23. Yang, Z. P., Kuo, C. C., and Grayston, J. T., 1995, Systemic dissemination of *Chlamydia pneumoniae* following intranasal inoculation in mice, *J. Infect. Dis.* **171:**736.
24. Kaukoranta-Tolvanen, S. S., Laurila, A. L., Saikku, P., Leinonen, M., Liesirova, L., and Laitinen, K., 1993, Experimental infection of *Chlamydia pneumoniae* in mice, *Microb. Pathog.* **15:**293.
25. Penttilä, J. M., Anttila, M., Puolakkainen, M., Laurila, A., Varkila, K., Sarvas, M., Mäkelä, P. H., and Rautonen, N., 1998, Local immune responses to *Chlamydia pneumoniae* in the lungs of BALB/c mice during primary infection and reinfection, *Infect. Immun.* **66:**5113.
26. Rottenberg, M. E., Gigliotti Rothfuchs, A. C., Gigliotti, D., Svanholm, C., Bandholtz, L., and Wigzell, H., 1999, Role of innate and adaptive immunity in the outcome of primary infection with *Chlamydia pneumoniae*, as analyzed in genetically modified mice, *J. Immunol.* **162:**2829.
27. Erkkilä, L., Laitinen, K., Laurila, A., Saikku, P., and Leinonen, M., 2002, Experimental *Chlamydia pneumoniae* infection in NIH/S mice: Effect of reinoculation with chlamydial or cell preparation on culture, PCR and histological findings of lung tissue, *Vaccine* **20:** 2318.
28. Moazed, T. C., Kuo, C., Grayston, J. T., and Campbell, L. A., 1997, Murine models of *Chlamydia pneumoniae* infection and atherosclerosis, *J. Infect. Dis.* **175:**883.
29. Lin, T. M., Campbell, L. A., Rosenfeld, M. E., and Kuo, C. C., 2000, Monocyte–endothelial cell coculture enhances infection of endothelial cells with *Chlamydia pneumoniae, J. Infect. Dis.* **181:**1096.
30. Puolakkainen, M., Campbell, L. A., Lin, T. M., Richards, T., Patton, D. L., and Kuo, C. C., 2003, Cell-to-cell contact of human monocytes with infected arterial smooth-muscle cells enhances growth of *Chlamydia pneumoniae, J. Infect. Dis.* **187:**435.
31. Kaukoranta-Tolvanen, S. E., Laurila, A. L., Saikku, P., Leinonen, M., and Laitinen, K., 1995, Experimental *Chlamydia pneumoniae* infection in mice: Effect of reinfection and passive immunization, *Microb. Pathog.* **18:**279.
32. Heinemann, M., Susa, M., Simnacher, U., Marre, R., and Essig, A., 1996, Growth of *Chlamydia pneumoniae* induces cytokine production and expression of CD14 in a human monocytic cell line, *Infect. Immun.* **64:**4872.
33. Kaukoranta-Tolvanen, S. S., Ronni, T., Leinonen, M., Saikku, P., and Laitinen, K., 1996, Expression of adhesion molecules on endothelial cells stimulated by *Chlamydia pneumoniae, Microb. Pathog.* **21:**407.
34. Halme, S., Latvala, J., Karttunen, R., Palatsi, I., Saikku, P., and Surcel, H. M., 2000, Cell-mediated immune response during primary *Chlamydia pneumoniae* infection, *Infect. Immun.* **68:**7156.
35. Maksymowych, W. P., and Kane, K. P., 2000, Bacterial modulation of antigen processing and presentation, *Microb. Infect.* **2:**199.
36. Halme, S., Saikku, P., and Surcel, H. M., 1997, Characterization of *Chlamydia pneumoniae* antigens using human T cell clones, *Scand. J. Immunol.* **45:**378.
37. Stephens, R. S., Kalman, S., Lammel, C., Fan, J., Marathe, R., Aravind, L., Mitchell, W., Olinger, L., Tatusov, R. L., Zhao, Q., Koonin, E. V., and Davis, R. W., 1998, Genome sequence of an obligate intracellular pathogen of humans: *Chlamydia trachomatis* [see comments], *Science* **282:**754.
38. Kalman, S., Mitchell, W., Marathe, R., Lammel, C., Fan, J., Hyman, R. W., Olinger, L., Grimwood, J., Davis, R. W., and Stephens, R. S., 1999, Comparative genomes of *Chlamydia pneumoniae* and *C. trachomatis, Nat. Genet.* **21:**385.
39. Fan, P., Dong, F., Huang, Y., and Zhong, G., 2002, *Chlamydia pneumoniae* secretion of a protease-like activity factor for degrading host cell transcription factors is required for major histocompatibility complex antigen expression, *Infect. Immun.* **70:**345.

40. Shaw, A. C., Vandahl, B. B., Larsen, M. R., Roepstorff, P., Gevaert, K., Vandekerckhove, J., Christiansen, G., and Birkelund, S., 2002, Characterization of a secreted *Chlamydia protease, Cell Microbiol.* **4:**411.
41. Nadareishvili, Z. G., Koziol, D. E., Szekely, B., Ruetzler, C., LaBiche, R., McCarron, R., and DeGraba, T. J., 2001, Increased CD8(+) T cells associated with *Chlamydia pneumoniae* in symptomatic carotid plaque, *Stroke* **32:**1966.
42. Bandholtz, L., Kreuger, M. R., Svanholm, C., Wigzell, H., and Rottengerg, M. E., Adjuvant modulation of the immune responses and the outcome of infection with *Chlamydia pneumoniae, Clin. Exp. Immunol.* **130:**393.
43. Svanholm, C., Bandholtz, L., Castanos-Velez, E., Wigzell, H., and Rottenberg, M. E., 2000, Protective DNA immunization against *Chlamydia pneumoniae, Scand. J. Immunol.* **51:** 345.
44. Yang, X., HayGlass, K. T., and Brunham, R. C., 1996, Genetically determined differences in IL-10 and IFN-gamma responses correlate with clearance of *Chlamydia trachomatis* mouse pneumonitis infection, *J. Immunol.* **156:**4338.
45. Perry, L. L., Feilzer, K., and Caldwell, H. D., 1997, Immunity to *Chlamydia trachomatis* is mediated by T helper 1 cells through IFN-gamma–dependent and –independent pathways, *J. Immunol.* **158:**3344.
46. Wang, S., Fan, Y., Brunham, R. C., and Yang, X., 1999, IFN-gamma knockout mice show Th2-associated delayed-type hypersensitivity and the inflammatory cells fail to localize and control chlamydial infection, *Eur. J. Immunol.* **29:**3782.
47. Vuola, J. M., Puurula, V., Anttila, M., Mäkelä, P. H., and Rautonen, N., 2000, Acquired immunity to *Chlamydia pneumoniae* is dependent on gamma interferon in two mouse strains that initially differ in this respect after primary challenge, *Infect. Immun.* **68:** 960.
48. Rottenberg, M. E., Gigliotti Rothfuchs, A., Gigliotti, D., Ceausu, M., Une, C., Levitsky, V., and Wigzell, H., 2000, Regulation and role of IFN-gamma in the innate resistance to infection with *Chlamydia pneumoniae, J. Immunol.* **164:**4812.
49. Geng, Y., Shane, R. B., Berencsi, K., Gonczol, E., Zaki, M. H., Margolis, D. J., Trinchieri, G., and Rook, A. H., 2000, *Chlamydia pneumoniae* inhibits apoptosis in human peripheral blood mononuclear cells through induction of IL-10, *J. Immunol.* **164:**5522.
50. Byrne, G. I., Grubbs, B., Marshall, T. J., Schachter, J., and Williams, D. M., 1988, Gamma interferon-mediated cytotoxicity related to murine *Chlamydia trachomatis* infection, *Infect. Immun.* **56:**2023.
51. Rottenberg, M. E., Gigliotti-Rothfuchs, A., and Wigzell, H., 2002, The role of IFN-gamma in the outcome of chlamydial infection, *Curr. Opin. Immunol.* **14:**444.
52. Yang, X., 2001, Distinct function of Th1 and Th2 type delayed type hypersensitivity: Protective and pathological reactions to chlamydial infection, *Microsc. Res. Tech.* **53:**273.
53. Holland, M. J., Bailey, R. L., Conway, D. J., Culley, F., Miranpuri, G., Byrne, G. I., Whittle, H. C., and Mabey, D. C., 1996, T helper type-1 (Th1)/Th2 profiles of peripheral blood mononuclear cells (PBMC): Responses to antigens of *Chlamydia trachomatis* in subjects with severe trachomatous scarring, *Clin. Exp. Immun.* **105:**429.
54. Kinnunen, A. H., Surcel, H. M., Lehtinen, M., Karhukorpi, J., Tiitinen, A., Halttunen, M., Bloigu, A., Morrison, R. P., Karttunen, R., and Paavonen, J., 2002, HLA DQ alleles and interleukin-10 polymorphism associated with *Chlamydia trachomatis*–related tubal factor infertility: A case–control study, *Hum. Reprod.* **17:**2073.
55. Mosmann, T. R., and Coffman, R. L., 1989, TH1 and TH2 cells: Different patterns of lymphokine secretion lead to different functional properties, *Ann. Rev. Immun.* **7:**145.
56. Romagnani, S., Parronchi, P., D'Elios, M. M., Romagnani, P., Annunziato, F., Piccinni, M. P., Manetti, R., Sampognaro, S., Mavilia, C., De Carli, M., Maggi, E., and Del Prete, G. F., 1997, An update on human Th1 and Th2 cells, *Int. Arch. Allerg. Immun.* **113:**153.
57. Tseng, C. T., and Rank, R. G., 1998, Role of NK cells in early host response to chlamydial genital infection, *Infect. Immun.* **66:**5867.

58. Fenton, M. J., Vermeulen, M. W., Kim, S., Burdick, M., Strieter, R. M., and Kornfeld, H., 1997, Induction of gamma interferon production in human alveolar macrophages by *Mycobacterium tuberculosis, Infect. Immun.* **65:**5149.
59. Rothfuchs, A. G., Gigliotti, D., Palmblad, K., Andersson, U., Wigzell, H., and Rottenberg, M. E., 2001, IFN-alpha beta-dependent, IFN-gamma secretion by bone marrow–derived macrophages controls an intracellular bacterial infection, *J. Immunol.* **167:**6453.
60. Frucht, D. M., Fukao, T., Bogdan, C., Schindler, H., O'Shea, J. J., and Koyasu, S., 2001, IFN-gamma production by antigen-presenting cells: Mechanisms emerge, *Trends Immunol.* **22:**556.
61. Stagg, A. J., Stackpoole, A., Elsley, W. J., and Knight, S. C., 1993, Acquisition of chlamydial antigen by dendritic cells and monocytes, *Adv. Exp. Med. Biol.* **329:**581.
62. Ojcius, D. M., Bravo de Alba, Y., Kanellopoulos, J. M., Hawkins, R. A., Kelly, K. A., Rank, R. G., and Dautry-Varsat, A., 1998, Internalization of Chlamydia by dendritic cells and stimulation of Chlamydia-specific T cells, *J. Immunol.* **160:**1297.
63. Airenne, S., Surcel, H. M., Bloigu, A., Laitinen, K., Saikku, P., and Laurila, A., 2000, The resistance of human monocyte–derived macrophages to *Chlamydia pneumoniae* infection is enhanced by interferon-gamma, *Apmis.* **108:**139.
64. Airenne, S., Kinnunen, A., Leinonen, M., Saikku, P., and Surcel, H.-M., 2002, Secretion of IL-10 and IL-12 in *Chlamydia pneumoniae* infected human monocytes, presented at *Chlamydial Infections Proceedings of the Tenth International Symposium on Human Chlamydial Infections*, Antalya, Turkey.
65. Nicod, L. P., Cochand, L., and Dreher, D., 2000, Antigen presentation in the lung: Dendritic cells and macrophages, *Sarc. Vasc. Diff. Lung Dis.* **17:**246.
66. Igietseme, J. U., Smith, K., Simmons, A., and Rayford, P. L., 1995, Effect of gamma-irradiation on the effector function of T lymphocytes in microbial control, *Int. J. Rad. Biol.* **67:**557.
67. Su, H., and Caldwell, H. D., 1995, CD4+ T cells play a significant role in adoptive immunity to *Chlamydia trachomatis* infection of the mouse genital tract, *Infect. Immun.* **63:** 3302.
68. Morrison, R. P., Feilzer, K., and Tumas, D. B., 1995, Gene knockout mice establish a primary protective role for major histocompatibility complex class II–restricted responses in *Chlamydia trachomatis* genital tract infection, *Infect. Immun.* **63:**4661.
69. Penttilä, J. M., Anttila, M., Varkila, K., Puolakkainen, M., Sarvas, M., Mäkelä, P. H., and Rautonen, N., 1999, Depletion of CD8+ cells abolishes memory in acquired immunity against *Chlamydia pneumoniae* in BALB/c mice, *Immunology* **97:**490.
70. Rank, R. G., 1999, Models of immunity, in: *Chlamydia: Intracellular Biology, Pathogenesis and Immunity* (R. S. Stephens, ed.), American society for Microbiology, Washington, p. 239.
71. Li, L., Sad, S., Kagi, D., and Mosmann, T. R., 1997, CD8Tc1 and Tc2 cells secrete distinct cytokine patterns *in vitro* and *in vivo* but induce similar inflammatory reactions, *J. Immunol.* **158:**4152.
72. Mosmann, T. R., Li, L., and Sad, S., 1997, Functions of CD8 T-cell subsets secreting different cytokine patterns, *Semin. Immunol.* **9:**87.
73. Mayer, L., 1997, Review article: Local and systemic regulation of mucosal immunity, *Aliment. Pharmacol. Ther.* **11**(Suppl. 3)**:**81.
74. Hingorani, M., Metz, D., and Lightman, S. L., 1997, Characterisation of the normal conjunctival leukocyte population, *Exp. Eye Res.* **64:**905.
75. Surcel, H. M., Syrjälä, H., Leinonen, M., Saikku, P., and Herva, E., 1993, Cell-mediated immunity to *Chlamydia pneumoniae* measured as lymphocyte blast transformation *in vitro, Infect. Immun.* **61:**2196.
76. Braun, J., Laitko, S., Treharne, J., Eggens, U., Wu, P., Distler, A., and Sieper, J., 1994, *Chlamydia pneumoniae*—A new causative agent of reactive arthritis and undifferentiated oligoarthritis, *Ann. Rheum. Dis.* **53:**100.

77. Halme, S., von Hertzen, L., Bloigu, A., Kaprio, J., Koskenvuo, M., Leinonen, M., Saikku, P., and Surcel, H. M., 1998, *Chlamydia pneumoniae*–specific cell-mediated and humoral immunity in healthy people, *Scand. J. Immun.* **47:**517.
78. Hanna, L., Schmidt, L., Sharp, M., Stites, D. P., and Jawetz, E., 1979, Human cell-mediated immune responses to chlamydial antigens, *Infect. Immun.* **23:**412.
79. Brunham, R. C., Martin, D. H., Kuo, C. C., Wang, S. P., Stevens, C. E., Hubbard, T., and Holmes, K. K., 1981, Cellular immune response during uncomplicated genital infection with *Chlamydia trachomatis* in humans, *Infect. Immun.* **34:**98.
80. Halme, S., Syrjälä, H., Bloigu, A., Saikku, P., Leinonen, M., Airaksinen, J., and Surcel, H. M., 1997, Lymphocyte responses to *Chlamydia antigens* in patients with coronary heart disease, *Eur. Heart J.* **18:**1095.
81. Mabey, D. C., Holland, M. J., Viswalingam, N. D., Goh, B. T., Estreich, S., Macfarlane, A., Dockrell, H. M., and Treharne, J. D., 1991, Lymphocyte proliferative responses to chlamydial antigens in human chlamydial eye infections, *Clin. Exp. Immunol.* **86:**37.
82. von Hertzen, L., Surcel, H. M., Kaprio, J., Koskenvuo, M., Bloigu, A., Leinonen, M., and Saikku, P., 1998, Immune responses to *Chlamydia pneumoniae* in twins in relation to gender and smoking, *J. Med. Microbiol.* **47:**441.
83. File, T. M., 2000, The epidemiology of respiratory tract infections, *Semin. Respir. Infect.* **15:**184.
84. Smieja, M., Leigh, R., Petrich, A., Chong, S., Kamada, D., Hargreave, F. E., Goldsmith, C. H., Chernesky, M., and Mahony, J. B., 2002, Smoking, season, and detection of *Chlamydia pneumoniae* DNA in clinically stable COPD patients, *BMC Infect. Dis.* **2:**12.
85. Sopori, M. L., and Kozak, W., 1998, Immunomodulatory effects of cigarette smoke, *J. Neuroimmunol.* **83:**148.
86. Caspar-Bauguil, S., Puissant, B., Nazzal, D., Lefevre, J. C., Thomsen, M., Salvayre, R., and Benoist, H., 2000, *Chlamydia pneumoniae* induces interleukin-10 production that down-regulates major histocompatibility complex class I expression, *J. Infect. Dis.* **182:** 1394.
87. Zhong, G., Fan, T., and Liu, L., 1999, *Chlamydia* inhibits interferon gamma–inducible major histocompatibility complex class II expression by degradation of upstream stimulatory factor 1, *J. Exp. Med.* **189:**1931.
88. Zhong, G., Liu, L., Fan, T., Fan, P., and Ji, H., 2000, Degradation of transcription factor RFX5 during the inhibition of both constitutive and interferon gamma–inducible major histocompatibility complex class I expression in chlamydia-infected cells, *J. Exp. Med.* **191:**1525.
89. Zhong, G., Fan, P., Ji, H., Dong, F., and Huang, Y., 2001, Identification of a chlamydial protease-like activity factor responsible for the degradation of host transcription factors, *J. Exp. Med.* **193:**935.
90. Rodel, J., Groh, A., Vogelsang, H., Lehmann, M., Hartmann, M., and Straube, E., 1998, Beta interferon is produced by *Chlamydia trachomatis*–infected fibroblast-like synoviocytes and inhibits gamma interferon–induced HLA-DR expression, *Infect. Immun.* **66:**4491.
91. Grayston, J. T., Kuo, C.-C., Campbell, L. A., and Wang, S.-P., 1989, *Chlamydia pneumoniae* sp. nov. for *Chlamydia* sp. strain TWAR, *Int. J. Syst. Bact.* **39:**88.
92. Saikku, P., Wang, S. P., Kleemola, M., Brander, E., Rusanen, E., and Grayston, J. T., 1985, An epidemic of mild pneumoniae due to an unusual strain of *Chlamydia psittaci*, *J. Infect. Dis.* **151:**832.
93. Grayston, J. T., Kuo, C.-C., Wang, S.-P., and Altman, J., 1986, A new *Chlamydia psittaci* strain, TWAR, isolated in acute respiratory tract infection, *New Eng. J. Med.* **315:**161.
94. Wang, S.-P., and Grayston, J. T., 1970, Immunologic relationship between genital TRIC, lymphogranuloma venereum, and organisms in a new microtiter indirect immunofluorescence test, *Am. J. Ophthal.* **70:**367.
95. Wang, S., 2000, The microimmunofluorescence test for *Chlamydia pneumoniae* infection: Technique and interpretation, *J. Infect. Dis.* **181:**S421.

96. Dowell, S. F., Peeling, R. W., Boman, J., Carlone, G. M., Fields, B. S., Guarner, J., Hammerschlag, M. R., Jackson, L. A., Kuo, C. C., Maass, M., Messmer, T. O., Talkington, D. F., Tondella, M. L., and Zaki, S. R., 2001, Standardizing *Chlamydia pneumoniae* assays: Recommendations from the Centers for Disease Control and Prevention (USA) and the Laboratory Centre for Disease Control (Canada), *Clin. Infect. Dis.* **33:**492.
97. Grayston, J. T., Campbell, L. A., Kuo, C. C., Mordhorst, C. H., Saikku, P., Thom, D. H., and Wang, S. P., 1990, A new respiratory tract pathogen: *Chlamydia pneumoniae* strain TWAR, *J. Infect. Dis.* **161:**618.
98. Ekman, M. R., Leinonen, M., Syrjälä, H., Linnanmäki, E., Kujala, P., and Saikku, P., 1993, Evaluation of serological methods in the diagnosis of *Chlamydia pneumoniae* pneumonia during an epidemic in Finland, *Eur. J. Clin. Microbiol. Infect. Dis.* **12:**756.
99. Ekman, M. R., Grayston, J. T., Visakorpi, R., Kleemola, M., Kuo, C. C., and Saikku, P., 1993, An epidemic of infections due to *Chlamydia pneumoniae* in military conscripts, *Clin. Infect. Dis.* **17:**420.
100. Kauppinen, M., and Saikku, P., 1995, Pneumonia due to *Chlamydia pneumoniae*: Prevalence, clinical features, diagnosis, and treatment, *Clin. Infect. Dis.* **21** (Suppl. 3)**:**5244.
101. Block, S., Hedrick, J., Hammerschlag, M. R., Cassell, G. H., and Craft, J. C., 1995, *Mycoplasma pneumoniae* and *Chlamydia pneumoniae* in pediatric community-acquired pneumonia: Comparative efficacy and safety of clarithromycin vs. erythromycin ethylsuccinate, *Pediatr. Infect. Dis. J.* **14:**471.
102. Harris, J. A., Kolokathis, A., Campbell, M., Cassell, G. H., and Hammerschlag, M. R., 1998, Safety and efficacy of azithromycin in the treatment of community-acquired pneumonia in children, *Pediatr. Infect. Dis. J.* **17:**865.
103. Kutlin, A., Roblin, P. M., and Hammerschlag, M. R., 1998, Antibody response to *Chlamydia pneumoniae* infection in children with respiratory illness, *J. Infect. Dis.* **177:**720.
104. Gnarpe, J., Gnarpe, H., and Sundelof, B., 1991, Endemic prevalence of *Chlamydia pneumoniae* in subjectively healthy persons, *Scand. J. Infect. Dis.* **23:**387.
105. Hyman, C. L., Roblin, P. M., Gaydos, C. A., Quinn, T. C., Schachter, J., and Hammerschlag, M. R., 1995, Prevalence of asymptomatic nasopharyngeal carriage of *Chlamydia pneumoniae* in subjectively healthy adults: Assessment by polymerase chain reaction–enzyme immunoassay and culture, *Clin. Infect. Dis.* **20:**1174.
106. Miyashita, N., Fukano, H., Niki, Y., Matsushima, T., and Okimoto, N., 2001, Etiology of community-acquired pneumonia requiring hospitalization in Japan, *Chest* **119:** 1295.
107. Dalhoff, K., and Maass, M., 1996, *Chlamydia pneumoniae* pneumonia in hospitalized patients. Clinical characteristics and diagnostic value of polymerase chain reaction detection in BAL, *Chest* **110:**351.
108. Thom, D. H., Grayston, J. T., Campbell, L. A., Kuo, C. C., Diwan, V. K., and Wang, S. P., 1994, Respiratory infection with *Chlamydia pneumoniae* in middle-aged and older adult outpatients, *Eur. J. Clin. Microbiol. Infect. Dis.* **13:**785.
109. Aldous, M. B., Grayston, J. T., Wang, S. P., and Foy, H. M., 1992, Seroepidemiology of *Chlamydia pneumoniae* TWAR infection in Seattle families, 1966–1979, *J. Infect. Dis.* **166:** 646.
110. Patnode, D., Wang, S.-P., and Grayston, J. T., 1990, *Persistence of* Chlamydia pneumoniae, *Strain TWAR Micro-Immunofluorescent Antibody*, Cambridge.
111. Grayston, J. T., Campbell, L. A., Kuo, C.-C., Mordhorst, C. H., Saikku, P., Thom, D. H., and Wang, S.-P., 1990, A new respiratory tract pathogen: *Chlamydia pneumoniae* strain TWAR, *J. Infect. Dis.* **161:**618.
112. Karvonen, M., Tuomilehto, J., Pitkäniemi, J., Naukkarinen, A., and Saikku, P., 1994, Importance of smoking for *Chlamydia pneumoniae* seropositivity, *Int. J. Epidemiol.* **23:** 1315.
113. Kawa, D. E., and Stephens, R. S., 2002, Antigenic topology of chlamydial PorB protein and identification of targets for immune neutralization of infectivity, *J. Immunol.* **168:**5184.

114. Wiedmann-Al-Ahmad, M., Schuessler, P., and Freidank, H. M., 1997, Reactions of polyclonal and neutralizing anti-p54 monoclonal antibodies with an isolated, species-specific 54-kilodalton protein of *Chlamydia pneumoniae, Clin. Diag. Lab. Immunol.* **4:**700.
115. Peterson, E. M., Cheng, X., Qu, Z., and de La Maza, L. M., 1996, Characterization of the murine antibody response to peptides representing the variable domains of the major outer membrane protein of *Chlamydia pneumoniae, Infect. Immun.* **64:**3354.
116. Donati, M., Rumpianesi, F., Pavan, G., D'Apote, L., and Cevenini, R., 1996, Detection of serum antibodies against *Chlamydia pneumoniae* by *in vitro* neutralization and microimmunofluorescence assays, *Zentralblatt fur Bakteriologie* **284:**52.
117. Qu, Z., Cheng, X., de la Maza, L. M., and Peterson, E. M., 1993, Characterization of a neutralizing monoclonal antibody directed at variable domain I of the major outer membrane protein of *Chlamydia trachomatis* C-complex serovars, *Infect. Immun.* **61:**1365.
118. Peterson, E. M., Cheng, X., Markoff, B. A., Fielder, T. J., and de la Maza, L. M., 1991, Functional and structural mapping of *Chlamydia trachomatis* species-specific major outer membrane protein epitopes by use of neutralizing monoclonal antibodies, *Infect. Immun.* **59:**4147.
119. Peterson, E. M., de La Maza, L. M., Brade, L., and Brade, H., 1998, Characterization of a neutralizing monoclonal antibody directed at the lipopolysaccharide of *Chlamydia pneumoniae, Infect. Immun.* **66:**3848.
120. Rank, R. G., and Batteiger, B. E., 1989, Protective role of serum antibody in immunity to chlamydial genital infection, *Infect. Immun.* **57:**299.
121. Cotter, T. W., Meng, Q., Shen, Z. L., Zhang, Y. X., Su, H., and Caldwell, H. D., 1995, Protective efficacy of major outer membrane protein–specific immunoglobulin A (IgA) and IgG monoclonal antibodies in a murine model of *Chlamydia trachomatis* genital tract infection, *Infect. Immun.* **63:**4704.
122. Su, H., Feilzer, K., Caldwell, H. D., and Morrison, R. P., 1997, *Chlamydia trachomatis* genital tract infection of antibody-deficient gene knockout mice, *Infect. Immun.* **65:**1993.
123. Morrison, S. G., Su, H., Caldwell, H. D., and Morrison, R. P., 2000, Immunity to murine *Chlamydia trachomatis* genital tract reinfection involves B cells and CD4(+) T cells but not CD8(+) T cells, *Infect. Immun.* **68:**6979.
124. Byrne, G. I., and Moulder, J. W., 1978, Parasite-specified phagocytosis of *Chlamydia psittaci* and *Chlamydia trachomatis* by L and HeLa cells, *Infect. Immun.* **19:**598.
125. Su, H., and Caldwell, H. D., 1991, *In vitro* neutralization of *Chlamydia trachomatis* by monovalent Fab antibody specific to the major outer membrane protein, *Infect. Immun.* **59:**2843.
126. Wyrick, P. B., Brownridge, E. A., and Ivins, B. E., 1978, Interaction of *Chlamydia psittaci* with mouse peritoneal macrophages, *Infect. Immun.* **19:**1061.
127. Caldwell, H. D., and Perry, L. J., 1982, Neutralization of *Chlamydia trachomatis* infectivity with antibodies to the major outer membrane protein, *Infect. Immun.* **38:**745.
128. Moore, T., Ananaba, G. A., Bolier, J., Bowers, S., Belay, T., Eko, F. O., and Igietseme, J. U., 2002, Fc receptor regulation of protective immunity against *Chlamydia trachomatis, Immunology* **105:**213.
129. Geng, Y., Berencsi, K., Gyulai, Z., Valyi-Nagy, T., Gonczol, E., and Trinchieri, G., 2000, Roles of interleukin-12 and gamma interferon in murine *Chlamydia pneumoniae* infection, *Infect. Immun.* **68:**2245.

8

Vaccines Against *Chlamydia pneumoniae*: Can They Be Made?

MARTIN E. ROTTENBERG, ANTONIO GIGLIOTTI ROTHFUCHS, and HANS WIGZELL

1. INTRODUCTION

The introduction of vaccines during the early part of the last century and the introduction of antibiotics and modern hygiene practices a few decades later contributed to the decline of infectious diseases responsible for much of the morbidity and mortality of humans during recorded history. Although immunization is the most cost-effective and efficient means to control microbial diseases, vaccines are not yet available to prevent many major bacterial infections. Examples include chlamydial infections, dysentery (shigellosis), gonorrhoea, gastric ulcers, and maybe even cancer (*Helicobacter pylori*). Moreover, while some vaccines against external bacteria or bacterial products have been successful, vaccines against many intracellular bacteria are more difficult or so far impossible to produce. Improved vaccines are also needed to combat some diseases such as TB for which current vaccines are inadequate. However, vaccine development appears not to be highly profitable: worldwide, vaccine potential sales are estimated to be approximately $6.5 billion, which represents only about 2% of the global pharmaceutical market, an amount roughly equivalent to the sales of one successful ulcer drug. The costs of vaccine development to the commercial sector are also huge and the process lengthy (5–10 years).[1]

Vaccines against chlamydial infections are also lacking. Development of chlamydial vaccines also implies scientific challenges of understanding the

MARTIN E. ROTTENBERG, ANTONIO GIGLIOTTI ROTHFUCHS, and HANS WIGZELL • Microbiology & Tumorbiology Center, Karolinska Institute, S 171 77 Stockholm, Sweden.

complex biology and antigenic structure of these bacteria. These issues will be analyzed here, in relation to *C. pneumoniae* in particular.

A heightened understanding of protective immunity to *C. pneumoniae* infection is emerging from studies using a mouse model of chlamydial respiratory infection. The insights are of considerable interest and offer promise for the development of an efficacious chlamydial vaccine. The availability of complete genome sequences, including that of *C. pneumoniae*,[2] has revolutionized the possibilities of developing vaccines. Never before has there been a catalogue of all vaccine candidate molecules from which to choose the one most likely to be effective. How to arrive at a judicious decision on which of the many candidates to take forward into the experimental and human trials is now where the challenge is shifted to.

2. DO WE NEED AN ANTI–*C. pneumoniae* VACCINE?

A better knowledge of the natural history of *C. pneumoniae* is required in order to identify the target populations to be vaccinated against *C. pneumoniae* and to decide whether prophylactic and/or therapeutic vaccines would be most desirable. The issue of chronic infection versus repeated infection; persistent quiescent infection (which has been unquestionably shown *in vitro*[3]) or persistent active infection; the presence of periods of reactivation and the clinical significance of super infection, need further studies.

Whether there are sufficient benefits to justify children's vaccination seems at present unlikely, since chlamydial respiratory tract infection in children is generally mild. However, transmission of infection to siblings or parents, and the consequences of infection in asthmatic pediatric patients may here be significant parameters to consider.[4] The prevention of arteriosclerosis by blocking *C. pneumoniae* infection is unlikely to provide a justification for a vaccine for children, until a truly causal role of *C. pneumoniae* in the development of atherosclerosis has been shown to exist. If *C. pneumoniae* is proven to be a cause of atherosclerosis, a therapeutic and effective vaccine would be of immense value. In fact, the development and use of an effective *C. pneumoniae* vaccine might provide the best experiment to investigate the role of *C. pneumoniae* in cardiovascular disease.

An improved understanding of the biology of *C. pneumoniae* in adults is also necessary to justify a vaccine for this group. The elderly might be a realistic target for vaccination since the impact of a respiratory tract disease caused by *C. pneumoniae* can here be serious. Likewise, there is increasing evidence of the involvement of *C. pneumoniae* infection in bronchial asthma, and a pathological role of this agent in immunocompromised patients has also begun to be appreciated.[5] Thus, adults suffering from chronic bronchitis, asthma, or chronic obstructive airway disease (COPD) might also benefit from vaccination.

3. *C. pneumoniae* VACCINE DEVELOPMENT

The complex biology and antigenic composition and the fact that infection may generate both protective and pathological immune responses present obstacles in achieving a *C. pneumoniae* vaccine. Chlamydial infections often recur or remain persistent, indicating absence of sterilizing immunity. However, studies in an experimental model of infection with the related organism *C. trachomatis* indicate a short-lived immunity after natural infection. The resistance to genital chlamydial diseases augments with age (and hence exposure), and vaccination with inactivated organisms produce a short-lived protection against ocular challenge.

C. pneumoniae provokes a similar lung pathology in humans and rodents: high infective doses in mice produced acute patchy pneumonia with PMN infiltration and alveolar and bronchiolar exudates, whereas lower infective doses produced more chronic inflammation, developing gradually with perivascular and peribronchial lymphocyte infiltration.[6,7] In the mouse pneumonitis model, bronchial ciliated epithelial cells and interstitial macrophages were shown to be infected after intranasal inoculation.

The immune response against chlamydia can mediate pathogenesis, as exemplified by the sensitization to a more severe disease in individuals vaccinated against *C. trachomatis*. It is speculated that this may be due to the presence of proinflammatory molecules such as LPS or cross-reactive antigens with host molecules. Protective or adverse effects may not only depend on specific antigen(s) but also on innate immune mechanisms that are mobilized by the infection. Diverse innate immune mechanisms are known not only to constitute a first barrier against pathogens but also dictate the quality of the clonal-dependent immune responses elicited. Thus, results from our laboratory suggest that the quality of specific responses against the same antigen, depending on the adjuvant used, would determine the outcome of infection in the direction of protection or increased susceptibility. Unequivocally, heat shock protein–60 (Hsp60) and the outer membrane protein (OMP-2) could both induce protection if inoculated as a DNA vaccine or in presence of CpG, whereas inoculation of these proteins in Freund's complete adjuvant resulted in increased bacterial proliferation and pathology.[8] This latter increased susceptibility was associated with enhanced Th2 immune responses. Thus, knowledge of innate immune mechanisms evoked are central in vaccine design.

Facts about immune mechanisms involved in human or experimental *C. pneumoniae* respiratory infections are scanty compared with what is known about diseases caused by *C. trachomatis*. Knowledge from the experimental *C. trachomatis* infections might at least in part be supportive for an anti–*C. pneumoniae* vaccine development, since they share a common biology, with more than 80% of their genes being orthologs.

3.1. Adaptive Immune Responses

Both CD8+ and CD4+ T cells produce IFN-γ in response to *C. pneumoniae* infection, and are probably complementary in warranting protective levels of this cytokine, which is a hallmark of Th1 immune responses. IFN-γ is needed for protection of mice against *C. pneumoniae*, as shown by the enhanced susceptibility of mice genetically deficient in IFN-γ, IFN-γR, or administered with neutralizing anti-IFN-γ antibodies.[9] IFN-γ enhances accumulation of the inducible nitric oxide synthase (iNOS), indoleamine-2,3-dioxygenase and the gp91 subunit of the phagocyte NADPH oxidase. iNOS activity partially accounts for the IFN-γ-mediated protection of mice against *C. pneumoniae* pulmonary infection.[10] The depletion of intracellular tryptophan pools by indoleamine-2,3-dioxygenase is inhibitory of chlamydial growth because of the auxotrophy of the parasite tryptophan. IFN-γ may however have a dual role in controlling the outcome of chlamydial infection *in vivo*, as exposure of infected cells to high IFN-γ concentration irreversibly inhibits chlamydial replication whereas lower concentrations induce the formation of persistent forms.[3] Importantly, IFN-γ is not only secreted by T cells but also by cells of the innate immune response, such as NK cells, macrophages, and dendritic cells.

Adaptive immune protection against *C. trachomatis* in mice can be demonstrated by transfer of bacteria-specific CD4+ or CD8+ T cells.[11,12] Both subsets are also recruited to sites of infection in nonhuman primate models. IFN-γ release by T cells has been associated to the protective ability of DNA vaccines against *C. trachomatis.*[13,14] Likewise, wild-type but not IFN-γ$R^{-/-}$ mice were partially protected against *C. pneumoniae* by a DNA vaccine expressing chlamydial Hsp60-inducing IFN-γ secretion by T cells.[15]

A role for CD8+ T cells in resistance to *C. pneumoniae* and *C. trachomatis* infection is likely, since mice depleted of CD8+ T cells by antibody administration and $CD8^-$, $ß2M^-$, or $ß2M^-/TAP1^-$ mice all exhibited exacerbated infection.[10,12,16] Specific CD8+ T cells recognize products of chlamydial antigens expressed in the cytoplasm of the infected cells or peptides derived from different outer membrane proteins, chaperones, the homologue to Inca, DNAk, and Hsp60, and a family of 43-kDa homologues.[17–21] CD8+ T cell lines generated against such peptides lysed *C. pneumoniae*–infected cells and synthesized IFN-γ. Lysis of cells containing the intracellular replicative reticulate bodies would be an attractive way to interfere with infection since this form is noninfectious. However, clearance of chlamydial infections *in vivo* is normal in the absence of perforin or Fas, suggesting that CD8+ T cell–mediated cytolysis plays no significant role in protection.[10,22] However, chlamydia-specific CD8+ T cells, derived from IFN-γ-deficient mice, failed to provide protection under conditions where the wild-type CD8+ T cells did, suggesting a role for IFN-γ in such CD8+ T cell resistance to chlamydia.[23] Thus, production of IFN-γ by *C. pneumoniae*–specific T cells may be critical to achieving local threshold levels, inhibiting *C. pneumoniae* growth and persistence. A major challenge here is to produce a strong Th1 response conferring protection without immunopathology.

Despite their intravacuolar localization, chlamydiae interact with multiple host cell processes to ensure that the inclusion is a safe niche for survival and replication. Presence of a type III secretion system allowing export of selected molecules from inclusion into the cytoplasm is necessary for chlamydial survival, but may also allow MHC class I presentation of chlamydial antigens.[24] However, of the currently identified MHC class I–restricted protein antigens, the majority have not been reported to be secreted into the cytoplasm. The route by which such peptides become presented by MHC class I molecules remains unknown.

The relative role of adaptive immune mechanisms in control of the primary infection is not necessarily similar to that creating resistance to a recall infection. Thus, the resolution of a primary *C. trachomatis* infection of μMT gene knockout mice (i.e., B cell–deficient) is equivalent to that of immunocompetent wild-type (WT) mice, whereas B cell–deficient mice are in contrast to WT mice uniformly susceptible to reinfection and exhibit delayed clearance of chlamydiae,[25] implicating a functional role for B cells/products in adaptive recall protective immunity. Antibodies binding to the elementary bodies may neutralize chlamydial infectivity and will reduce infection by blocking chlamydial attachment to epithelial cells. This protective mechanism is supported by *in vitro* studies showing that antibodies to the chlamydial major outer membrane protein (MOMP) block attachment, and subsequent infectivity, of chlamydiae to epithelial cells.[26–28]

Whether antibodies can play a protective role once Chlamydia has infected the target cell is doubtful, and even more if a persistent infection is established, during which assembly of extracellular forms is low or absent. However, chlamydial antigens on the surfaces of infected cells could function as ADCC targets.

One approach to identify the elements of acquired immunity important in resistance to *C. trachomatis* is to observe how the immune system responds to natural infections or reinfections. However, immunity after a natural *C. pneumoniae* infection is not completely protective. Thus, another equally important approach is to develop immunization strategies that prime those arms of the immune response that are most effective in reducing the replicative capacity of the organism, as such responses may well have been selected against during natural infections.

3.2. Escape Mechanisms

Chlamydia has developed a numbers of ways to evade aggressive immune responses and secure survival. Chlamydial molecules involved in such escapes are probably relevant in establishing persistent infections, and relevant potential targets for a therapeutic vaccine.

Apoptosis of the infected cells (induced by the pathogen itself or by cytokines, NK cells, or T cells) is a common component of the antimicrobial response. It is thus not surprising that many pathogens have evolved inhibitory mechanisms against apoptosis. Chlamydia display strong antiapoptotic activity to evade the killing mechanisms of CD8+ T cells.[29–31] However, depending on the stage of development, chlamydia has also been indicated to do the opposite, to induce apoptosis of infected cells.[32] However, without killing, T cell recognition of infected cells may lead to local secretion of cytokines, restricting intracellular pathogen growth. In this respect, chlamydia also may seem to have evolved strategies for evading CD4+ and CD8+ T cell cognition. Thus, chlamydia can suppress IFN-γ-inducible MHC class I and II expression. One way to do this is by secretion of a chlamydial proteosome-like activity molecule into host cell cytoplasm that degrades host transcription factors RFX5 and USF-1 necessary for transcription of class I and II molecules.[33,34] Chlamydia-induced IFN-γ has also been suggested to inhibit IFN-γ-dependent expression of MHC class II proteins.

4. ANTIGENS

Like other intracellular pathogens, chlamydial biological complexity underscores the likely need of multiple target antigens in a vaccine. Use of whole inactivated chlamydial agents appears to be because of the presence of immunopathogenic components, plus the fact that early trials indicated that trachoma was exacerbated following episodes of *C. trachomatis* infection. On the other hand, a live attenuated strain of *C. psittaci* is used in veterinary medicine as a safe way to prevent abortion in ewes, suggesting that there might be hope for an attenuated vaccine. A live vaccine replicates like the target pathogen and promotes processing and presentation of antigens most similar to a natural infection. Furthermore, while replicating, a live vaccine presumably expresses all or most of its important target immunogens. This may be important for chlamydiae, which exist in (at least) two *developmental* forms. Live attenuated vaccines could also stimulate mucosal immune responses and are capable of inducing systemic humoral and cell-mediated responses. Major obstacles in generating and isolating attenuated *C. trachomatis* strains are the inability to genetically manipulate the organism and to isolate and propagate clonal lineages. Plaque cloning techniques have been recently developed and will facilitate the isolation of clonal lineages of chlamydiae.[35] However, efficient genetic transformation systems for chlamydiae still need development.

Dendritic cells (DC) pulsed with chlamydial antigens or infected with *C. trachomatis* can induce protection when transferred to naïve animals.[36] This method is based on the use of DCs that will be activated and present chlamydial epitopes on their MHC class I and II. Such cells are then used in adoptive cell transfers to autologous recipients. DC vaccines could be used as tools to unravel protective antigens, co-stimulatory molecules, and homing requirements. The ability of dendritic cells to produce IL-12 upon chlamydial-organism stimulation is required for the induction of DC vaccine protection against chlamydial infection. Murine DCs undergo phenotypic maturation to become activated upon exposure to type I interferons (type I IFNs) *in vivo* or *in vitro* and can also be matured by different adjuvants and used as such to stimulate adaptive responses.[37] Although unlikely to constitute an alternative to other types of immunotherapy because of methodological constraints, protection conferred by DC vaccines constitute a way to obtain the proof of principle for possibilities to develop a chlamydial vaccine. Moreover, induction of protective immunity occurred following chlamydial vaccination when coinoculated with a transgene-based GM-CSF adjuvant, probably allowing DCs to migrate to the site of immunization.[38]

To date there has been little progress in the identification of promising candidate chlamydia vaccine antigens. The most studied antigen here is MOMP. *C. trachomatis* MOMP is a predominant disulfide cross-linked surface protein and an immunodominant B cell antigen. MOMP is also the primary serotyping antigen. Antibodies specific to MOMP neutralize infectivity by blocking chlamydial attachment to host cells, suggesting a role of MOMP as a chlamydial adhesin (reviewed in ref. 39). MOMP has thus been the focus of many vaccination studies because of these important immunological properties and implication in chlamydial pathogenesis. Recombinant MOMP, MOMP synthetic peptides, DNA vaccines encoding MOMP, and the passive transfer of MOMP-specific monoclonal antibodies have all been evaluated for protective efficacy.(40) However, they have all yielded disappointing results with no or at best partial protective immunity. This is also true with chlamydial vaccines using other antigens such as OMP-2, Hsp60, Cap-1, ADP/ATP translocase.(8,15,30,41,42) The reason for the ineffectiveness of MOMP as a vaccine is not known, but it may result from the use of MOMP immunogens that do not mimic the native structure of the protein. Studies of the *C. pneumoniae* MOMP antigenic structure have produced a portrayal of exposure and immunogenicity different from the MOMP of *C. trachomatis.* Whereas *C. trachomatis* MOMP varies antigenically among many strains, the *C. pneumoniae* MOMP is antigenically constant, suggesting that the selective pressure responsible for antigenic variation seen in *C. trachomatis* is not active for *C. pneumoniae. C. pneumoniae* probably behave like a more time-persistent intracellular organism being less exposed to anti-MOMP-neutralizing antibodies. Like the *C. trachomatis* homologue, *C. pneumoniae* MOMP is exposed on the surface of the bacteria and conformational epitopes can be recognized by species-specific neutralizing monoclonal antibodies as well as human sera.(26)

The complete genome sequence of *C. pneumoniae* provides an inventory of all proteins potentially produced by this bacteria. Thus, no potential vaccine candidates need slip through the net through ignorance of its existence. However, the genome revolution has brought to the forefront immense difficulties in evaluating the large number of potential vaccine candidates. The challenge is now shifted to which on the many candidates should be taken forward. The information from the bacterial genome sequences allows those skilled in bioinformatics to classify genes and their products and identify those most potentially interesting for vaccine development. Software packages will assign function to genes, predict their key functions such as their cellular locations, MW, pI, etc. However experimental biology should always supplement this information. Stephens has classified the genome encoded proteins into four categories(43): (1) those present in chlamydiae and other organisms, (2) those unique to chlamydiae, (3) those unique to one chlamydia species but related to those in other species, and (4) those unique to *C. pneumoniae* lacking identifiable relatives in other organisms.

Proteins present in chlamydiae and other organisms (type 1) would typically have metabolic functions common to most organisms, including proteins important in translation, transcription, replication, and metabolic pathways. Although generally not good candidates for vaccine—there are exceptions such as Hsp60—type III secretion structural proteins constitute attractive candidates. Those shared by different species of chlamydia but not by other species (type 2) are interesting candidates for vaccination, since, in general, they would participate in chlamydial structure or biochemistry. This category includes many outer membrane proteins, inclusion membrane proteins, and type III secretion effectors. Antigens unique to one chlamydial species but present in other bacterial species (type 3) are few and probably involved in the metabolism or virulence of the particular species. This set is not promising for vaccination. Those unique to

C. pneumoniae without relatives in other species (type 4) characterize the species and reflect exclusive biology and virulence traits. Their function is not known and thereby cannot be ruled out as vaccine candidates.

Using immune responses to indicate genes expressed *in vivo* which could also be used for vaccine design is logical, powerful, and specific. A direct and innovative approach is based on the finding that amplified PCR fragments can be rendered transcriptionally active when inoculated into animals (i.e., DNA immunization). Because all candidate antigens are encoded in the genome, a systematic search could be realized through construction of expression libraries. Animals will be immunized with a representative library of pathogen DNA prepared in such a way that the encoded genes are expressed and secreted, the animal will be subsequently challenged and assayed for protection.[44] Expression libraries capable of inducing protection could be subfractionated in order to identify the relevant protective DNA immunogens. Availability of the genome should facilitate selection since only few nucleotides of the insert DNA responsible for inducing protection need to be sequenced, and then matched to the whole genome sequence. The ability to identify epitopes by expression cloning is particularly useful when applied to organisms such as *C. pneumoniae*, where large numbers of organisms are difficult to obtain, and where techniques for genetic manipulation of the organism are unavailable. Expression libraries have been used to characterize antigens reactive with antibodies directed against chlamydia. Recently prokaryotic and eukaryotic bacterial expression libraries have been used to identify *C. trachomatis* antigens recognized by MHC class I and II restricted T cells.[30,45] In brief, chlamydial proteins expressed by *E. coli* were processed by antigen-presenting cells (APC), or macrophage cell lines infected with retroviral expression libraries and the resulting peptide/MHC complexes screened by antigen-specific T cells. Using these approaches the antigens recognized by CD4+ and CD8+ T cell lines were identified.[30,45] Selection of unknown antigens screened with immune T cells or a panel of T cell hybridomas from infected mice selected on the basis of the ability of these cells to secrete high levels of IFN-γ if in contact with proper immunogens can be useful in vaccine design. This tactic could complement the approaches in which cDNA-encoding chlamydia open reading frames are used to vaccinate experimental animals to determine antigens associated with protection.

Alternative approaches involving purification and sequencing of peptides eluted from MHC molecules have been used to identify pathogen-derived antigens. This technique is more applicable for identifying antigens recognized by CD8+ T cells, where peptides of relatively restricted defined length (8–10 amino acids) are recognized, although this could also be used for identification of antigens eluted from MHC class II.

Although this technique has still not been used in identification of chlamydial epitopes, search for chlamydial sequences encoding potential target epitopes of murine anti–*C. pneumoniae* CD8+ T cells, among *C. pneumoniae* proteins likely to reach the cytosol of infected cells, has been performed.[20,21] Sequences were tested for their ability to bind MHC molecules and, when coincubated with APC, these were recognized and lysed by *C. pneumoniae*–specific CTLs.

5. ADJUVANTS

5.1. Activation of Innate Immunity

Besides physical properties of the antigen additional factors affect the immune response to a vaccine. These include immune status of the host, route of administration, delivery vehicle, and adjuvants. Even the most promising vaccines may fail to establish protection against the pathogen because of inadequate delivery (adjuvants or vectors). Although safety concerns are precluded by usage of subunit or epitope candidate vaccines, they are usually immunogenically meagre, and need efficient adjuvants to induce protective immunity. Incorrect delivery might not only fail to induce protective responses, but can in fact induce a detrimental immune profile that might enhance the susceptibility of the host to the infection. Adjuvants (and delivery vectors) normally act by triggering innate immune responses, and these in turn will allow development of specific adaptive responses.

The innate immune responses will to a large extent determine the quality of the adaptive immune system and thereby the result of infection. The direct importance of the innate immune responses is further illustrated by the life threatening conditions observed in humans and experimental animals showing genetic defects in this system. For recognition of pathogens, plants, insects, and vertebrates have relied upon a system of receptors that share a characteristic cytoplasmic domain termed TIR (toll/interleukin-1 receptor domain). Toll-like receptors (TLR) discriminate "self" from pathogen-derived ligands also termed pathogen-associated molecular patterns (PAMPs).[9] Different TLRs can discriminate between different PAMPs. Thus, the innate immune system may have evolved to express different TLR (or combinations thereof) within distinct subsets of immunocompetent cells, such as phagocytes and dendritic cells. These receptors generate both shared and unique intracellular signals, which is reflected at the level of gene expression, resulting in distinct proinflammatory patterns. Various response patterns that microbes induce in dendritic cells will accordingly affect the differentiation of the adaptive immune responses. For example, signalling through different TLRs stimulates secretion of IL-1 and TNF-γ, whereas genes such as iNOS, IL-12, and type I IFN will only be induced via certain TLRs. Type I IFNs have been reported to be induced in response to TLR 3, 4, and 7, which are expressed in different cell types. Coincubation of dendritic cells, endothelial cells, human vascular smooth muscle cells, or peripheral-blood mononuclear cells with chlamydia or chlamydial components has recently been shown to induce TLR-mediated activation.[46,47] Whether TLR recognition plays a role in the outcome of chlamydial infection *in vivo* remains to be explored but is most likely. In other intracellular bacterial infections such as *L. monocytogense*, absence of TLR signalling thus results in increased susceptibility to infection. Innate recognition of the pathogens will determine the release of various cytokines that in turn will activate common or different resistance mechanisms. Among cytokines secreted by innate immune cells, IFN-γ can control chlamydial infections, as shown by the increased susceptibility of (T and B cell and IFN-γ signalling deficient) RAG-$1^{-/-}$/IFN-$\gamma R^{-/-}$ mice compared to (T and B cell deficient) RAG-$1^{-/-}$ mice.[48] IL-12 as well as other cytokines can also regulate synthesis of IFN-γ. Unpublished data from our laboratory show that type I IFNs (usually induced by viral infections), are induced in a TLR-dependent way by infection with *C. pneumoniae*. Type I IFNs will in turn induce IFN-γ, which in an autocrine/paracrine way will control the infection of the infected cells.[49]

5.2. Adjuvants and Delivery Systems

Various strategies have been used to deliver immune chlamydial antigens to enhance protective immunity. Nonionic detergents, lipophilic immune stimulating complexes (ISCOMS) have been used with some success in the *C. trachomatis* model.[50,51] Experimental mucosal adjuvants such as cholera or heat-labile enterotoxins are strategies to be further explored.[52] Vector-mediated immunization with naked DNA has recently received the most attention, and has been successful in murine *C. trachomatis* and *C. pneumoniae* infections, and has also been used to screen for vaccine candidates.[42] DNA vectors used for immunization usually contain CpG motifs. These motifs will stimulate innate immunity through activation of TLR9-mediated signalling potent Th1 responses. In fact, immunization of chlamydial protein antigens with CpG motifs alone has also induced a Th1-protective antichlamydial immunity.[8,53] We have recently succeeded in preparing a fusion construct of CTLA4, the ligand for the costimulatory molecule B7 on APCs with the OMP-2 gene of *C. pneumoniae*. Such DNA fusion gene vaccine did improve protection as compared to a DNA vaccine encoding the microbial gene alone.[8] Plasmids encoding cytokine genes or coinoculation of cytokines such as IL-12 or IFN-γ are strategies also to be analyzed for chlamydial vaccines. Use of recombinant viral and bacterial vectors does also provide promising delivery systems, which have been relatively less explored. However, Salmonella transfected with plasmids containing chlamydial MOMP has been used with success as an experimental anti–*C. trachomatis* mucosal vaccine.[54]

6. CONCLUSIONS

We are in the early phases of establishing a *C. pneumoniae* vaccine. The new knowledge arising from the sequencing of the complete genome of the pathogen now allows for a rational analysis of a selection of vaccine candidate antigens. Likewise, the availability of a murine model of infection and the selective-gene-deficient mouse strains add further form to the vaccine research in the very next few years.

ACKNOWLEDGMENTS. This work was supported by the European Community QLK2-CT-2002-00846 grant, the Karolinska Institute, Sweden, The Swedish Cancer Society, and The Swedish Research Council, Sweden.

REFERENCES

1. Rappuoli, R., Miller, H. I., and Falkow, S., 2002, Medicine. The intangible value of vaccination, *Science* **297:**937–939.
2. Kalman, S., Mitchell, W., Marathe, R., Lammel, C., Fan, J., Hyman, R. W., Olinger, L., Grimwood, J., Davis, R. W., and Stephens, R. S., 1999, Comparative genomes of *Chlamydia pneumoniae* and *C. trachomatis, Nat. Genet.* **21:**385–389.
3. Beatty, W. L., Morrison, R. P., and Byrne, G. I., 1994, Persistent chlamydiae: From cell culture to a paradigm for chlamydial pathogenesis, *Microbiol. Rev.* **58:**686–699.
4. Murdin, A. D., Gellin, B., Brunham, R. C., Campbell, L. A., Christiansen, G., Deal, C. D., Jenson, H. B., Metcalf, B., Sankaran, B., Stephens, R. S., and Wilfert, C., 2000, Collaborative multidisciplinary workshop report: Progress toward a *Chlamydia pneumoniae* vaccine, *J. Infect. Dis.* **181**(Suppl. 3):S552–S557.
5. Blasi, F., Cosentini, R., and Tarsia, P., 2000, *Chlamydia pneumoniae* respiratory infections, *Curr. Opin. Infect. Dis.* **13:**161–164.
6. Yang, Z. P., Kuo, C. C., and Grayston, J. T., 1993, A mouse model of *Chlamydia pneumoniae* strain TWAR pneumonitis, *Infect. Immun.* **61:**2037–2040.
7. Kaukoranta-Tolvannen, S., Laurila, A., Saikku, P., Leinonen, M., and Laitinen, K., 1995, Experimental *Chlamydia pneumoniae* infection in mice: Effect of reinfection and passive immunization, *Microb. Pathog.* **18:**279–288.
8. Bandholtz, L., Kreuger, M. R., Svanholm, C., Wigzell, H., and Rottengerg, M. E., 2002, Adjuvant modulation of the immune responses and the outcome of infection with *Chlamydia pneumoniae, Clin. Exp. Immunol.* **130:**393–403 .
9. Aderemand, A., and Ulevitch, R. J., 2000, Toll-like receptors in the induction of the innate immune response, *Nature* **406:**782–787.
10. Rottenberg, M. E., Gigliotti Rothfuchs, A. C., Gigliotti, D., Svanholm, C., Bandholtz, L., and Wigzell, H., 1999, Role of innate and adaptive immunity in the outcome of primary infection with *Chlamydia pneumoniae*, as analyzed in genetically modified mice, *J. Immunol.* **162:**2829–2836.
11. Suand, H., and Caldwell, H. D., 1995, CD4+ T cells play a significant role in adoptive immunity to *Chlamydia trachomatis* infection of the mouse genital tract, *Infect. Immun.* **63:**3302–3308.
12. Starnbach, M. N., Bevan, M. J., and Lampe, M. F., 1994, Protective cytotoxic T lymphocytes are induced during murine infection with *Chlamydia trachomatis, J. Immunol.* **153:**5183–5189.
13. Zhang, D. J., Yang, X., Shen, C., and Brunham, R. C., 1999, Characterization of immune responses following intramuscular DNA immunization with the MOMP gene of *Chlamydia trachomatis* mouse pneumonitis strain, *Immunology* **96:**314–321.
14. Zhang, D., Yang, X., Berry, J., Shen, C., McClarty, G., and Brunham, R. C., 1997, DNA vaccination with the major outer-membrane protein gene induces acquired immunity to *Chlamydia trachomatis* (mouse pneumonitis) infection, *J. Infect. Dis.* **176:**1035–1040.
15. Svanholm, C., Bandholtz, L., Castanos-Velez, E., Wigzell, H., and Rottenberg, M. E., 2000, Protective DNA immunization against *Chlamydia pneumoniae, Scand. J. Immunol.* **51:**345–353.
16. Magee, D. M., Williams, D. M., Smith, J. G., Bleicker, C. A., Grubbs, B. G., Schachter, J., and Rank, R. G., 1995, Role of CD8 T cells in primary Chlamydia infection, *Infect. Immun.* **63:**516–521.
17. Kim, S. K., Angevine, M., Demick, K., Ortiz, L., Rudersdorf, R., Watkins, D., and DeMars, R., 1999, Induction of HLA class I-restricted CD8+ CTLs specific for the major outer membrane protein of *Chlamydia trachomatis* in human genital tract infections, *J. Immunol.* **162:**6855–6866.

18. Kim, S. K., Devine, L., Angevine, M., DeMars, R., and Kavathas, P. B., 2000, Direct detection and magnetic isolation of *Chlamydia trachomatis* major outer membrane protein–specific CD8+ CTLs with HLA class I tetramers, *J. Immunol.* **165:**7285–7292.
19. Kuon, W., Holzhutter, H. G., Appel, H., Grolms, M., Kollnberger, S., Traeder, A., Henklein, P., Weiss, E., Thiel, A., Lauster, R., Bowness, P., Radbruch, A., Kloetzel, P. M., and Sieper, J., 2001, Identification of HLA-B27-restricted peptides from the *Chlamydia trachomatis* proteome with possible relevance to HLA-B27-associated diseases, *J. Immunol.* **167:**4738–4746.
20. Saren, A., Pascolo, S., Stevanovic, S., Dumrese, T., Puolakkainen, M., Sarvas, M., Rammensee, H. G., and Vuola, J. M., 2002, Identification of *Chlamydia pneumoniae*–derived mouse CD8 epitopes, *Infect. Immun.* **70:**3336–3343.
21. Wizel, B., Starcher, B. C., Samten, B., Chroneos, Z., Barnes, P. F., Dzuris, J., Higashimoto, Y., Appella, E., and Sette, A., 2002, Multiple *Chlamydia pneumoniae* antigens prime CD8+ Tc1 responses that inhibit intracellular growth of this vacuolar pathogen, *J. Immunol.* **169:**2524–2535.
22. Perry, L. L., Feilzer, K., Hughes, S., and Caldwell, H. D., 1999, Clearance of *Chlamydia trachomatis* from the murine genital mucosa does not require perforin-mediated cytolysis or Fas-mediated apoptosis, *Infect. Immun.* **67:**1379–1385.
23. Lampe, M. F., Wilson, C. B., Bevan, M. J., and Starnbach, M. N., 1998, Gamma interferon production by cytotoxic T lymphocytes is required for resolution of *Chlamydia trachomatis* infection, *Infect. Immun.* **66:**5457–5461.
24. Bavoil, P. M., Hsia, R., and Ojcius, D. M., 2000, Closing in on Chlamydia and its intracellular bag of tricks, *Microbiology* **146**(Part 11)**:**2723–2731.
25. Morrison, S. G., Su, H., Caldwell, H. D., and Morrison, R. P., 2000, Immunity to murine *Chlamydia trachomatis* genital tract reinfection involves B cells and CD4(+) T cells but not CD8(+) T cells, *Infect. Immun.* **68:**6979–6987.
26. Wolf, K., Fischer, E., Mead, D., Zhong, G., Peeling, R., Whitmire, B., and Caldwell, H. D., 2001, *Chlamydia pneumoniae* major outer membrane protein is a surface-exposed antigen that elicits antibodies primarily directed against conformation-dependent determinants, *Infect. Immun.* **69:**3082–3091.
27. Pal, S., Theodor, I., Peterson, E. M., and de la Maza, L. M., 1997, Monoclonal immunoglobulin A antibody to the major outer membrane protein of the *Chlamydia trachomatis* mouse pneumonitis biovar protects mice against a chlamydial genital challenge, *Vaccine* **15:**575–582.
28. Peterson, E. M., Cheng, X., Motin, V. L., and de la Maza, L. M., 1997, Effect of immunoglobulin G isotype on the infectivity of *Chlamydia trachomatis* in a mouse model of intravaginal infection, *Infect. Immun.* **65:**2693–2699.
29. Fan, T., Lu, H., Hu, H., Shi, L., McClarty, G. A., Nance, D. M., Greenberg, A. H., and Zhong, G., 1998, Inhibition of apoptosis in chlamydia-infected cells: Blockade of mitochondrial cytochrome c release and caspase activation, *J. Exp. Med.* **187:**487–496.
30. Fling, S. P., Sutherland, R. A., Steele, L. N., Hess, B., D'Orazio, S. E., Maisonneuve, J., Lampe, M. F., Probst, P., and Starnbach, M. N., 2001, CD8+ T cells recognize an inclusion membrane–associated protein from the vacuolar pathogen *Chlamydia trachomatis, Proc. Natl. Acad. Sci. U.S.A.* **98:**1160–1165.
31. Geng, Y., Shane, R. B., Berencsi, K., Gonczol, E., Zaki, M. H., Margolis, D. J., Trinchieri, G., and Rook, A. H., 2000, *Chlamydia pneumoniae* inhibits apoptosis in human peripheral blood mononuclear cells through induction of IL-10, *J. Immunol.* **164:**5522–5529.
32. Ojcius, D. M., Souque, P., Perfettini, J. L., and Dautry-Varsat, A., 1998, Apoptosis of epithelial cells and macrophages due to infection with the obligate intracellular pathogen *Chlamydia psittaci*, *J. Immunol.* **161:**4220–4226.
33. Zhong, G., Liu, L., Fan, T., Fan, P., and Ji, H., 2000, Degradation of transcription factor RFX5 during the inhibition of both constitutive and interferon gamma–inducible major histocompatibility complex class I expression in chlamydia-infected cells, *J. Exp. Med.* **191:**1525–1534.

34. Zhong, G., Fan, T., and Liu, L., 1999, Chlamydia inhibits interferon gamma–inducible major histocompatibility complex class II expression by degradation of "Upstream Stimulatory Factor 1," *J. Exp. Med.* **189:**1931–1938.
35. Gieffers, J., Belland, R. J., Whitmire, W., Ouellette, S., Crane, D., Maass, M., Byrne, G. I., and Caldwell, H. D., 2002, Isolation of *Chlamydia pneumoniae* clonal variants by a focus-forming assay, *Infect. Immun.* **70:**5827–5834.
36. Su, H., Messer, R., Whitmire, W., Fischer, E., Portis, J. C., and Caldwell, H. D., 1998, Vaccination against chlamydial genital tract infection after immunization with dendritic cells pulsed *ex vivo* with nonviable Chlamydiae, *J. Exp. Med.* **188:**809–818.
37. Le Bon, A., Schiavoni, G., D'Agostino, G., Gresser, I., Belardelli, F., and Tough, D. F., 2001, Type I interferons potently enhance humoral immunity and can promote isotype switching by stimulating dendritic cells *in vivo*, *Immunity* **14:**461–470.
38. Lu, H., Xing, Z., and Brunham, R. C., 2002, GM-CSF transgene-based adjuvant allows the establishment of protective mucosal immunity following vaccination with inactivated *Chlamydia trachomatis*, *J. Immunol.* **169:**6324–6331.
39. Igietseme, J. U., Black, C. M., and Caldwell, H. D., 2002, Chlamydia vaccines: Strategies and status, *BioDrugs* **16:**19–35.
40. Morrison, R. P., and Caldwell, H. D., 2002, Immunity to murine chlamydial genital infection, *Infect. Immun.* **70:**2741–2751.
41. Penttila, T., Vuola, J. M., Puurula, V., Anttila, M., Sarvas, M., Rautonen, N., Makela, P. H., and Puolakkainen, M., 2000, Immunity to *Chlamydia pneumoniae* induced by vaccination with DNA vectors expressing a cytoplasmic protein (Hsp60) or outer membrane proteins (MOMP and Omp2), *Vaccine* **19:**1256–1265.
42. Murdin, A. D., Dunn, P., Sodoyer, R., Wang, J., Caterini, J., Brunham, R. C., Aujame, L., and Oomen, R., 2000, Use of a mouse lung challenge model to identify antigens protective against *Chlamydia pneumoniae* lung infection, *J. Infect. Dis.* **181** (Suppl. 3)**:**S544–S551.
43. Stephens, R. S., 2000, Chlamydial genomics and vaccine antigen discovery, *J. Infect. Dis.* **181** (Suppl. 3)**:**S521–S523.
44. Barry, M. A., Lai, W. C., and Johnston, S. A., 1995, Protection against mycoplasma infection using expression-library immunization, *Nature* **377:**632–635.
45. Goodall, J. C., Yeo, G., Huang, M., Raggiaschi, R., and Gaston, J. S., 2001, Identification of *Chlamydia trachomatis* antigens recognized by human CD4+ T lymphocytes by screening an expression library, *Eur. J. Immunol.* **31:**1513–1522.
46. Prebeck, S., Kirschning, C., Durr, S., da Costa, C., Donath, B., Brand, K., Redecke, V., Wagner, H., and Miethke, T., 2001, Predominant role of toll-like receptor 2 versus 4 in *Chlamydia pneumoniae*–induced activation of dendritic cells, *J. Immunol.* **167:**3316–3323.
47. Bulut, Y., Faure, E., Thomas, L., Karahashi, H., Michelsen, K. S., Equils, O., Morrison, S. G., Morrison, R. P., and Arditi, M., 2002, Chlamydial heat shock protein 60 activates macrophages and endothelial cells through toll-like receptor 4 and MD2 in a MyD88-dependent pathway, *J. Immunol.* **168:**1435–1440.
48. Rottenberg, M. E., Gigliotti Rothfuchs, A., Gigliotti, D., Ceausu, M., Une, C., Levitsky, V., and Wigzell, H., 2000, Regulation and role of IFN-gamma in the innate resistance to infection with *Chlamydia pneumoniae*, *J. Immunol.* **164:**4812–4818.
49. Rothfuchs, A. G., Gigliotti, D., Palmblad, K., Andersson, U., Wigzell, H., and Rottenberg, M. E., 2001, IFN-alphabeta-dependent, IFN-gamma secretion by bone marrow–derived macrophages controls an intracellular bacterial infection, *J. Immunol.* **167:**6453–6461.
50. Dong-Ji, Z., Yang, X., Shen, C., Lu, H., Murdin, A., and Brunham, R. C., 2000, Priming with *Chlamydia trachomatis* major outer membrane protein (MOMP) DNA followed by MOMP ISCOM boosting enhances protection and is associated with increased immunoglobulin A and Th1 cellular immune responses, *Infect. Immun.* **68:**3074–3078.
51. Igietseme, J. U., and Murdin, A., 2000, Induction of protective immunity against *Chlamydia trachomatis* genital infection by a vaccine based on major outer membrane protein-lipophilic immune response-stimulating complexes, *Infect. Immun.* **68:**6798–6806.

52. Taylor, H. R., Whittum-Hudson, J., Schachter, J., Caldwell, H. D., and Prendergast, R. A., 1988, Oral immunization with chlamydial major outer membrane protein (MOMP), *Invest. Ophthalmol. Vis. Sci.* **29:**1847–1853.
53. Pal, S., Davis, H. L., Peterson, E. M., and de la Maza, L. M., 2002, Immunization with the *Chlamydia trachomatis* mouse pneumonitis major outer membrane protein by use of CpG oligodeoxynucleotides as an adjuvant induces a protective immune response against an intranasal chlamydial challenge, *Infect. Immun.* **70:**4812–4817.
54. Brunham, R. C., and Zhang, D., 1999, Transgene as vaccine for chlamydia, *Am. Heart J.* **138:**S519–S522.

9

Chlamydia pneumoniae and Atherosclerosis—An Overview of the Association

JOSEPH NGEH and SANDEEP GUPTA

1. INTRODUCTION

Atherosclerosis is now acknowledged as an inflammatory disease.[1–7] The major clinical manifestations of atherosclerotic disease include coronary heart disease (CHD), ischemic stroke, abdominal aortic aneurysm (AAA), and peripheral arterial disease (PAD). These in turn are among the leading causes of death and disability in the world. However, only about 50% of the occurrences of these conditions can be attributed to classical vascular risk factors such as hypertension, smoking, diabetes mellitus, and dyslipidemia.[3,4,7] "Novel" risk factors such as infection have emerged as potentially important, linking inflammation and the pathogenesis of atherosclerosis.[1–7] Specifically, *Chlamydia pneumoniae* (*C. pneumoniae*) is the microorganism most implicated in this "infectious" hypothesis of atherosclerosis, and will be the focus of discussion.

1.1. Historical Perspectives on the Association between Infection and Atherosclerosis

The idea that infection could play an important role in atherosclerosis is not entirely new. In fact the concept of an "infectious" hypothesis was proposed by several prominent historical figures such as Virchow in 1859 and Osler in 1908, followed by Frothingham in 1911, and Ophuls in 1921.[7] Indeed, over 20 specific microorganisms (bacteria and viruses) have been named to associate

JOSEPH NGEH • Department of Clinical Geratology, Radcliffe Infirmary, Oxford OX2 6HE, United Kingdom. SANDEEP GUPTA • Consultant Cardiologist, Whipps Cross and St Bartholomew's Hospitals, London E 11 1NR, United Kingdom.

TABLE I
Microorganisms Implicated to Associate Infection with Atherosclerotic Vascular Disease[6–11]

Microorganism	Year and author of publication
Bacillus typhosus	1889, Gilbert and Lion
Streptococci	1931, Benson *et al.*
Coxsackie B virus	1968, Sohal *et al.*
Adenovirus	1973, Fabricant *et al.*
Mycoplasma gallisepticum	1973, Clyde and Thomas
Marek's disease virus	1978, Fabricant *et al.*
Cytomegalovirus	1987, Petrie *et al.*
Herpes simplex virus	1987, Hajjar *et al.*
Chlamydia pneumoniae	1988, Saikku *et al.*
Measles virus	1990, Csonka *et al.*
Epstein–Barr virus	1993, Straka *et al.*
Human immunodeficiency virus	1993, Paton *et al.*
Helicobacter pylori	1994, Mendall *et al.*
Mycoplasma fermentans	1996, Ong *et al.*
Enteroviruses	1998, Roivainen *et al.*[9]
Coxiella burnetti	1999, Lovey *et al.*
Porphyromonas gingivalis	1999, Chiu *et al.*
Streptococcus sanguis	1999, Chiu *et al.*
Actinobacillus actinomycetemcomitans	2000, Haraszthy *et al.*
Bacteroides forsythus	2000, Haraszthy *et al.*
Hepatitis A virus	2000, Zhu *et al.*
Influenza virus	2000, Naghavi *et al.*
Mycoplasma pneumoniae	2000, Higuchi *et al.*[10]
Prevotella intermedia	2000, Haraszthy *et al.*
Varicella-zoster virus	2000, Moriuchi and Rodriguez[11]

Note: Adapted and updated from Ngeh J., *Chlamydia Pneumoniae* in Elderly Patients with Stroke Study (CPEPS): A case-control study on the seroprevalence of *Chlamydia pneumoniae* in patients aged over 65 years admitted with acute stroke or transient ischaemic attack, MSc dissertation, University of Keele, UK, 2000).[8]

infection and atherosclerosis (Table I).[6–11] Among these, *Cytomegalovirus* (CMV), *C. pneumoniae, Helicobacter pylori*, and dental pathogens are the microorganisms most commonly studied.[3,5–8] By far, the evidence linking *C. pneumoniae* infection and atherosclerosis is most abundant and appear to be the strongest.[1–8,12]

C. pneumoniae was first isolated from the conjunctiva of a child in Taiwan in 1965 and designated as TW-183.[4,6–8] In 1983, it was isolated from the respiratory tract of a university student suffering from pharyngitis in Seattle, and designated as AR-39.[4,6–8] *C. pneumoniae*, thought to be a new strain of *Chlamydia psitacci* causing acute respiratory tract infection, was described by Grayston and his colleagues in 1986, and given the acronym TWAR (TaiWan Acute Respiratory).[6–8] However, it was Saikku and his colleagues who first suggested that *Chlamydia* TWAR was associated with CHD or atherosclerosis in 1988.[5–8]

TWAR was subsequently renamed as *Chlamydia pneumoniae*, and recognised as a separate species in the genus *Chlamydia* in 1989.[(6–8)]

1.2. Microbiology and Epidemiology of *C. pneumoniae*

Apart from *Chlamydia trachomatis* and *C. psitacci*, *C. pneumoniae* is the only other species that causes human diseases within the genus of *Chlamydia*.[(6–8)] *C. pneumoniae* is a gram-negative, obligate intracellular bacterium that depends on its host cell for growth and survival.[(2)] The outer membrane of *C. pneumoniae* is composed of lipopolysaccharides (LPS) and heat-shock proteins (HSP) that are genus specific.[(7,8)] However, several major outer membrane proteins (MOMP) identified are detectable by monoclonal antibodies and are species specific.[(7,8)]

C. pneumoniae exists in three forms in its unique life cycle: (1) infectious, nonreplicating elementary body (EB); (2) noninfectious but actively replicating reticulate body (RB); and (3) noninfectious, nonreplicating persistent body (PB) (Fig. 1).[(2,4,7,8)] For example, the EB spore attacks the host's cell by endocytosis and differentiates into RBs by binary fission in enlarging vacuoles called inclusion bodies.[(2,4,7,8)] After inclusion maturation, the RBs redifferentiate into EBs, which are released by cytolysis, and start another cycle of infection.[(2,4,7,8)] However, the EB may transform into PB under conditions of immune stress, such as in the presence of gamma interferon.[(2,7,8)] The PB may remain metabolically

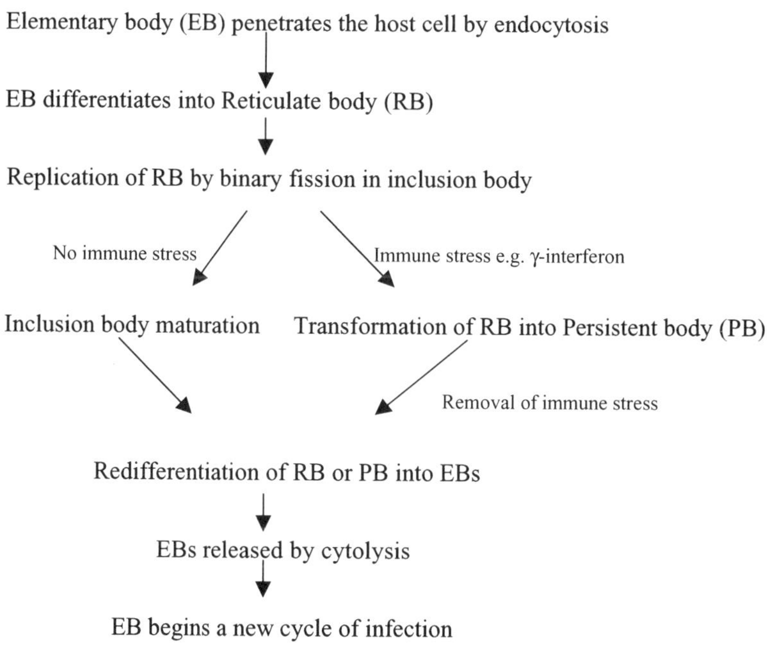

FIGURE 1. Life cycle of *C. pneumoniae*.[(2,7)]

inactive and undetected by the host's immune system, and remain unresponsive to antibiotics.[2,7,8] When the stress is removed, the PB may redifferentiate into infectious EB and released by cell lysis to begin a new life cycle.[2]

Clinically, *C. pneumoniae* is primarily a respiratory pathogen, although it also contributes to extrapulmonary manifestations.[2,5–8] The infection is common worldwide and accounts for 5–20% of all cases of community-acquired pneumonia.[6] However, *C. pneumoniae* infection can also be asymptomatic.[4] The incubation period of the infection is around 3–4 weeks.[7] Cyclical epidemics due to *C. pneumoniae* have been observed to occur every 4–7 years in western European countries.[7] Most people are infected several times in their lifetime.[4,7] The seroprevalence of *C. pneumoniae* is about 50% in middle-aged adults and rises to 70–80% in older populations.[1,2,4,6–8]

2. EVIDENCE LINKING *C. pneumoniae* AND ATHEROSCLEROSIS

The evidence linking *C. pneumoniae* infection and atherosclerosis is derived from five areas of research: seroepidemiological, pathological, animal, immunological, and antibiotic treatment studies.[1–8,12]

2.1. Seroepidemiological Observations

Seroepidemiological studies are credited as the earliest area of research linking *C. pneumoniae* and atherosclerotic vascular diseases. Since Saikku and his colleagues first showed a positive association between CHD and *C. pneumoniae* antibodies in a case–control study in 1988, numerous (probably 100) seroepidemiological studies of various designs, i.e. retrospective, case–control, cross-sectional, or prospective studies, have suggested a positive association.[1–8] More recently, seroepidemiological studies have also showed a positive association between *C. pneumoniae* serological markers and other atherosclerotic vascular diseases such as stroke, abdominal aortic aneurysm, and peripheral arterial diseases.[4,6,8]

Most of these positive seroepidemiological studies reported odds ratios (ORs) of 2 or higher.[1,2,7,8] However, a number of negative studies have also emerged.[1–4,6–8] Recent large-scale prospective seroepidemiological studies and meta-analyses involving over 5,000 cases have suggested a modest or weak association between *C. pneumoniae* serological markers and CHD, with ORs between 1.15 and 1.25.[2,3,7,8] Although some studies do not fully adjust for the influence of risk factors associated with *C. pneumoniae* infection, others have been criticized for overadjustment.[3]

The current gold standard in *C. pneumoniae* serology is microimmunofluorescence (MIF), a test originally developed for the diagnosis of clinical *C. pneumoniae* infection, as opposed to that for an epidemiological study of the role of *C. pneumoniae* in atherosclerosis.[5] The MIF is a subjective test and requires an experienced microscopist.[7] Different seroepidemiological studies have used different cutoff titers, making comparison of data difficult.[1,2,6–8] Further efforts are required to standardize *C. pneumoniae* serological assays and promote uniformity in laboratory practice.[2,6]

Newer serological methods such as the enzyme-linked immunosorbent assays (ELISA) have been developed and are readily available as commercial kits.[6–8,13] These ELISAs give qualitative results that can be read by machine rapidly.[13] Because ELISAs can afford a relatively high throughput and objectivity, they are increasingly used in seroepidemiological studies.[6–8,13] Preliminary data suggest that ELISA has a good sensitivity, specificity, and reproducibility when compared to MIF.[6–8,13] However, like the MIF, ELISA is still far from perfect as a diagnostic tool, and will require further standardization and improvement.[7,13]

Serological markers such as immunoglobulin titers tend to fluctuate over time, and there is variability both within individuals and across populations.[2,5,7] Moreover, *C. pneumoniae* infection is more common in the autumn or winter, and during cyclical epidemics.[4,7,8] The time of sampling will therefore affect the results of serological studies. In older age groups, the seroprevalence of *C. pneumoniae* is over 70%, making comparison of the actual, though small, difference in seropositivity between the cases and the controls difficult.[5,8] There were also reports suggesting a poor correlation between endovascular *C. pneumoniae* infection and serological positivity.[5,6]

Nevertheless, obtaining patients' serum and other blood components remains a relatively convenient and noninvasive method to study for *C. pneumoniae* infection. Currently, serological methods alone may not reliably detect chronic or persistent *C. pneumoniae* infection in a subgroup of patients that might benefit from antibiotic interventional therapy. Indeed, investigators are studying other blood markers such as *C. pneumoniae* immune complexes and circulating leukocytes, e.g. macrophages with detectable *C. pneumoniae* DNA and mRNA as markers of underlying infection.[2,6,7] Seroepidemiological studies have progressed to examine for other blood markers of infections implicated in atherosclerosis, to correlate inflammatory markers such as C-reactive protein (CRP) in conjunction with *C. pneumoniae* serology, and have shown association between "infectious" and "inflammatory" burdens and atherosclerosis.[5]

2.2. Pathological Specimen Findings

Shor and colleagues provided the first evidence that *C. pneumoniae* can be detected within atherosclerotic or CHD specimen and visualized with electron microscopy (EM) in 1992 (Fig. 2).[6–7,14] Since then, over 55 published reports have demonstrated the presence of *C. pneumoniae* not only in coronary arteries, but also in cerebral, carotid, internal mammary, pulmonary, aorta, renal, iliac, femoral, and popliteal arteries, and occluded bypass grafts, obtained from postmortem and surgical atheromatous tissues.[5–7,15] Different techniques such as immunocytochemistry (ICC), polymerase chain reaction (PCR), *in situ* hybridization (ISH), and EM were used in these studies.[1–7] The rate of detection is about 50% (range varies from 0 to 100%) in atheromatous tissues, but less than 1% in "healthy" arteries free from obvious atheroma.[1–7] *C. pneumoniae*

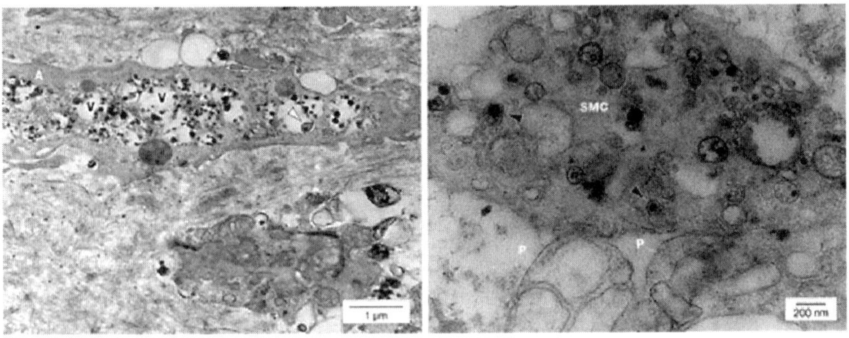

FIGURE 2. Transmission electron micrograph of smooth muscle cells in an early atherosclerotic lesion of the aorta positive for *C. pneumoniae* by polymerase chain reaction.[14] On the left, a smooth muscle cell containing vacuoles (V) and *C. pneumoniae* elementary bodies (arrowhead) are demonstrated. The other fragmenting cell and actin (A) filaments are also shown. On the right, macrophage pseudopodia (P) is shown in contact with a fragment of smooth muscle cell (SMC) containing *C. pneumoniae* (arrowheads). (Reproduced from Shor, A., and Phillips, J., 2000, Histological and ultrastructural findings suggesting an initiating role for *Chlamydia pneumoniae* in the pathogenesis of atherosclerosis: A study of fifty cases, *Cardiovasc. J. S. Afr.* **11**:16–23, with permission.)[14]

can also be detected in about 10% of noncardiovascular and granulomatous tissues, showing a nonspecific, ubiquitous distribution in the human body.[6,7] However, the distribution of *C. pneumoniae* is known to correlate well with the distribution of atherosclerosis within the same individual.[6]

The presence of *C. pneumoniae* in atherosclerotic lesions is not unique. It is known that other microorganisms such as *H. pylori, CMV, Herpes simplex virus,* periodontal pathogens, and *Mycoplasma pneumoniae*[10] have also been detected in human atherosclerotic tissues.[3,5] However, *C. pneumoniae* is, to date, the only microorganism that can be cultured alive, albeit with difficulty, from atheromatous plaques[5] (Table II).[2,6,16]

The methods used in histopathological studies, i.e. EM, ICC, PCR, and ISH, do not correlate with one another consistently, making comparison of studies in this area and others difficult.[2,5,7] PCR technique alone tends to give the lowest rate of *C. pneumoniae* detection, whereas the highest rate of detection is by ICC or a combination of methods.[2] Further efforts are therefore required to improve and standardize these methods.[7]

In spite of methodological limitations, pathological examinations have shown that *C. pneumoniae* has a tropism for arterial tissues and atherosclerotic lesions.[6,7] Although the mere presence of *C. pneumoniae* or its antigens in atheroma do not prove a causal role, pathological studies nevertheless support the rejection of the notion that *C. pneumoniae* is simply an innocent "bystander."[6,7] Increasingly, histopathological methods are used in conjunction with animal, immunological, and antibiotic studies to examine for causal, mechanistic, and treatment susceptibility effects of *C. pneumoniae* in atherosclerosis.

TABLE II
Pathological Studies That Have Detected or Cultured Viable *C. pneumoniae* from Human Atherosclerotic Tissues[2,6,16]

Year	Author	Specimen	Detection method	Positive (%)	Viable (%)
1996	Ramirez	Coronary	Culture/EM/ ICC/ISH/PCR	7/10 patients (70%)	1/10 patients (10%), by culture
1997	Jackson *et al.*	Carotid	Culture/EM/ ICC/ISH/PCR	12/16 specimens (75%), other than by culture	1/25 specimens (4%), by culture
1998	Maass *et al.*	Coronary, CABG	Culture/PCR	21/70 patients (30%)	11/70 patients (16%), by culture
1999	Bartels *et al.*	Saphenous vein CABG	Culture/PCR	8/32 specimens (25%)	5/32 specimens (16%), by culture
1999	Esposito *et al.*	Carotid	PCR/RT-PCR	18/30 patients (60%)	10/30 patients (33%), presence of chlamydial mRNA by RT-PCR
2000	Karlsson *et al.*	AAA	Culture/IHC	21/26 patients (81%)	10/25 patients (40%), by culture
2000	Apfalter *et al.*	Carotid, coronary, AAA, iliac, femoral, AV	Culture/DIF/ PCR	-	3/38 specimens (7.9%)
2001	Johnston *et al.*[16]	Carotid	PCR/RT-PCR	19/48 specimens (40%)	18/48 specimens (38%), presence of chlamydial mRNA by RT-PCR

Note: AAA = Abdominal aortic aneurysm, AV = Aortic valve, CABG = Coronary artery bypass graft, DIF = Direct, genus specific immunofluorescence staining, EM = Electron microscopy, ICC = Immunocytochemistry, IHC = Immunohistochemistry, ISH = *In situ* hybridization, PCR = Polymerase chain reaction, RT-PCR= Reverse transcriptase-PCR.

2.3. Animal Experimental Models

Historical precedents of animal models used to demonstrate infection-induced atherosclerosis include Benson's *Streptococcus*–rabbit and Fabricant's *Herpesvirus*–chicken experimental models in 1931 and 1978, respectively.[5,7] In 1997, Fong, Laitinen, and coworkers were among the first to report that repeated *C. pneumoniae* infection through the respiratory tract could induce not only pneumonia, but also inflammatory and atherosclerotic changes in the aortas of rabbits.[5–7] *M. pneumoniae*, another atypical respiratory pathogen used as a control in the study, was not found to induce any atherosclerotic lesions.[5] Further animal experimentation demonstrated that treatment using azithromycin, a macrolide antibiotic active against *C. pneumoniae*, could—though not invariably—counteract or reduce the atherogenic effect of *C. pneumoniae* in the aortas of infected animals.[4–6]

Animal models are invaluable laboratory tools used to study the synergistic effects of some of the well-established cardiovascular risk factors, e.g. atherogenic diet and genetic predisposition, in the context of *C. pneumoniae*–induced atherosclerosis. Dietary supplementations with cholesterol in *C. pneumoniae*–infected rabbits have been shown to increase the animals' vascular intimal wall thickening.[2,3,5–7] Genetically modified mouse such as ApoE-deficient mouse are known to develop atherosclerosis in the absence of a fatty diet.[2,3,5,7] Repeated *C. pneumoniae* infection has been shown to accelerate atherosclerotic lesion progression in this model, although a short 2-week course of azithromycin treatment did not reduce the size of the lesion.[7] Specifically, *C. pneumoniae* infection was shown to increase the T lymphocyte influx in the atherosclerotic plaques and to accelerate the formation of more advanced atherosclerotic lesions in ApoE3-Leiden mice.[17]

Further evidence suggests that the atherogenic effects of *C. pneumoniae* infection were dependent on cholesterol and species specific to *C. pneumoniae* as opposed to *C trachomatis*, in LDL receptor knockout transgenic mouse model.[7] On the other hand, the C57BL/6J mouse that does not develop atherosclerosis when fed a normal diet was shown to produce intimal thickening and carditis on repeated intranasal inoculation with *C. pneumoniae.*[7] In addition, when fed a high-fat diet, the infected C57BL/6J mice developed significantly larger atherosclerotic lesion areas compared with control mice.[18]

Recently, larger mammals such as dogs and pigs were used in animal studies.[19–21] Using EM, ICC, and PCR methods, a group of Japanese researchers have reported the presence of viable *C. pneumoniae* in canine atheromatous tissues.[19] Report from Sweden has demonstrated that acute *C. pneumoniae* infection of pigs through the respiratory tract could cause profound endothelial dysfunction in coronary arteries, thus supporting the role of *C. pneumoniae* in atherogenesis.[20] More recently, a report using a pig model has shown that intracoronary and intrapulmonary *C. pneumoniae* inoculation could induce coronary intimal proliferation in the absence of a lipid-rich diet.[21] However, direct administration of macrophages infected with *C. pneumoniae* into the pigs' coronary arterial wall was not associated with the development of coronary lesions.[21]

Animal models are useful to demonstrate a causal and temporal effect of *C. pneumoniae* infection and subsequent development of atherosclerosis.[6,7] They have shown that *C. pneumoniae* infection could initiate and accelerate atherosclerotic process.[2,5] Moreover, the induced atherogenic process appears to correlate with increasing *C. pneumoniae* infective dose and time from exposure, and accelerate with other atherogenic risk factors.[5–7] The models could also help to define the role of antibiotic treatment in atherosclerosis.[2,6] However, exactly to what extent could laboratory-induced atherogenesis in animal models reflect human atherosclerosis (which may take years to develop) remains an important issue to be resolved.[6]

2.4. Molecular and Immunological Mechanisms

Studies in this area have focused on the mechanistic roles of *C. pneumoniae* in the inflammatory processes of atherosclerosis. A variety of studies have examined the interactions between *C. pneumoniae* infections or antigens and components of atherothrombotic tissues: (1) cells such as leukocytes, platelets, monocytes, endothelium, smooth muscle cells (SMCs), and fibroblasts; (2) extracellular matrix and substrates such as collagens, elastin, gelatin, and fibronectin.[1–7,22]

A range of molecules expressed by these cells have been identified:[1–7]

1. Cytokines and chemokines: interleukins (e.g. IL-1, IL-2, IL-6, IL-8), tumor necrosis factors (e.g. TNF-α), gamma interferon, monocyte integrins, monocyte chemotactic protein-1 (MCP-1), colony stimulating factor, growth factors (e.g. platelet-derived growth factor or PDGF, basic fibroblast growth factor or bFGF), tissue factor.
2. Adhesion molecules: vascular cell adhesion molecule-1 (VCAM-1), intercellular adhesion molecule-1 (ICAM-1), E-selectin, P-selectin.

These inflammatory cells and mediators are known to participate in complicated immunological cross-talks between *C. pneumoniae* antigens and atherogenic/atherothrombotic processes.[1–7]

C. pneumoniae is capable of infecting macrophages/monocytes, endothelial and smooth muscle cells.[1–3,5–7] These are the key cells involved in atherogenesis that have been shown to support the growth and proliferation of *C. pneumoniae* in laboratory studies.[6,7] From pathological, animal, and clinical observations, it is postulated that *C. pneumoniae* can infect macrophages in human respiratory tract and be carried by circulatory monocytes to invade arterial sites that are either developing or prone to develop atherosclerosis.[1,2,4,6,7] At the endothelial surfaces of these sites, *C. pneumoniae* antigens could cause endothelial dysfunction and monocyte activation, and initiate or exacerbate a series of endovascular immunological reactions (Fig. 3).[2,6,7,12]

Experimental data has shown that *C. pneumoniae* A-03 (a coronary strain) could stimulate the production of MCP-1, IL-8, and ICAM-1 by human endothelial cells *in vitro.*[7] Circulating leukocytes, e.g. monocytes and T lymphocytes, attracted by various mediators to the activated endothelial surface, could themselves be activated and express further inflammatory molecules (e.g. ILs and TNFs), which contribute to further cellular recruitment and chronic endothelial cell damage.[7] Indeed, the "response to injury" hypothesis refers to the fact that endothelial cell damage is the crucial initial stage in atherogenesis.[7] Moreover, a number of traditional cardiovascular risk factors and infections are known to cause endothelial injury, dysfunction, and activation.[1–7]

Monocytes and macrophages are the key transporters of LDL cholesterol from systemic circulation into subendothelial space.[2,6,7] A study has identified

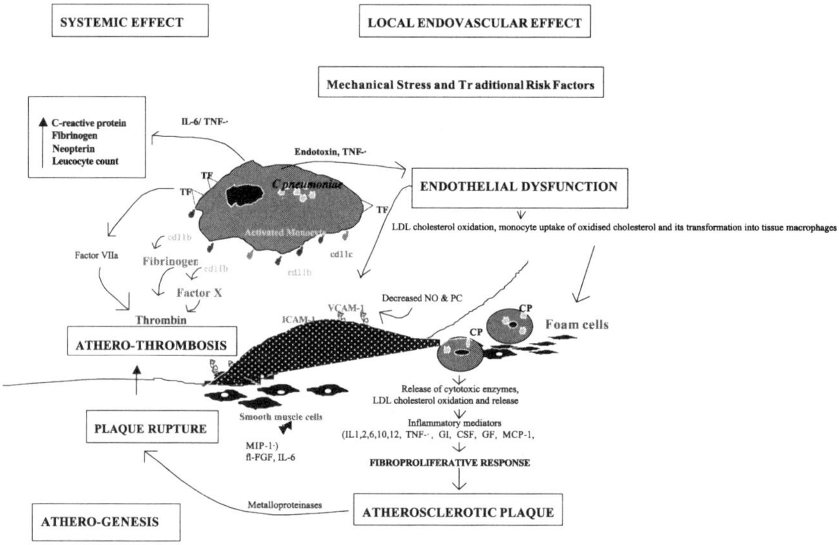

FIGURE 3. The roles of *C. pneumoniae*, monocytes, and inflammatory markers in atherogenesis and atherothrombosis: interplay between molecular and immunological mechanisms.[2,6,7,12] In conjunction with traditional atherosclerotic risk factors, *C. pneumoniae*–infected monocyte may be activated, which contributes toward endothelial dysfunction and subsequent foam cell transformation and inflammatory response in the subendothelial space. Monocyte activation and endothelial dysfunction also lead to secretion of cytokines and acute-phase proteins, expression of adhesion molecules, and upregulation of tissue factor and monocyte integrins. These processes may either act independently or interact, leading to atherogenesis and atherothrombosis. Cd11b/c = monocyte integrins, Cp = *C. pneumoniae*, CSF = colony stimulating factor, ß-FGF = ß-fibroblast growth factor, GF = growth factor, GI = gamma interferon, ICAM/VCAM = intercellular/vascular cell adhesion molecules, IL = interleukin, LDL = low density lipoprotein, MCP = monocyte chemoattractant protein, MIP = macrophage inflammatory protein, NO = nitric oxide, PG = prostacyclin, TF = tissue factor, TNF = tumor necrosis factor. (Adapted and modified from Gupta, S., 1999, *Chlamydia pneumoniae*, monocyte activation and antimicrobial therapy in coronary heart disease. MD Thesis, University of London, with permission)[12]

C. pneumoniae LPS as the antigen that could enhance LDL cholesterol uptake and downregulate cholesterol efflux in monocytes and macrophages.[2,5,6] Once the monocytes and leukocytes (T lymphocytes and neutrophils) are anchored by adhesion molecules onto the dysfunctional endothelial surface, transendothelial penetration into the subendothelial space occurs.[7] Many of these monocytes loaded with oxidised LDL cholesterol then transform into tissue macrophages known as foam cells.[2,3,5–7] Indeed, Kalayoglu and Byrne in 1998 demonstrated that *C. pneumoniae* LPS could induce not only LDL cholesterol oxidation but also macrophage transformation into foam cells, a key atherogenic process.[2] Furthermore, *C. pneumoniae* heat shock protein–60 (cHsp60) was found to induce cellular oxidation of LDL cholesterol.[1–3,5–7] The foam cells are known to participate in inflammatory and immunological reactions that lead to a fibroproliferative response from the arterial SMCs and the formation of atherosclerotic plaque.[2,6,7] Moreover, *C. pneumoniae*–infected human SMCs have been shown to stimulate the production of IL-6 and bFGF, which contribute to atherosclerotic fibrous plaque formation.[2,6]

Chlamydial cHsp60 and human hHsp60 were frequently found together in human atherosclerotic plaques.[7] Because they share sequence homology, cHsp60 could cross-react with hHsp60, causing antibody-mediated cytotoxicity.[1,2,4,5,7] In contrast, *C. pneumoniae*–reactive T lymphocytes in human atherosclerotic plaques represent specific cell-mediated immunity.[2,5,7] These lymphocytes can produce gamma interferon and convert *C. pneumoniae* into PB or persistent infection.[2] *C. pneumoniae* can also inhibit apoptosis in infected host cells through inhibition of mitochondrial cytochrome c release and caspase-3 activation, mediating a chronic infection.[23]

Hsp60 could activate macrophages to release TNF-α and matrix metalloproteinases that can degrade connective tissue and cause plaque rupture leading to acute atherothrombotic events.[2,3,6,7] These adverse events are enhanced by an increased vascular tone resulting from decreased nitric oxide and prostacyclin production by dysfunctional endothelium, in a procoagulant environment due to tissue factor expression by activated monocytes.[1–7] Apart from its local endovascular effects, *C. pneumoniae* can also systemically interact with circulating blood cells; stimulate hepatic synthesis of inflammatory markers, e.g. CRP, fibrinogen, and neopterin; disturb lipoprotein metabolism; enhance clotting cascade; and promote atherothrombotic events (Fig. 3).[1–7,12]

Recent studies have reported specific signalling pathways involved in *C. pneumoniae* pathogenic mechanisms. For example, *C. pneumoniae* and cHsp60 were shown to stimulate human vascular SMC proliferation through the activation of toll-like receptor 4 that acts as an antigenic sensor for *C. pneumoniae,* with signalling through the p44/p42 mitogen-activated protein kinase (MAPK) pathway.[6] The activation of nuclear transcription factors such as NF-κB and Activator Protein-1 (proliferative intracellular signals) by *C. pneumoniae* was found to induce vascular SMC proliferation, an effect that can be abolished by the antibiotic azithromycin.[6] In addition, NF-κB activation in *C. pneumoniae*–infected cells was shown to induce the production of proinflammatory and procoagulant substances.[6]

In summary, various mechanistic, molecular, and immunological studies have demonstrated a pathogenic role of *C. pneumoniae* infection in different stages of atherosclerosis. Further understanding and progress in this area of research may provide important "laboratory based" causal evidence of *C. pneumoniae* in atherosclerosis and stimulate "clinical" interventional and therapeutic strategies.

2.5. Antibiotic Treatment Studies

Most of the investigations in this area involved clinical antibiotic interventional trials in the secondary prevention of CHD. Others have examined the impact of antibiotic treatment on CHD in retrospective epidemiological studies. Apart from CHD, antibiotic treatment studies in the clinical context of AAA, PAD, and carotid artery stenosis or stroke are emerging (Table III).[1,4,6,24–26,29,30] The results of these studies are inconclusive, although

TABLE III
Antibiotic Treatment Studies and Atherosclerotic Vascular Diseases [1,4,6,24–26,29,30]

Trial, Year	Population, number	Antibiotic, course	Trial duration	Results
Antibiotic Treatment and Coronary Heart Disease				
St George's Hospital, London, UK. RCT, 1997[1,24]	Post MI, 220 males	Azithromycin, 3 or 6 days	1.5 years	Positive
ROXIS, Argentina. RCT, 1997[1,24]	ACS, 202 (76% men)	Roxithromycin, 30 days	6 months	Positive at 1 month, negative at 3 and 6 months
University of Washington, USA. RCT, 1998[4]	Post-PTCA, 88	Azithromycin, 28 days	6 months	Positive
ACADEMIC, USA. RCT, 1999[1,24]	Stable CAD, 302 (89% men)	Azithromycin, 3 month	2 years	Negative
ISAR-3, Germany. RCT, 2001[6,24,25]	Post–coronary stenting, 1010 (78% men)	Roxithromycin, 28 days	1 year	Negative, restenosis reduced in patients with titres 1:512
Siriraj Hospital, Bangkok, Thailand. RCT, 2001[1,24]	ACS, 84 (63% men)	Roxithromycin, 30 days	90 days	Negative
STAMINA, UK. RCT, 2002[1,24]	ACS, 325 (69% men)	Azithromycin or amoxicillin, and metronidazole, 7 days	1 year	Positive
WIZARD, International multicentre trial. RCT, 2002[1,24]	Stable CAD, 7,724 (83% men)	Azithromycin, 3 months	1–4 years (mean 2.5 years)	7% event reduction (hazard ratio 0.93, $P = 0.23$)
CLARIFY, Finland. RCT, 2002[1,24]	ACS, 148 (70% men)	Clarithromycin, 3 months	138–924 days (mean 555 days or 1.5 years)	Positive
St George's Hospital, London, UK. RCT, 2002[30]	Stable CAD, 40 males	Azithromycin, 3 days, then weekly for 4 weeks	5 weeks	Improvement in FMD of brachial artery, $P < 0.005$

TABLE III
(*Cont.*)

Trial, Year	Population, number	Antibiotic, course	Trial duration	Results
AZACS, USA. RCT, 2003[25]	ACS, 1,439 (74% men)	Azithromycin, 5 days	6 months	Negative
ANTIBIO, Germany. RCT, 2003[24]	ACS, 872 (79% men)	Roxithromycin, 6 weeks	1 year	Negative
MARBLE, UK. RCT[1,6]	CAD on waiting list for CABG surgery, 1240	Azithromycin, 3 months	2 years	Expected in 2004(?)
ACES, USA. RCT[1,6,24]	Stable CAD, 4000	Azithromycin, weekly for 1 year	4 years	Expected in 2004(?)
PROVE IT, International Multicentre trial. RCT[1,6,24]	ACS, 4000+	Gatifloxacin, 2 weeks, then 10-day course monthly for 2 years+	2 years+	Expected in 2004(?)
CROAATS, Croatia. RCT[1]	Documented MI, 340	Azithromycin, 3-day course at days 1, 10, 20	1.5 years	Expected in 2004(?)
CLARICOR, Denmark. RCT[26]	Stable CAD, 4600	Clarithromycin, 14 days	2 years	Expected in 2004(?)
CLAINF, Italy. RCT[6]	Acute MI, 800	Clarithromycin, 3 weeks	1 year	Expected in 2004(?)
Antibiotic Treatment and Carotid Atherosclerosis (CA), Abdominal Aortic Aneurysm (AAA), and Peripheral Arterial Disease (PAD)				
Italy RT, 1999[29]	*C. pneumoniae* burden in CA, 32 (16 males)	Roxithromycin, 17–35 days (mean 26 days)	17–35 days (mean 26 days)	Positive, $P = 0.034$
Finland RCT, 2001[24]	AAA growth, 32 (29 males)	Doxycycline, 3 months	1.5 years	Positive, $P = 0.03$
Denmark RCT, 2001[24]	AAA growth, 92 males	Roxithromycin, 28 days	1.5 years	Positive, $P = 0.02$
Switzerland RCT, 2002[24]	PAD progression, 40 males	Roxithromycin, 28 days	2.7 years	Positive, $P < 0.05$
Germany RCT, 2002[24]	CA progression, 272 (56% men)	Roxithromycin, 30 days	2 years	Positive, $P < 0.01$

Note: ACS = Acute coronary syndrome, CAD = Coronary artery disease, MI = Myocardial infarction, PTCA = Percutaneous transluminal coronary angioplasty, RCT = Randomized controlled trial.

the reports of some large-scale antibiotic interventional studies currently ongoing are still awaited.

In 1997, the positive results of two pilot antibiotic treatment studies were reported independently by Gupta and colleagues in the United Kingdom and Gurfinkel and colleagues in Argentina.[1,24] In the U.K. study, 60 male survivors of acute myocardial infarction (MI) with persistently raised *C. pneumoniae* IgG antibody titers ≥1:64 were randomized to receive placebo or a macrolide antibiotic azithromycin (500 mg per day for 3 or 6 days). Apart from a reduction in cardiovascular events at 18 months in the treatment group, there was also a reduction in several inflammatory/monocyte activation markers and *C. pneumoniae* IgG antibody titers, in comparison with the placebo group and nonrandomized group with high antibody titers.

The Argentinian ROXIS (Roxithromycin in Ischaemic Syndromes) study randomized hospitalized patients with unstable angina or non–Q-wave MI ($n =$ 202) to receive a placebo or another macrolide antibiotic roxithromycin (150 mg twice daily) for 30 days.[1,24] At 1 month, a reduction in the combined triple endpoint (severe recurrent angina, acute MI, and ischemic death) in the treatment group was reported. However, at 3 and 6 months, the beneficial effect of antibiotic treatment became nonsignificant, although inflammatory markers such as the CRP level decreased more significantly in the treatment group. This implied that a longer duration of antibiotic treatment trial might have been necessary to confer more lasting protective effects.

A third pilot antibiotic treatment study was reported in 1998.[4] Eighty-eight patients who had had percutaneous transluminal coronary angioplasty were randomized to receive azithromycin 500 mg daily or placebo for 2 days, followed by azithromycin 250 mg daily or placebo for 28 days. The investigators found that patients in the treatment group had less restenosis (9%) than those in the placebo group (16%), and also less recurrent angina (40 vs. 60%, respectively). However, the positive results were not sustained in a larger follow-up study.

In 1999, the findings of the Azithromycin in Coronary Artery Disease Elimination of Myocardial Infection with Chlamydia (ACADEMIC) study were reported from the United States.[1,24] The ACADEMIC study randomized 302 subjects with elevated *C. pneumoniae* antibody titers and CHD to receive a placebo or a 3-month course of azithromycin (500 mg per day for 3 days, then 500 mg per week). The investigators found no difference in clinical outcome at 6 months and at 2 years in both placebo and treatment groups. However, a reduction in inflammatory markers such as CRP, IL-1, IL-6, and TNF-α was found at 6 months.

The ISAR-3 (Intracoronary Stenting and Antibiotic Regime trial-3) study published in 2001 investigated whether roxithromycin could prevent *C. pneumoniae*–associated restenosis after coronary stent placement.[6,24] Consecutive patients were randomized in a double-blind fashion to receive a placebo ($n = 504$) or 300 mg daily of roxithromycin ($n = 506$) for 28 days. The investigators reported no real difference in outcome as the rate of angiographic stenosis at follow-up was 31% in the treatment group versus 29% in the placebo group. The combined 1-year mortality and MI rates were also similar in both groups, i.e. 7% in the treatment group versus 6% in the placebo group. However, in those patients with high *C. pneumoniae* antibody titers, i.e. 1:152, the investigators found that roxithromycin treatment significantly reduced the rate of restenosis after coronary stenting. This suggested that selective usage of roxithromycin in some patients with *C. pneumoniae* infection might reduce post–coronary-stenting restenosis.

In a small Thai study, 84 patients with acute coronary syndrome were randomized to receive roxithromycin (150 mg BD) or placebo for 30 days.[1,24] After a follow-up period of 90 days, the investigators found no significant difference in adverse cardiovascular events between the roxithromycin and placebo groups (17 vs. 16).

Several antibiotic treatment studies were reported in 2002. The South Thames Trial of Antibiotics in Myocardial Infarction and Unstable Angina (STAMINA) is another small U.K. study.[1,24] In this study, 325 patients with acute coronary syndromes (unstable angina or first MI) were randomized to receive a 1-week course of either placebo; amoxicillin (500 mg BD), metronidazole (400 mg BD), and omeprazole (20 mg BD); or azithromycin (500 mg OD), metronidazole (400 mg BD), and omeprazole (20 mg BD). The investigators found that antibiotic treatment significantly reduced adverse cardiac events at 12 months. This favorable treatment effect appeared to be independent of seropositivity to *C. pneumoniae* and *H. pylori*, and there was no difference found between the azithromycin and amoxicillin groups. The CRP and fibrinogen levels were found to decrease in the antibiotic treatment groups.

The WIZARD (Weekly Intervention with Zithromax for Atherosclerosis and Its Related Disorders) study is a large randomized, placebo-controlled study that has enrolled more than 7,000 patients who had an MI at least 6 weeks before enrolment and had *C. pneumoniae* IgG titers >1:16.[1,24] Of these, 3,868 patients were randomized to receive azithromycin 600 mg per day for 3 days, then 600 mg weekly for 11 weeks; and 3,856 patients received placebo. At the 2-year follow-up, the investigators found a 7% reduction in the composite primary endpoint of all-cause mortality, recurrent MI, hospitalization, and revisualization (hazard ratio $= 0.93$, $p =$ NS).

The Clarithromycin in Acute Coronary Syndrome Patients in Finland (CLARIFY) study reported favorable results with antibiotic treatment.[1,24] In this small study, 148 patients with unstable angina or acute non–Q-wave coronary events were randomized to receive either clarithromycin 500 mg or a placebo once daily for 85 days. During the initial 3-month treatment period, there were 11 patients in the treatment group compared with 19 patients in the placebo group, which met the combined primary endpoint of death, MI, or unstable angina ($p = 0.1$). Over an average period of 555 days, there were also fewer patients experiencing cardiovascular events in the treatment group (16 vs. 27 in the placebo group, $p = 0.03$).

In the AZACS (Azithromycin in Acute Coronary Syndromes) study, 1,439 patients with unstable angina or acute MI were randomly assigned to receive azithromycin (500 mg for 1 day, then 250 mg for 4 more days) or placebo.[25] Patients were followed up for 6 months. There was no difference between the two groups in terms of adverse outcome such as the incidence of death, recurrent MI, and revascularization.

The ANTIBIO was a German study reported in 2003.[24] A total of 872 patients with an acute MI were randomized to receive either roxithromycin 300 mg or placebo daily for 6 weeks. After a 12-month follow-up period, the investigators found that roxithromycin treatment did not reduce death and adverse cardiac events when compared with the placebo group.

The results of several antibiotic treatment studies currently underway are due to report over the next few years (Table III). These include ACES (Azithromycin and Coronary Events Study),[1,6,24] PROVE-IT (Pravastatin or Atorvastatin Evaluation and Infection Therapy),[1,6,24] CLARICOR (Intervention with Clarithromycin in patients with stable Coronary heart disease),[26] CLAINF (*Chlamydia pneumoniae* and Myocardial Infarction),[6] and CROAATS (Croatian Azithromycin in Atherosclerosis Study)[1]. The ACES, PROVE-IT, and CLARICOR are large-scale, randomized controlled studies, and have each recruited more than 4,000 patients.

Similar contradictory results have been reported in several large-scale retrospective, antibiotic treatment case–control studies. In the first study involving 3,315 cases and 13,139 controls, Meier and colleagues provided indirect evidence for a positive association between past usage of antibiotics such as tetracycline or quinolones (but not macrolides) and the reduction in risk of first-time MI.[6,7] However, the results of another similar study (1,796 cases vs. 4,882 controls) suggested that past usage of tetracycline, doxycycline, or erythromycin (a macrolide) was not associated with the risk of first-time MI.[6] Further study on the usage of a variety of antibiotics and their relationship to first-time MI in a population involving 354,258 patients was also reported with negative results.[27] Yet another study involving a database of 26,195 patients concluded that exposure to antichlamydial antibiotics (tetracyclines, macrolides, quinolones) during the 3 months after an acute MI was associated with a small survival benefit, whereas exposure during the 6 months before an acute MI did not affect survival.[28]

A positive treatment effect of roxithromycin on small AAA expansion rate particularly in patients with *C. pneumoniae* IgA seropositivity has been reported in 2001.[24] The investigators in this study randomized 92 male subjects to receive either roxithromycin 300 mg per day or placebo for 28 days, and patients were followed up for a mean period of 1.5 years. They found that the expansion rate of AAA was reduced by 44% i.e. 1.56 mm per year in the treatment group versus 2.80 mm per year in the placebo group ($p = 0.02$). During the second year, the difference was only 5%. The results remained significant even after multiple linear and logistic regression statistical analyses. A similar positive result of a significant reduction in the AAA expansion rate was reported in another randomized, double-blind, placebo-controlled pilot study (32 patients followed up for 18 months) using the antibiotic doxycycline.[24]

In the clinical context of PAD, a pilot study that randomized 40 *C. pneumoniae*–seropositive men suffering from peripheral arterial occlusive disease to receive either roxithromycin 300 mg daily or placebo for 28 days reported its results in 2002.[24] During a follow-up period of 2.7 years, the investigators found that there were significantly fewer numbers of invasive revascularization per patient in the treatment group (5 interventions performed on 4 patients) in comparison with the placebo group (29 interventions on 9 patients), even after multiple regression analyses. They also reported that limitation of walking distance to 200 m or less occurred in fewer patients in the treatment group than in the placebo group (4 vs. 13). Interestingly, they also observed a significant regression of the size of soft carotid plaques in the roxithromycin-treated patients.

In 1999, an Italian randomized study reported that treatment with a course of roxithromycin (150 mg twice daily for a mean of 26 days) before carotid endarterectomy appeared effective in reducing the bacterial burden of *C. pneumoniae* within carotid atherosclerotic specimens using PCR detection method.[(29)] In 2002, a group of investigators in Germany reported that roxithromycin treatment reduced progression of early carotid atherosclerosis in *C. pneumoniae*–seropositive patients with ischemic stroke.[(24)] They randomized 272 patients with ischemic stroke aged over 55 years to receive either roxithromycin 150 mg twice daily or placebo for 30 days. In the 62 *C. pneumoniae*–seropositive patients that received roxithromycin, there was a significantly decreased common carotid artery intima-to-media thickness (IMT) progression after 2 years, when compared with *C. pneumoniae*–seropositive patients ($n = 63$) that received placebo. They observed no significant difference in IMT progression between *C. pneumoniae*–seronegative patients that received either roxithromycin ($n = 74$) and those that received placebo ($n = 73$). In addition, no significant difference in the occurrence of future cardiovascular events (stroke, MI, or vascular death) was observed in either the treatment or placebo groups. The CRP levels decreased after roxithromycin treatment more significantly in the *C. pneumoniae*–seropositive than in the –seronegative patients, when compared with their respective pretreatment values. However, when the CRP levels in these two treatment groups were compared with their respective placebo groups, a significant difference was only observed in the *C. pneumoniae*–positive group. These results implied that roxithromycin might have greater antichlamydial than anti-inflammatory effects.

Rather than focusing on adverse clinical events as outcome measures, a recent U.K. study assessed the treatment effect of azithromycin on endothelial function in patients with CHD that were seropositive to *C. pneumoniae.*[(30)] The investigators randomized 40 male patients with CHD and *C. pneumoniae* IgG antibody to receive either azithromycin (500 mg per day for 3 days, then 500 mg once weekly for 4 weeks) or placebo. They found that patients in the treatment group, irrespective of their initial *C. pneumoniae* antibody titers, had a significant improvement in flow-mediated dilation (FMD) of the brachial artery when compared with those in the placebo group. Azithromycin treatment also resulted in a significant decrease in E-selectin and von Willebrand factor (both markers of endothelial dysfunction) but not CRP levels. They concluded that azithromycin treatment had a favorable effect on endothelial function in patients with CHD and evidence of *C. pneumoniae* infection.

Several important issues need to be addressed in future antibiotic treatment studies. The macrolides and tetracyclines used in the treatment studies are known to possess anti-inflammatory and immunomodulatory effects, in addition to their antichlamydial activities.[(1,5–7)] The relative contribution of these independent pharmacological properties in treatment studies and their concurrent antimicrobial effects on other microorganisms need to be delineated. Furthermore, the optimal dose and dosing regime, duration, and even cycles of antibiotic treatment will need to be tested in future studies. Apart from its bacteriostatic or bacteriocidal property, whether an antibiotic can reliably eradicate different forms of *C. pneumoniae* (including PB) will also need to be explored and determined.

It is also important to develop reliable methods that can detect true markers reflecting chronic/persistent *C. pneumoniae* infection or high infectious burden in individuals targeted in future antibiotic treatment studies. As *C. pneumoniae* is known to associate with different stages of atherosclerosis, it will be important to target homogenous populations with similar atherosclerotic burden and age to verify treatment effects. Hence, antibiotic treatment studies in the setting of primary prevention in high risk populations with silent/subclinical atherosclerosis should be considered. Future antibiotic treatment studies should increasingly utilize and incorporate other surrogate (e.g. radiological, pathological, blood) markers of plaque activities (which may be clinically silent) to assess treatment effects, rather than using clinical vascular events as the sole outcome measure.

3. *C. pneumoniae* ATHEROSCLEROTIC RISK FACTORS AND KOCH'S POSTULATES

Because atherosclerosis is a complicated multifactorial disease with many clinical variables, it is important to assess the role of *C. pneumoniae* in the context of other well-established atherosclerotic risk factors. Interestingly, *C. pneumoniae* infection is known to be associated with classical vascular risk factors such as smoking and those of the metabolic syndrome: hypertension, dyslipidemia, diabetes, obesity.[2,3,5,7] In addition, *C. pneumoniae* infection is also more commonly associated with other nonmodifiable vascular risk factors such as male sex, age, and certain genetic predisposition (HLA DRII genotype 13a or 17).[5,7] Furthermore, *C. pneumoniae* infection is also associated with novel vascular risk factors such as hyperhomocysteinaemia[31] and elevated CRP.[2,5–7] As all major vascular risk factors are known to correlate with one another in a synergistic way, it is conceivable that chronic *C. pneumoniae* infection may represent a forceful inflammatory stimulus that will interact with all these major vascular risk factors in the pathogenesis and clinical manifestations of atherosclerosis. Indeed, studies that examined specific mechanistic interactions (e.g., iron and lipid metabolisms, immunological expressions) between *C. pneumoniae* and specific vascular risk factors (e.g. gender, dyslipidemia, genetic polymorphisms) are emerging.[5]

C. pneumoniae is unlikely to be the sole cause of atherosclerosis, a complicated disease with many clinical manifestations. This is in contrast to the *H.pylori*–peptic ulcer disease causal link, where the latter is a more straightforward disease with fewer risk factors and clinical variables. Traditionalists may demand a strict fulfillment of Koch's postulates in order to establish that *C. pneumoniae* infection is a cause of atherosclerosis.[7] Although this is not always possible in the case of the *C. pneumoniae*–atherosclerosis association, it must be remembered that the accepted *H. pylori*–peptic ulcer causal link does not always satisfy Koch's postulates either.[2,6,7]

4. CONCLUDING COMMENTS

Seroepidemiological studies have provided weak, circumstantial evidence linking *C. pneumoniae* infection and various atherosclerotic vascular diseases. The challenge in this area of study is to identify other (blood) surrogate markers that are accessible and reliable in reflecting underlying endovascular *C. pneumoniae* infection, applicable in large-scale population (antibiotic treatment) studies. The development of reliable methods to detect surrogate markers of *C. pneumoniae* infection is a priority.

Similarly, the various pathological methods used to detect *C. pneumoniae* and its antigens in atherosclerotic specimens will need further improvement and standardization. In conjunction with studies in other areas, pathological specimen studies will help to clarify a possible atherogenic role of *C. pneumoniae* infection. Animal models will continue to be useful in the investigations of *C. pneumoniae*–induced atherogenic mechanisms in conjunction with other atherosclerotic risk factors and antibiotic treatment studies. A mechanistic understanding of the interaction between *C. pneumoniae* infection and atherosclerosis at the molecular or immunological levels will provide possible therapeutic opportunities and surrogates of treatment effects. Indeed, *C. pneumoniae* has been shown to be involved in key events of atherogenesis and atherothrombosis in *in vitro* studies.

Current antibiotic treatment studies could not prove or disprove a causal role of *C. pneumoniae* in atherogenesis, since the process in its early or stable stages may not always manifest itself clinically as adverse vascular events. Conversely, most of these treatment studies focused on antibiotic treatment benefits on patients with preexisting (advanced) atheromas and secondary prevention of acute atherothrombotic events. Future antibiotic treatment studies should incorporate specific surrogate parameters of atherosclerosis as outcome measures. Studies should be conducted in different populations exposed to *C. pneumoniae* that may have different atherosclerotic burdens, in both primary and secondary prevention settings. In addition, different antibiotics active against *C. pneumoniae* and treatment regimes will need to be developed and standardized in these studies. This is of particular relevance if and when a causal association between *C. pneumoniae* and atherosclerosis is proven, since widespread antibiotic misuse and resistance may then become a serious problem. A minority (4%) of physicians (mostly cardiologists) in the United States have already used antibiotics to treat cardiovascular disease as if it were an infectious disease, although this practice is currently considered inappropriate and premature.[6]

C. pneumoniae on its own is an important pathogen that causes a wide spectrum of respiratory tract infections and extrapulmonary diseases worldwide. The recognition of an association between *C. pneumoniae* infection and atherosclerotic vascular diseases has generated an exponential interest in this area of research over the last decade. The body of evidence as a whole is supportive of *C. pneumoniae* as a plausible and modifiable risk factor in atherosclerosis. With the recent discovery of *C. pneumoniae*'s genome, the development of an effective vaccination[1] program in future is becoming a pressing reality that will not only prevent *C. pneumoniae* infection but also clarify if *C. pneumoniae* has a causal relationship with atherosclerosis. Meanwhile, we rely on a progressive, multidisciplinary collaborative approach among scientists and clinicians to finally reach a conclusion beyond reasonable doubt, to either confirm or refute *C. pneumoniae* as a causal factor in atherosclerosis.

REFERENCES

1. Higgins, J. P., 2003, *Chlamydia pneumoniae* and coronary artery disease: The antibiotic trials, *Mayo Clin. Proc.* **78:**321–332.
2. Kalayoglu, M. V., Libby, P., and Byrne, G. I., 2002, *Chlamydia pneumoniae* as an emerging risk factor in cardiovascular disease, *JAMA* **288:**2724–2731.
3. Fong, I. W., 2002, Infections and their role in atherosclerotic vascular disease, *JADA* **133:**7S–13S.
4. Dugan, J. P., Feuge, R. R., and Burgess, D. S., 2002, Review of evidence for a connection between *Chlamydia pneumoniae* and atherosclerotic disease, *Clin. Ther.* **24:**719–735.
5. Leinonen, M., and Saikku, P., 2002, Evidence for infectious agents in cardiovascular disease and atherosclerosis, *Lancet Infect. Dis.* **2:**11–17.
6. Ngeh, J., Anand, V., and Gupta, S., 2002, *Chlamydia pneumoniae* and atherosclerosis—what we know and what we don't, *Clin. Microbiol. Infect.* **8:**2–13.
7. Ngeh, J., and Gupta, S., 2002, Inflammation and infection in coronary artery disease, in: *Cardiology: Current Perspectives* (G. Jackson, ed.), Martin Dunitz Ltd., London., 125–144.
8. Ngeh, J., 2000, *Chlamydia pneumoniae* in Elderly Patients with Stroke study (CPEPS): A case–control study on the seroprevalence of *Chlamydia pneumoniae* in patients aged over 65 years admitted with acute stroke or transient ischaemic attack, MSc Dissertation, University of Keele, UK.
9. Roivainen, M., Alfthan, G., Jousilahti, P., Kimpimäki, M., Hovi, T., and Tuomilehto, J., 1998, Enterovirus infections as a possible risk factor for myocardial infarction, *Circulation* **98:**2534–2537.
10. Higuchi, M. L., Sambiase, N., Palomino, S., Gutierrez, P., Demarchi, L. M., Aiello, V. D., and Ramires, J. A. F., 2000, Detection of *Mycoplasma pneumoniae* and *Chlamydia pneumoniae* in ruptured atherosclerotic plaques, *Braz. J. Med. Biol. Res.* **33:**1023–1026.
11. Moriuchi, H., and Rodriguez, W., 2000, Role of varicella-zoster virus in stroke syndromes, *Pediatr. Infect. Dis. J.* **19:**648–653.
12. Gupta, S., 1999, *Chlamydia pneumoniae*, monocyte activation and antimicrobial therapy in coronary heart disease. MD Thesis, University of London, London.
13. Ngeh, J., Gupta, S., and Goodbourn, C., 2004, The reproducibility of an enzyme-linked immunosorbent assay for detection of *Chlamydia pneumoniae*–specific antibodies, *Clin. Microbiol. Infect.* **10:**171–174.
14. Shor, A., and Phillips, J., 2000, Histological and ultrastructural findings suggesting an initiating role for *Chlamydia pneumoniae* in the pathogenesis of atherosclerosis: A study of fifty cases, *Cardiovasc. J. S. Afr.* **11:**16–23.
15. Rassu, M., Cazzavillan, S., Scagnelli, M., Peron, A., Bevilacqua, P. A., Facco, M., Bertoloni, G., Lauro, F. M., Zambello, R., and Bonoldi, E., 2001, Demonstration of *Chlamydia pneumoniae* in atherosclerotic arteries from various vascular regions, *Atherosclerosis* **158:**73–79.
16. Johnston, S. C., Messina, L. M., Browner, W. S., Lawton, M. T., Morris, C., and Dean, D., 2001, C-reactive protein levels and viable *Chlamydia pneumoniae* in carotid artery atherosclerosis, *Stroke* **32:**2748–2752.
17. Ezzahiri, R., Nelissen-Vrancken, H. J. M. G., Kurvers, H. A. J. M., Stassen, F. R. M., Vliegen, I., Grauls, G. E. L. M., van Pul, M. M. L., Kitslaar, P. J. E. H. M., and Bruggeman, C. A., 2002, Chlamydophila pneumoniae (*Chlamydia pneumoniae*) accelerates the formation of complex atherosclerotic lesions in Apo E3-Leiden mice, *Cardiovasc. Res.* **56:**269–276.

18. Blessing, E., Campbell, L. A., Rosenfeld, M. E., Chough, N., and Kuo, C.-C., 2001, *Chlamydia pneumoniae* infection accelerates hyperlipidemia induced atherosclerotic lesion development in C57BL/6J mice, *Atherosclerosis* **158:**13–17.
19. Sako, T., Takahashi, T., Takehana, K., Uchida, E., Nakade, T., Umemura, T., and Taniyama, H., 2002, Chlamydial infection in canine atherosclerotic lesions, *Atherosclerosis* **162:**253–259.
20. Liuba, P., Pesonen, E., Paakkari, I., Forslid, A., and Sandstrom, S., 2002, Acute *Chlamydia pneumoniae* infection causes diffuse coronary endothelial dysfunction in pigs, *JACC* **39**(Suppl. 2)**:**S277.
21. Pislaru, S. V., Ranst, M. V., Pislaru, C., Szelid, Z., Theilmeier, G., Ossewaarde, J. M., Holvoet, P., Janssens, S., Verbeken, E., and Van de Werf, F. J., 2003, *Chlamydia pneumoniae* induces neointima formation in coronary arteries of normal pigs, *Cardiovasc. Res.* **57:**834–842.
22. Loftus, I. M., Naylor, A. R., Bell, P. R. F., and Thompson, M. M., 2002, Matrix metalloproteinases and atherosclerotic plaque instability, *Br. J. Surg.* **89:**680–694.
23. Airenne, S., Surcel, H. M., Tuukkanen, J., Leinonen, M., and Saikku, P., 2002, *Chlamydia pneumoniae* inhibits apoptosis in human epithelial and monocyte cell lines, *Scand. J. Immunol.* **55:**390–398.
24. Grayston, J. T., 2003, Antibiotic treatment of atherosclerotic cardiovascular disease, *Circulation* **107:**1228.
25. Cercek, B., Shah, P. K., Noc, M., Zahger, D., Zeymer, U., Matetzky, S., Maurer, G., and Mahrer, P., 2003, Effect of short-term treatment with azithromycin on recurrent ischemic events in patients with acute coronary syndrome in the Azithromycin in Acute Coronary Syndrome (AZACS) trial: A randomised controlled trial, *Lancet* **361:**809–13.
26. Hansen, S., Als-Nielsen, B., Damgaard, M., Helø, O. H., Petersen, L., and Jespersen, C. M., 2001, Intervention with clarithromycin in patients with stable coronary heart disease, *Heart Drug* **1:**14–19.
27. Luchsinger, J. A., Pablos-Méndez, A., Knirsch, C., Rabinowitz, D., and Shea, S., 2002, Relation of antibiotic use to risk of myocardial infarction in the general population, *Am. J. Cardiol.* **89:**18–21.
28. Pilote, L., Green, L., Joseph, L., Richard, H., Eisenberg, M. J., 2002, Antibiotics against *Chlamydia pneumoniae* and prognosis after myocardial infarction, *Am. Heart J.* **143:**294–300.
29. Melissano, G., Blasi, F., Esposito, G., Tarsia, P., Dordoni, L., Arosio, C., Tshomba, Y., Fagetti, L., Allegra, L., and Chiesa, R., 1999, Chlamydia pneumoniae eradication from carotid plaques. Results of an open randomized treatment study, *Eur. J. Vas. Endovasc. Surg.* **18:**355–359
30. Parchure, N., Zouridakis, E. G., and Kaski, J. C., 2002, Effect of azithromycin treatment on endothelial function in patients with coronary artery disease and evidence of *Chlamydia pneumoniae* infection, *Circulation* **105:**1298–1303.
31. Stanger, O. H., Semmelrock, H. J., Rehak, P., Tiran, B., Meinitzer, A., Rigler, B., and Tiran, A., 2002, Hyperhomocyst(e)inemia and *Chlamydia pneumoniae* IgG seropositivity in patients with coronary artery disease, *Atherosclerosis* **162:**157–162.

10

The Biology of *Chlamydia pneumoniae* in Cardiovascular Disease Pathogenesis

SCOT P. OUELLETTE, ROBERT J. BELLAND, JENS GIEFFERS, and GERALD I. BYRNE

1. INTRODUCTION

In 1988, Saikku *et al.* published their observations linking *Chlamydia pneumoniae* serology to coronary heart disease in a population composed of Finnish men under the age of 50.[1] The notion that infections cause heart disease is not new,[2] but this remarkable finding proved to be a bellwether for the reemergence of the idea that a specific infectious agent may be causally involved in heart disease. The Saikku *et al.* report led to a series of studies to epidemiologically corroborate or refute those findings. Concurrent with the publication of epidemiologic reports, molecular diagnostic tests were being adapted for the identification of this microorganism to determine if its presence could be detected in diseased tissue. Indeed, immunohistochemical and *C. pneumoniae*–specific PCR methods demonstrated that these bacteria are found associated with about half of atheromatous tissue whereas the organism was rarely found in normal tissues. These clinical studies, although lacking causal implications, clearly document that *C. pneumoniae* is present within atherosclerotic lesions. Animal models have also been developed to study the involvement of *C. pneumoniae* in cardiovascular disease. Infections of atherogenesis-prone mice have extended the information gleaned from clinical observations to include data showing that *C. pneumoniae* localizes to the site of lesion development and causes much more rapid lesion

---

SCOT P. OUELLETTE, ROBERT J. BELLAND and GERALD I. BYRNE • University of Tennessee Health Science Center, Memphis, Tennessee. JENS GIEFFERS • Medical University of Luebeck, Luebeck, Germany.

progression than occurs in either uninfected animals or in animals infected with other pathogens (see ref. 3 for review).

These *in situ* and *in vivo* studies have prompted more basic work focused on the assessment of the role that the organism and its putative virulence factors may play in atherogenesis and cardiovascular disease. Virulence factors for most microbial pathogens are traditionally defined by deleting the gene of interest and assessing if an avirulent phenotype causes disease or infection in appropriate model systems. Unfortunately, there is no stable DNA transformation system available for chlamydiae, making it more difficult to establish a causal role for any given virulence factor in chlamydial disease pathogenesis. Although more descriptive approaches have been used to study chlamydial pathogenic mechanisms, some progress has been made. Chlamydial heat shock protein–60 (cHsp60) localizes with human atheromas and regulates macrophage tumor necrosis factor-alpha (TNFα) and matrix metalloproteinase (MMP) expression.[(4)] cHsp60 is also involved in LDL oxidation[(5)] whereas chlamydial LPS has been linked to foam cell formation.[(6)] A key component of the atherogenic process is cellular lipid accumulation in the form of cholesteryl esters derived from modified LDL. Lipid accumulation can occur in a variety of cell types, but lipid-laden macrophages, or foam cells, are a hallmark of early lesion development. Several factors that induce lipid accumulation and may be involved in atherosclerosis include the presence of IL-1β, lipopolysaccharide (LPS), cytomegalovirus (CMV), and *C. pneumoniae.*[(7)] Of these, *C. pneumoniae* is the only bacterium that has been repeatedly associated with atherosclerosis and localized to atheromas.

Chlamydia pneumoniae initially infects a variety of cell types within the lung, including epithelial cells and leukocytes. Figure 1 illustrates how an initial acute or asymptomatic infection can lead to dissemination of the organism by macrophages and lymphocytes to sites of inflammation within the circulation, thereby contributing to lesion progression and instability. In this model, *C. pneumoniae* would be a contributing factor to the atheroma but not necessarily a direct causal factor.

Recently, the complete genome sequencing of three *C. pneumoniae* isolates was carried out, which has greatly expanded our ability to understand and study this organism.[(8–10)] Although the three isolates were more than 99.9% identical, there were also several key differences that suggest a few surprises. For example, Read *et al.* found that there was roughly the same level of gene polymorphisms between different *C. pneumoniae* strains as there were *within* the AR39 strain,[(9)] thus giving rise to the concept of intra- as well as interstrain heterogeneity in *C. pneumoniae* populations. Intrastrain heterogeneity may not be an unexpected finding considering that clonal isolates of *C. pneumoniae* have not been routinely obtained. This is because *Chlamydia* is an obligate intracellular pathogen, and clonal isolates can only be collected by plaque purification methods similar to those used for viruses. This procedure is not commonly done for chlamydial isolates because it is highly labor intensive and takes 1 to 2 weeks to collect a small sample, which must then be propagated several times before a useful quantity of chlamydiae is obtained. In addition, some chlamydial species plaque more readily than others, and *C. pneumoniae* has been refractory to standard plaquing

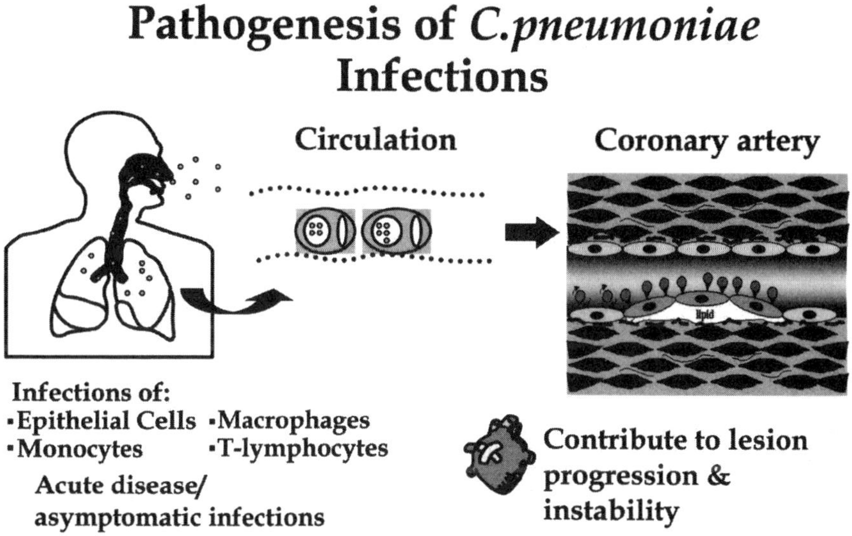

FIGURE 1. *C. pneumoniae* initially infects cells within the lung, resulting in acute disease or asymptomatic infections. Infected macrophages and lymphocytes subsequently enter the circulatory system and localize to sites of inflammation within the vasculature, thereby contributing to lesion progression and instability.

techniques. However, improved methods have recently been developed,(11) and the production and use of clonal isolates will allow the study of homogeneous strains for the first time. Consequences of strain heterogeneity to actual *C. pneumoniae* infections, particularly as this relates to atherosclerosis, will require more extensive analysis, but it is probable that if intrastrain heterogeneity is common for *C. pneumoniae*, then variants may be selected based on the variety of tissue-specific environments that the organism encounters as it moves from the lung to the site of lesion development.

2. CHLAMYDIAL LIFE CYCLE AND PERSISTENCE

Chlamydiae exhibit a unique obligate intracellular life cycle that is essential for understanding their virulence in both acute and chronic disease. These bacteria have a biphasic life cycle that alternates between a metabolically inert, infectious elementary body (EB) and a noninfectious, metabolically active reticulate body (RB) (see ref. 12 for review). The molecular events of chlamydial growth and differentiation have yet to be defined, but their intracellular life cycle has been sufficiently described to document several important fundamental processes essential for chlamydial survival. An EB begins an infection by attaching to a susceptible host cell. This primary interaction likely involves the chlamydial major outer membrane protein (MOMP), although roles have been postulated for glycosaminoglycans and other putative adhesins.(13) After

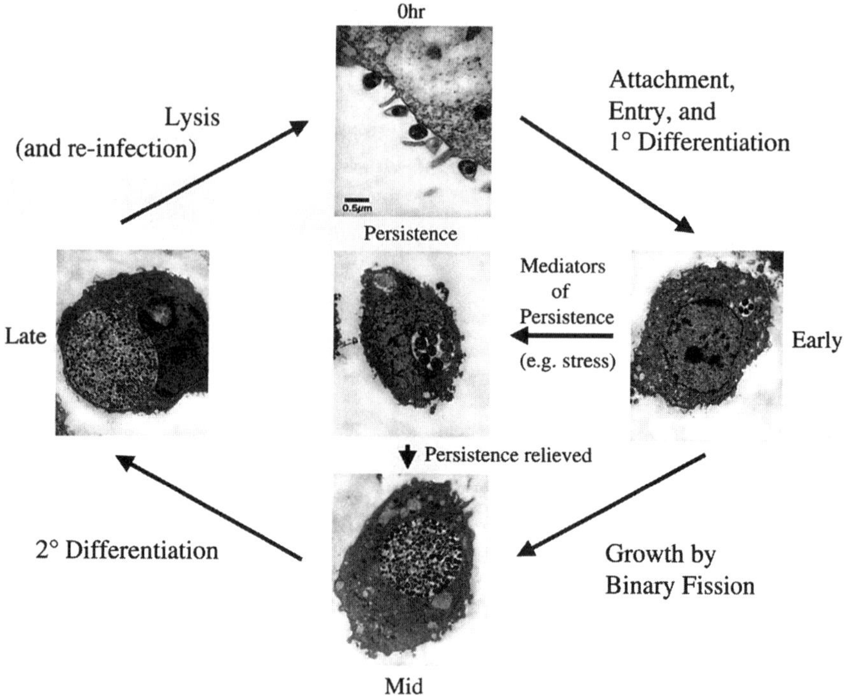

FIGURE 2. After attachment and internalization, *C. pneumoniae* EBs (smaller, electron-dense objects) differentiate into RBs (larger forms) and grow within the inclusion. At later time points, RBs begin to reorganize into EBs to be released from the cell. Host immune stress, such as IFN-γ activation of host cells, can induce persistence, resulting in abnormally large, aberrant morphological forms. Removal of the stress allows the organism to reenter the productive growth cycle.

attachment, the EB is internalized into a vesicle that avoids the endosomal pathway and subsequent lysosome fusion.[14] Sphingolipid trafficking and recruitment of post-Golgi vesicles to the chlamydia-containing vacuole indicate that this structure actually resides in the exocytic pathway.[15] These postentry trafficking activities suggest dynamic changes imposed on the host cell by chlamydiae. The next steps in chlamydial growth involve morphogenic changes. Each EB differentiates into an RB, which divides by binary fission within what is now termed the inclusion. At the completion of the life cycle, the RBs reorganize to EBs, which are then released from the host cell. Figure 2 illustrates some of these developmental steps ultrastructurally.

Defining the microbiology associated with productive intracellular chlamydial growth and development is an important first step in understanding the pathogenesis of acute infections caused by these organisms. Chronic infections, however, likely involve altered chlamydial characteristics because the intracellular organisms are not in a productive growth mode. For example, host immune responses, antibiotic treatment, or nutrient deprivation cause chlamydiae to enter a nonproductive, persistent state that has been used as a model for chronic chlamydial infections (see ref. 12 for review). Under these persistence-inducing conditions, the chlamydial Hsp60 appears to be highly upregulated. Chlamydial Hsp60 has proinflammatory effects by directly activating mononuclear leukocytes. Chronic chlamydial infections are inflammatory because of the presence of such cells at the site of infection. Chlamydial persistence is characterized by morphologically aberrant RBs that do not undergo cytokinesis. This was shown ultrastructurally in electron micrographs and molecularly by the absence of cell division transcripts during nonproductive growth.[16] Persistent chlamydiae, however, reenter the productive life cycle upon removal of the stress, suggesting that this chronic model is not a bactericidal pathway and may have *in vivo* relevance. For example, in genital tract infections of mice lacking inducible nitric oxide synthase, an enzyme that produces the potent antimicrobial nitric oxide, it is not possible to culture chlamydiae from the genital tract after the initial disease symptoms. However, treatment of the mouse with immunosuppressive drugs now makes possible chlamydial reactivation with subsequent culture.[17]

3. *C. pneumoniae* IN THE PATHOGENESIS OF CARDIOVASCULAR DISEASE

The current paradigm for atherosclerosis suggests that it is an inflammatory disease. Evidence for this is provided by the presence of monocyte-derived macrophages and T lymphocytes at the site of early lesions. Cytokines released by these immune cells, such as IL-6, induce acute phase response proteins, and elevated CRP levels, (a marker for atherosclerosis) are an indication of chronic inflammation. The inflammatory nature of the early atheroma causes modifications to LDL, which allow the molecule to bypass the usual regulatory mechanisms for entry into cells, thus facilitating foam cell formation. Furthermore, the effect of inflammation on endothelial cells allows for greater arterial permeability to LDL and increased adhesion molecule expression and diapedesis of leukocytes. Although the cause for inflammation remains controversial, it is likely multifactorial and may involve infectious agents. It remains unclear if atherosclerosis is due to continued, chronic inflammation at the site or episodic inflammation from other sources that exacerbates the condition at the lesion. If *C. pneumoniae* infection of the cell types associated with the atheroma is connected to modification and/or dysregulation of those cells in ways that promote or exacerbate inflammation, then a strong case exists for this bacterium to contribute to the disease state.

3.1. *C. pneumoniae* and Host Cell Signaling

Vascular remodeling is a characteristic of proatherogenic functions, and vascular smooth muscle and endothelial cells are integral to this process. As such, inflammatory mediators released from infected cells that induce proliferation likely aggravate the atherosclerotic condition. *Chlamydia pneumoniae* may play a role in this process. When this organism infects vascular cells, signal transduction pathway activation occurs that results in changes that are characteristically atherogenic. Because signal transduction involves complex pathways with redundant interactions between them, there is great potential for activation to be achieved via limited mechanisms. For example, a single ligand may bind a receptor, which results in the downstream activation of a single mediator that in turn can elicit pleiotropic effects. Consequently, a handful of chlamydial components may elicit activation even in the absence of the organism itself. These soluble mediators may be released by persistent or actively growing organisms in a low percentage of host cells; thus failure to detect the organism by culture or PCR may not always eliminate a chlamydial aspect to this disease.

Vascular smooth muscle cells (VSMC) are critical components of the atherosclerotic plaque, and unprogrammed growth can lead to the progression of disease toward enlarged plaques. If *C. pneumoniae* can induce vascular smooth muscle cell proliferation, either directly or indirectly, then this would be further evidence that this organism can contribute to the progression of lesions within the vasculature. Miller *et al.* demonstrated that infection of this cell type with *C. pneumoniae* can lead to the activation of transcriptional regulators that are known to induce proliferation.[(18)] In these experiments, smooth muscle cells were infected or uninfected in the absence or presence of azithromycin, an antibiotic with activity against chlamydiae. Activation of the transcriptional regulators NF-κB and AP-1 was monitored via electrophoretic mobility shift assays of nuclear extracts shortly after infection. In all cases, VSMCs infected with *C. pneumoniae* in the absence of azithromycin were able to elicit activation of NF-κB and AP-1. These changes were accompanied by an increase in cell growth compared to controls and infected cells treated with azithromycin. NF-κB and AP-1 are important proinflammatory transcriptional mediators that initiate the production of various cytokines and growth factors. These data suggest a mechanism whereby *C. pneumoniae* is able to modify cell growth in a way that promotes atherosclerotic lesion development through continued inflammation and cell proliferation.

Dechend *et al.* analyzed the effect of *C. pneumoniae* infection on the activation of NF-κB and PAI-1, plasminogen activator inhibitor 1, in VSMCs and endothelial cells.[(19)] Here, immunoblots were used to assess protein expression of prothrombic and immunomodulatory factors. Infected VSMCs showed significantly more procoagulant protein than mock-infected controls. Infected endothelial cells responded similarly but also produced significant amounts of IL-6, a cytokine that is associated with atherosclerotic inflammation by virtue of its ability to induce the acute phase response. In both cell types, these changes were accompanied by an increase in NF-κB activation with a subsequent decrease in the levels of its inhibitor IκBα. A previous report from Fryer *et al.* had established that *Chlamydia*-infected endothelial cells stimulate expression of tissue factor in a time- and dose-dependent manner independent of chlamydial viability.[(20)] These data provide a possible causal role for *C. pneumoniae* in plaque rupture and coagulant activity by its ability to both infect cells associated with the plaque and to elicit prothrombic and proinflammatory mediators.

Elaboration of cytokines from dysregulated smooth muscle cells affects endothelial cell function in concert with the inflammatory effect on this cell type. Endothelial cells comprise the luminal surface of the vasculature and thus are intimately involved in any event that leads to vascular remodeling as occurs in atherosclerosis. Several studies have analyzed the effect of *C. pneumoniae* on endothelial cells. This work has focused on activating signal transduction pathways and adhesion molecule expression. However, endothelial cell infection can affect other cell types within the vicinity of the atheroma. Coombes and Mahony showed that an infected endothelial cell secretes a factor(s) that can induce proliferation of smooth muscle cells.[(21)] In these experiments, culture medium from infected endothelial cells was transferred at various time points to smooth muscle cell monolayers. Proliferation was monitored by tritiated thymidine uptake and increase in cell number. These data demonstrated that *C. pneumoniae* infection of endothelial cells induces secretion of a smooth muscle cell stimulatory factor(s) in a dose- and time-dependent manner. This property did not appear to rely on live organisms but solely on cell contact or entry, as heat-killed chlamydiae and organisms grown in the presence of chloramphenicol, which prevents protein synthesis, did not abolish the activity. Further studies will be required to discern which components of the chlamydia activate endothelial cell signal transduction pathways, as well as the secreted factor responsible for inducing smooth muscle cell proliferation.

Diapedesis of leukocytes into the extravascular space of the lesion exacerbates the inflammatory condition of this site. Indeed, accretion of leukocytes, particularly monocyte-derived macrophages that are subsequently activated by cytokines, in the atherosclerotic intima contributes to plaque rupture and myocardial infarction. Molestina *et al.* provide an interesting study on the effect *C. pneumoniae* infection of endothelial cells has on this process.[(22)] Human umbilical vein endothelial cells were infected with heart, respiratory, or laboratory isolates of *C. pneumoniae* or different serovars of *C. trachomatis*, and subsequent production of chemotactic products were measured. Neutrophil and monocyte transendothelial migration were assessed in response to endothelial cell infection with live, heat-killed, and UV-treated *C. pneumoniae* in an *in vitro* model system. In most cases, live organisms were better able to induce diapedesis across the endothelial cell layer. Organisms that had been passaged greater than 40 times in HEp-2 cells resulted in greater induction. High-passaged organisms were also able to better induce IL-8 and MCP-1 secretion from endothelial cells, with little difference in that ability between UV-treated and live bacteria. These proteins are chemotactic for neutrophils and monocytes, thus providing a mechanism for transmigration of these cell types. Infection with *C. trachomatis* serovars failed to induce secretion of either chemotactic protein but was able to productively grow within endothelial cells. A slight but significant increase in transmigration of monocytes was observed for endothelial cell infection with *C. trachomatis* but not to the extent of infection with *C. pneumoniae.* The lack of secreted IL-8 and MCP-1 in response to heat treatment implies a heat-sensitive antigen in this process such as cHsp60. Because *C. trachomatis* failed to induce secretion of these proteins when it has a highly orthologous Hsp60, another antigen that is unique to *C. pneumoniae* is likely responsible for this effect.

In an earlier study, Molestina *et al.* had demonstrated heterogeneity in the major outer membrane protein among *C. pneumoniae* strains and in their ability to induce chemokine secretion and adhesion molecule expression.[23] In this more recent report (Molestina, 1999), the authors comment that serial passage through HEp-2 cells may have resulted in the acquisition/upregulation or loss/downregulation of key components that favor activation of endothelial cells in such a way as to elicit transmigration across the cell layer. This finding was interesting because low-passage heart isolates, more recently removed from the atheroma, were less efficient at activating endothelial cells than high-passage heart isolates, which have likely become adapted to the cell culture system. These data suggest that particular clonal variants are being selected during serial passage through the standard epithelial cell culture system. Furthermore, because the low-passage heart isolates are poorer inducers of chemotactic proteins, it becomes difficult to explain a role for this organism in the pathogenicity of atherosclerosis as this relates to endothelial cell activation and recruitment of leukocytes unless one posits that the heart isolates were initially potent activators that have downregulated expression of key antigens involved in this process.

Endothelial cell activation in response to infection with *C. pneumoniae* has been characterized by cDNA microarray analysis. Coombes and Mahony isolated mRNA from uninfected and infected endothelial cells 18 h postinfection and generated cDNA probes to an array of 268 human genes, including cytokines, growth factors, and receptors.[24] Data obtained from the array support earlier findings of IL-8 and MCP-1 activation. In addition, many growth factors that may induce proliferation of smooth muscle cells were upregulated in response to infection. Interestingly, Kothe *et al.* found that pretreatment of macrophages and endothelial cells with HMG-CoA inhibitors, cholesterol lowering agents, resulted in less secretion of IL-8 and MCP-1.[25] Krull *et al.* showed a marked increase in protein tyrosine phosphorylation, MAPK activation, and NF-κB activation in endothelial cells shortly after infection with a respiratory isolate of *C. pneumoniae.*[26] Activation of these signal transduction cascades resulted in an upregulation of transcription for adhesion molecules, including VCAM-1, ICAM-1, and E-selectin. Together, these data provide a mechanistic rationale for this pathogenic organism to recruit leukocytes to an area of potential inflammation.

3.2. Chlamydiae–Macrophage Interactions

Two components that have been implicated in the pathogenesis of a number of chronic chlamydial diseases, such as trachoma and pelvic inflammatory disease, are the organism's LPS and heat shock protein–60 (cHsp60).[27] LPS molecules in Gram-negative bacteria are potent stimulators of immune cells and induce the release of inflammatory mediators such as TNFα. Serological associations between heat shock antigens and chlamydial diseases have been established and suggest that these are possible virulence factors.[28] Furthermore, Kol *et al.* demonstrated the presence of cHsp60 in atheromatous tissue collected from humans.[4] Together, these observations support a role for chlamydial antigens in atherogenic events.

LDL uptake by macrophages is a tightly regulated process that prevents lipid accumulation within the cell. Under nonatherogenic conditions, LDL uptake causes the transcriptional downregulation of its receptor. Scavenger receptors, however, may bypass this control by allowing the endocytosis of modified LDL. Oxidized LDL is one such form that has been shown to bind to these receptors and to accumulate within macrophages. An initial report on the pathogenic capabilities of *C. pneumoniae* in atherosclerosis studied the interaction of the bacterium with macrophages in the presence of native LDL.[6] These data showed that *C. pneumoniae*–infected macrophages were much more likely to accumulate lipids, stored in the form of cholesteryl esters, and consequently, to form foam cells. Furthermore, this ability was not inhibited through blocking scavenger receptors, which proved that modified LDL uptake was not responsible. Interestingly, treatment with heparin, a competitive inhibitor for the LDL receptor, reduced foam cell formation in the presence of native LDL but not oxidized LDL. These experiments were the first to implicate *C. pneumoniae* as a causative agent of atherosclerosis by showing that the pathogen may dysregulate native LDL uptake or metabolism.

Subsequent experiments determined that the chlamydial component responsible for inducing foam cell formation was cLPS.[29] Heat-killed EBs maintained this ability whereas periodate-treated EBs abolished it. Purified cLPS was able to reproduce these results, and cholesteryl ester accumulation was blocked by an LPS inhibitor, lipid X. There is no data to suggest that *C. pneumoniae* clones produce differing amounts of cLPS. However, because of the low biologic activity of the cLPS compared with most Gram-negative LPS molecules, it is tempting to speculate that clonal variants selected in atherosclerotic plaques may upregulate cLPS synthesis genes. An analysis of laboratory-strain versus heart-isolate gene expression of the cLPS synthesis genes under productive and nonproductive growth conditions could begin to answer this question.

Oxidation of LDL is another important atherogenic event that promotes leukocyte diapedesis, smooth muscle cell migration and proliferation, and injury to vessels and can be linked to *C. pneumoniae*. Kalayoglu *et al.* showed that infected monocytes, cultured in the presence of LDL, generated oxidized LDL compared with mock-infected cells.[5] This occurred in a time- and dose-dependent mechanism and was inhibited by the antioxidant vitamin E. Since LDL oxidation is thought to occur by superoxide-dependent mechanisms, experiments were performed to determine what role this pathway played in the modification of LDL. No increase in superoxide anions was seen in infected monocytes compared with appropriate controls. Blockage of the superoxide pathway with inhibitors failed to decrease the oxidation of LDL. Heat- or UV-treated chlamydiae were assayed for their ability to induce lipoprotein modifications. Heat treatment severely abrogated LDL oxidation whereas UV treatment had little effect. To further evaluate chlamydial antigens that might be responsible for this effect, cHsp60 and cLPS were examined. cHsp60 consistently reproduced the previous observations in a dose-dependent fashion. The chlamydial Hsp60 has also been shown to regulate TNFα and matrix metalloproteinase expression in macrophages.[4]

These data provide a mechanism for chlamydial antigens to promote plaque destabilization with subsequent coronary events since matrix metalloproteinases can contribute to the break down of the atherosclerotic plaque.

Because of the importance of chlamydial heat shock proteins in diseases caused by this pathogen, it was particularly intriguing to find that there are three paralogous *Hsp60* genes in both *C. pneumoniae* and *C. trachomatis.*[8] Whereas each paralogue is only about 25% similar, the same orthologues between species are more than 90% identical. The cHsp60 used in the experiments performed by Kalayoglu *et al.* was cHsp60 cloned from *C. trachomatis.*[5] This in no way invalidates the results obtained because only *C. pneumoniae* whole organisms have been shown to elicit these proatherogenic responses. The possibility that clonal variants isolated from atheromatous tissue might preferentially express one or more of these *Hsp60* paralogues as compared to respiratory isolates is extremely interesting. Alternately, some strains may produce large amounts of cHsp60 protein when growing within macrophages. Indeed, it is documented that chlamydiae grown *in vitro* under conditions that induce persistence, a condition that is thought to mimic *in vivo* chronic infections, Hsp60 production is greatly upregulated.[30] Growth in macrophages is usually nonproductive, thus supporting this as a persistent-like growth state. The particular isoform of Hsp60 was not elaborated from these studies, so it remains to be seen whether one or more paralogues are expressed.

4. *C. pneumoniae* STRAIN HETEROGENEITY IN CARDIOVASCULAR DISEASE

Because *C. pneumoniae* has such broad host tropism and is able to cause both acute and chronic diseases, the possibility exists for specific clonal isolates to be associated with each disease or disease site. From the three isolates sequenced, it is clear that both interstrain and intrastrain heterogeneity is present in *C. pneumoniae* populations. Vascular isolates of this organism will likely have subtle genetic differences, which indicate a propensity for growth and/or survival in the vasculature, when compared with respiratory isolates. Figure 3 provides a model for how specific clonal isolates may be associated with disease sites by their propensity to infect and survive within specific cell types. Many isolates may be present during the initial infection of the host. However, only a subset of that population will have the ability, for example, to infect and survive within macrophages, which may subsequently localize to atheromas.

4.1. The *tyrP* Polymorphism and SNPs

There are several mechanisms for inducing persistence *in vitro.* These include activation of host cells with interferon gamma (IFNγ), treatment with antibiotics such as penicillin, and amino acid starvation.[12] The amino acid tryptophan has been shown to be critical for normal chlamydial growth and development and, indeed, IFNγ activates the enzyme indoleamine-2,3-dioxygenase

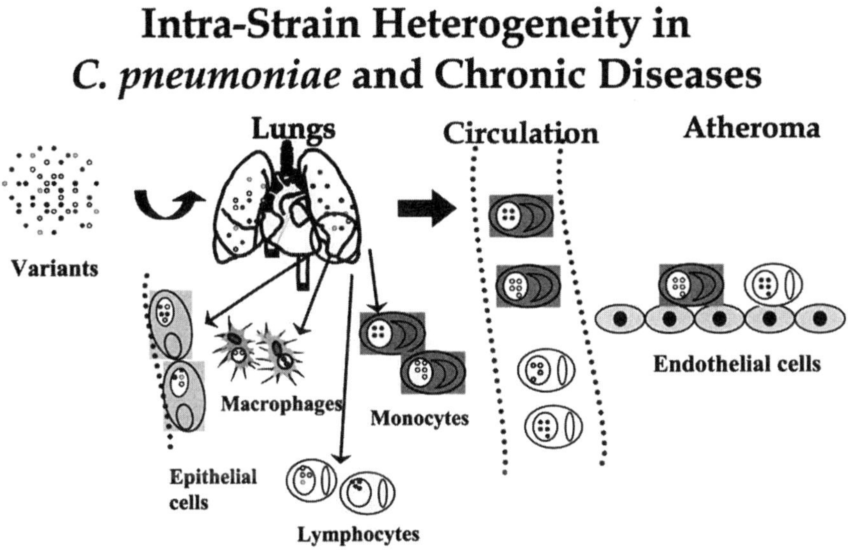

FIGURE 3. When *C. pneumoniae* is transmitted from host to host, multiple variants can infect the cells present in the lung. However, only a subset of variants has the ability to infect and to survive within macrophages and lymphocytes, which are subsequently disseminated through the circulation to the atheroma. Therefore, variants of a specific genotype that exist within atheromas should be identifiable.

(IDO), which catalyzes the formation of kynurenine from tryptophan.[31] This limits intracellular tryptophan for use by chlamydiae, thus rendering them persistent. The *tyrP* gene is classified as a tyrosine permease and is thought to be an aromatic amino acid transporter, making it one possible mechanism for importing tryptophan into the bacteria. The *tyrP* region of the chromosome was identified by Read *et al.* as a noticeable polymorphism, as in some readthroughs they found two copies present whereas in others there was only one.[9] The physiologic significance of this observation, if any, is not fully understood. One can envision a scenario where chlamydiae with multiple copies of this gene are better able to withstand tryptophan limitation within a host cell. Alternately, organisms that have only one copy may be predisposed to persistent growth, such as in macrophages.

Gieffers *et al.* developed a unique method for isolating clonal populations of *C. pneumoniae* using the *tyrP* polymorphism as an indicator of homogeneity.[11] They showed that in many of the routinely used laboratory strains, both one and two copies of the polymorphism were present as shown in the AR39 sequence. In a clinical respiratory isolate, these findings were also replicated. When the respiratory isolate was clonally purified however, strains containing either one *or* two copies were found, thus proving their ability to generate genetically homogeneous populations as assessed by this polymorphism. In a follow-up study examining *tyrP* gene copy number in clinical isolates, Gieffers *et al.* (submitted) found that respiratory isolates usually had multiple copies of *tyrP*, with some isolates showing a mixture of populations with single and multiple copies. Interestingly, all vascular isolates examined had only one copy of the gene. Copy number was also related to increased tyrosine transport as demonstrated by radiolabeled uptake assays. These data imply that vascular strains may indeed be predisposed toward persistent growth and thus be better able to maintain the chronic infection at the site of atheroma.

The vast majority of polymorphisms within the AR39 strain, as reported by Read *et al.*, were single-nucleotide polymorphisms or SNPs.(9) It has been documented in other pathogenic bacteria that certain strains, identified by their SNP patterns, are more virulent than others. Most of the SNPs within the AR39 strain, however, occur in the intergenic region, implicating that most of these changes are not phenotypically apparent. Analyses will have to be performed to assess what role these SNPs have on virulence and if they can be used to differentiate strains in a meaningful way. Another potentially intriguing polymorphism is the upstream promoter region for the uridine kinase gene. The AR39 region was in a reverse orientation compared with CWL029. Uridine kinase is an important enzyme in nucleoside metabolism, and the ability to regulate its expression suggests another means for either resisting or inducing persistent growth in otherwise unfavorable conditions. Again, analysis of various clinical strains may resolve this finding. The relative position of some of these more interesting polymorphisms as well as the genes related to the type III secretion system is indicated in Fig. 4. Many of the *pmp* genes are localized to a distinct region of the chromosome although some remain separate. There is no clear explanation for the apparent scattering of these interesting polymorphisms.

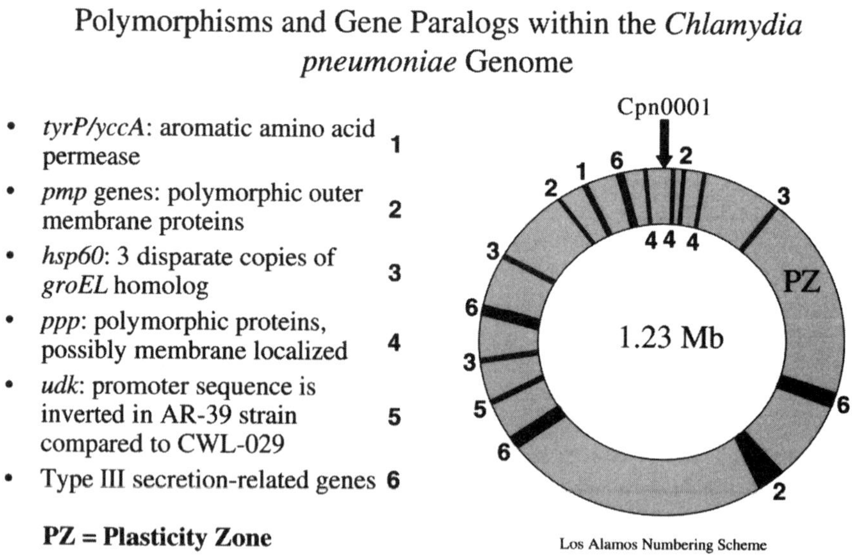

FIGURE 4. Listed is a set of polymorphisms within the *C. pneumoniae* genome and their relative positioning in the chromosome. These genes are likely involved in the pathogenicity of the microorganism.

4.2. Host Tropism and Polymorphic Outer Membrane Proteins

Chlamydia species collectively have the ability to infect a wide range of hosts, from birds to mammals, and host cells, from epithelial cells to macrophages. *C. pneumoniae*, in particular, has a broad tropism within its human host and can infect epithelial cells, endothelial cells, smooth muscle cells, lymphocytes, and phagocytes.[16,32] Atherosclerotic lesions contain both macrophages and T lymphocytes that actively secrete cytokines that may eventually lead to plaque rupture and infarcts. Haranaga *et al.* demonstrated the capacity of *C. pneumoniae* to infect and multiply within lymphocytes *in vitro*, particularly T lymphocytes.[33] Thus, macrophages and T lymphocytes at the site of atherosclerotic lesions may be activated by infection with this pathogen to accelerate the disease state. Alternately, infected lymphocytes may disseminate the organism from the lung to vascular sites. Moazed *et al.* demonstrated that *C. pneumoniae* infection was disseminated to the blood via peripheral blood mononuclear cells in experimental mice.[34] Monocytic cells were posited as the responsible cell type, but the work performed by Haranaga *et al.* strongly implicate lymphocytes as the cellular transporter for this bacteria.[33] Further work by Yamaguchi *et al.* suggests that *C. pneumoniae* infection may cause monocytes to differentiate into macrophages.[35] Data collection involved infecting a monocytic cell line as well as peripheral blood monocytes and looking for characteristic differentiation. A study by Kalayoglu *et al.* suggests that *C. pneumoniae*–infected monocytes exhibit increased adherence to endothelial cells and that this activity may be modulated by cLPS since adherence was inhibited by the LPS antagonist lipid X.[36] Together these data suggest a mechanism whereby chlamydiae that enter the lung might infect lymphocytes that subsequently enter the bloodstream, are localized to atherosclerotic lesions, transmit infection to circulating monocytes that differentiate into macrophages, and are able to attach to endothelial cells, and finally elicit cytokine secretion by all cellular components of the lesion.

Host specificity and tropism, at least for intracellular pathogens, are mediated by outer membrane proteins. For many pathogens, the ligand that attaches to host receptors can be determined by gene knockout and replacement analyses. For chlamydiae, these studies are more difficult. Prior to the availability of genomic sequences, the major outer membrane protein (MOMP) was thought to be the likely ligand for host cell attachment considering its prominence in the outer envelope as well as its immunodominance. Recent studies suggest that MOMP may provide an initial electrostatic interaction with the host cell whereas secondary, more specific interactions strengthen the attachment and facilitate entry.[13] The genome sequences for *C. pneumoniae* revealed a family of 21 polymorphic outer membrane proteins or Pmps.[8–10] Considering the small size of the genome (1.23 Mb), this was an interesting finding, as this gene family accounts for almost 2% of all open reading frames. Grimwood *et al.* assessed expression patterns of Pmps over the course of infection and found that all *pmp* genes were transcribed.[37] Antisera generated against each Pmp showed that 8 Pmps were expressed. Vandahl *et al.* detected 10 Pmps among other Omps in elementary bodies using mass spectrometry.[38] For all of these proteins, several spots were associated with each, indicating that more than one isoform may be present. What significance this may have for *C. pneumoniae* attachment and entry remains to be investigated. Finally, another study by Montigiani *et al.* expanded these observations by detecting 15 Pmps in EBs using mass spectrometry and flow cytometry.[39] Some studies have suggested that MOMP and Pmp variations do not account for host cell tropism; however, these studies indubitably relied on nonclonal populations of the bacteria as no technique was available at the time to isolate clonal variants. It is quite possible that, among the likely several strains in the sequenced population, any obvious differences that would imply host cell tropism are masked. Further investigation with clonal populations may help to explain the broad host cell tropism that is characteristic of *C. pneumoniae.*

Ppps, or *C. pneumoniae* polymorphic proteins, have recently been characterized by Rocha *et al.*[40] The function of *ppps* is not known, but the presence of a signal sequence and a transmembrane domain, and the composition of the protein itself, suggests that these genes are likely outer membrane components. This subset of genes contains repetitive elements (tandem and nontandem, of various sizes) that would be conducive to homologous or illegitimate recombination, allowing the authors to claim that "*C. pneumoniae* has the highest potential for recombination among fully sequenced *Chlamydiaceae.*"[40] Furthermore, these *ppp* genes appear unique to this organism. Most of the *ppp* genes are transcribed, prompting Rocha *et al.* to hypothesize that variation may occur from differential gene silencing and protein production.[40] Other mechanisms may include gene conversions and duplication/deletions between elements that are adjacent within the chromosome. Gene conversions may have occurred between *ppp2* and *ppp6* in the laboratory strains AR39 and CWL029, rendering these genes highly similar in both strains. Interestingly, strains TW-183 and IOL-207, which were originally isolated from the conjunctiva, had deleted certain *ppp* genes. Any significance these findings have *in vivo* will require further examination. Again, these strains were not clonal, so it is possible that within each "strain," there were clones where genetic events had occurred and others where they had not. This might point to, though not entirely accurately, the conclusions drawn in this report since only one sequence can be published even if there are multiple polymorphisms within a gene. Alternately, if *C. pneumoniae* is truly prone to recombination, then this is interesting as it differs from related organisms. These recombinational hotspots could be targets for introducing genes into the bacteria. The polymorphic potential of these genes indicate a strong probability that they are important in pathogenesis. Rocha *et al.* believe that Ppps may direct the broad tropism that is characteristic of the pathogen, although further studies will be required to justify this claim.[40]

5. CONCLUDING REMARKS

Given the plethora of studies associating *C. pneumoniae* with cardiovascular disease, there is a high probability that this microorganism contributes to the pathology in some way. The summary of experimental systems described here provide several mechanisms whereby *C. pneumoniae* can affect lesion development and progression through direct (via infecting cells) or indirect (via its antigens and the host's response to them) interactions with the components of the lesion. Clearly, this is an important pathogen, and further studies are necessary to determine not only causal roles in atherosclerosis but treatment strategies.

ACKNOWLEDGMENTS. GIB & RJB are supported in part by HL71735 and AI42790 NIH grants.

REFERENCES

1. Saikku, P., Leinonen, M., Matilla, K., *et al.*, 1988, Serologic evidence of an association of a novel Chlamydia, TWAR, with chronic coronary heart disease and acute myocardial infarction, *Lancet* **2:**983–986.
2. Frothingham, C., 1911, The relationship between acute infectious diseases and arterial lesions, *Arch. Intern. Med.* **8:**153–162.
3. Kalayoglu, M. V., Libby, P., and Byrne, G. I., 2002, *Chlamydia pneumoniae* as an emerging risk factor in cardiovascular disease, *JAMA* **288:**2724–2731.
4. Kol, A., Sukhova, G. K., Lichtman, A. H., and Libby, P., 1998, Chlamydial heat shock protein 60 localizes in human atheroma and regulates macrophage tumor necrosis factor-alpha and matrix metalloproteinase expression, *Circulation* **98:**300–307.
5. Kalayoglu, M. V., Hoerneman, B., LaVerda, D., Morrison, S. G., Morrison, R. P., and Byrne, G. I., 1999, Cellular oxidation of low-density lipoprotein by *Chlamydia pneumoniae*, *J. Infect. Dis.* **180:**780–790.
6. Kalayoglu, M. V., and Byrne, G. I., 1998, Induction of macrophage foam cell formation by *Chlamydia pneumoniae*, *J. Infect. Dis.* **177:**725–729.
7. Byrne, G. I., and Kalayoglu, M. V., 1999, *Chlamydia pneumoniae* and atherosclerosis: Links to the disease process, *Am. Heart J.* **138:**S488–S490.
8. Kalman, S., Mitchell, W., Marathe, R., Lammel, C., Fan. J., Hyman, R. W., Olinger, L., Grimwood, J., Davis, R. W., and Stephens, R. S., 1999, Comparative genomes of *Chlamydia pneumoniae* and *C. trachomatis*, *Nat. Genet.* **21:**385–389.
9. Read, T. D., Brunham, R. C., Shen, C., Gill, S. R., Heidelberg, J. F., White, O., Hickey, E. K., Peterson, J., Utterback, T., Berry, K., Bass, S., Linher, K., Weidman, J., Khouri, H., Craven, B., Bowman, C., Dodson, R., Gwinn, M., Nelson, W., DeBoy, R., Kolonay, J., McClarty, G., Salzberg, S. L., Eisen, J., and Fraser, C. M., 2000, Genome sequences of *Chlamydia trachomatis* MoPn and *Chlamydia pneumoniae* AR39, *Nucleic Acids Res.* **28:**1397–1406.
10. Shirai, M., Hirakawa, H., Kimoto, M., Tabuchi, M., Kishi, F., Ouchi, K., Shiba, T., Ishii, K., Hattori, M., Kuhara, S., and Nakazawa, T., 2000, Comparison of whole genome sequences of *Chlamydia pneumoniae* J138 from Japan and CWL029 from USA, *Nucleic Acids Res.* **28:**2311–2314.
11. Gieffers, J., Belland, R. J., Whitmire, W., Ouellette, S., Crane, D., Maass, M., Byrne, G. I., and Caldwell, H. D., 2002, Isolation of *Chlamydia pneumoniae* clonal variants by a focus-forming assay, *Infect. Immun.* **70:**5827–5834.
12. Beatty, W. L., Morrison, R. P., and Byrne, G. I., 1994, Persistent Chlamydiae: from cell culture to a paradigm for chlamydial pathogenesis, *Microbiol. Rev.* **58:**686–699.
13. Carabeo, R. A., and Hackstadt, T., 2001, Isolation and characterization of a mutant Chinese hamster ovary cell line that is resistant to *Chlamydia trachomatis* infection at a novel step in the attachment process, *Infect. Immun.* **69:**5899–5904.

14. Van Ooij, C., Apodaca, G., and Engel, J., 1997, Characterization of *Chlamydia trachomatis* vacuole and its interaction with the host endocytic pathway in HeLa cells, *Infect. Immun.* **65:**758–766.
15. Wolf, K., and Hackstadt, T., 2001, Sphingomyelin trafficking in *Chlamydia pneumoniae*-infected cells, *Cell. Microbiol.* **3:**145–152.
16. Byrne, G. I., Ouellette, S. P., Wang, Z., Rao, J. P., Lu, L., Beatty, W. L., and Hudson, A. P., 2001, *Chlamydia pneumoniae* expresses genes required for DNA replication but not cytokinesis during persistent infection of HEp-2 cells, *Infect. Immun.* **69:**5423–5429.
17. Ramsey, K. H., Miranpuri, G. S., Sigar, I. M., Ouellette, S., and Byrne, G. I., 2001, *Chlamydia trachomatis* persistence in the female mouse genital tract: Inducible nitric oxide synthase and infection outcome, *Infect. Immun.* **69:**5131–5137.
18. Miller, S. A., Selzman, C. H., Shames, B. D., Barton, H. A., Johnson, S. M., and Harken, A. H., 2000, *Chlamydia pneumoniae* activates nuclear factor kB and activator protein 1 in human vascular smooth muscle and induces cellular proliferation, *J. Surg. Res.* **90:**76–81.
19. Dechend, R., Maass, M., Gieffers, J., Dietz, R., Scheidereit, C., Leutz, A., and Gulba, D. C., 1999, *Chlamydia pneumoniae* infection of vascular smooth muscle and endothelial cells activates NF-kB and induces tissue factor and PAI-1 expression, *Circulation* **100:**1369–1373.
20. Fryer, R. H., Schwobe, E. P., Woods, M. L., and Rodgers, G. M., 1997, *Chlamydia* species infect human vascular endothelial cells and induce procoagulant activity, *J. Invest. Med.* **45:**168–174.
21. Coombes, B. K., and Mahony, J. B., 1999, *Chlamydia pneumoniae* infection of human endothelial cells induces proliferation of smooth muscle cells via an endothelial cell-derived soluble factor(s), *Infect. Immun.* **67:**2909–2915.
22. Molestina, R. E., Miller, R. D., Ramirez, J. A., and Summersgill, J. T., 1999, Infection of human endothelial cells with *Chlamydia pneumoniae* stimulates transendothelial migration of neutrophils and monocytes, *Infect. Immun.* **67:**1323–1330.
23. Molestina, R. E., Dean, D., Miller, R. D., Ramirez, J. A., and Summersgill, J. T., 1998, Characterization of a strain of *Chlamydia pneumoniae* isolated from a coronary atheroma by analysis of the *omp1* gene and biological activity in human endothelial cells, *Infect. Immun.* **66:**1370–1376.
24. Coombes, B. K., and Mahony, J. B., 2001, cDNA array analysis of altered gene expression in human endothelial cells in response to *Chlamydia pneumoniae* infection, *Infect. Immun.* **69:**1420–1427.
25. Kothe, H., Dalhoff, K., Rupp, J., Muller, A., Kreuzer, J., Maass, M., and Katus, H. A., 2000, Hydroxymethylglutaryl coenzyme A reductase inhibitors modify the inflammatory response of human macrophages and endothelial cells infected with *Chlamydia pneumoniae, Circulation* **101:**1760–1763.
26. Krull, M., Klucken, A. C., Wuppermann, F. N., Fuhrmann, O., Magerl, C., Seybold, J., Hippenstiel, S., Hegemann, J. H., Jantos, C. A., and Suttorp, N., 1999, Signal transduction pathways activated in endothelial cells following infection with *Chlamydia pneumoniae, J. Immunol.* **162:**4834–4841.
27. Kalayoglu, M. V., Indrawati, Morrison, R. P., Morrison, S. G., Yuan, Y., and Byrne, G. I., 2000, Chlamydial virulence determinants in atherogenesis: The role of chlamydial lipopolysaccharide and heat shock protein 60 in macrophage–lipoprotein interactions, *J. Infect. Dis.* **181** (Suppl. 3)**:**S483–S489.
28. LaVerda, D., Albanese, L. N., Ruther, P. E., Morrison, S. G., Morrison, R. P., Ault, K. A., and Byrne, G. I., 2000, Seroreactivity to *Chlamydia trachomatis* Hsp10 correlates with severity of human genital tract disease, *Infect. Immun.* **68:**303–309.
29. Kalayoglu, M. V., and Byrne, G. I., 1998, A *Chlamydia pneumoniae* component that induces macrophage foam cell formation is chlamydial lipopolysaccharide, *Infect. Immun.* **66:**5067–5072.
30. Beatty, W. L., Byrne, G. I., and Morrison, R. P., 1993, Morphological and antigenic characterization of interferon-gamma mediated persistent *Chlamydia trachomatis* infection *in vitro, Proc. Natl. Acad. Sci. U.S.A.* **90:**3998–4002.

31. Byrne, G. I., Lehmann, L. K., and Landry, G. J., 1986, Induction of tryptophan catabolism is the mechanism for gamma-interferon-mediated inhibition of intracellular *Chlamydia psittaci* replication in T24 cells, *Infect. Immun.* **53:**347–351.
32. Gaydos, C. A., Summersgill, J. T., Sahney, N. N., Ramirez, J. A., and Quinn, T. C., 1996, Replication of *Chlamydia pneumoniae in vitro* in human macrophages, endothelial cells, and aortic artery smooth muscle cells, *Infect. Immun.* **64:**1614–1620.
33. Haranaga, S., Yamaguchi, H., Friedman, H., Izumi, S.-I., and Yamamoto, Y., 2001, *Chlamydia pneumoniae* infects and multiplies in lymphocytes *in vitro*, *Infect. Immun.* **69:**7753–7759.
34. Moazed, T. C., Kuo, C., Grayston, J. T., and Campbell, L. A., 1997, Murine models of *Chlamydia pneumoniae* infection and atherosclerosis, *J. Infect. Dis.* **175:**883–890.
35. Yamaguchi, H., Haranaga, S., Widen, R., Friedman, H., and Yamamoto, Y., 2002, *Chlamydia pneumoniae* infection induces differentiation of monocytes into macrophages, *Infect. Immun.* **70:**2392–2398.
36. Kalayoglu, M. V., Perkins, B. N., and Byrne, G. I., 2001, *Chlamydia pneumoniae*–infected monocytes exhibit increased adherence to human aortic endothelial cells, *Microb. Infect.* **3:**963–969.
37. Grimwood, J., Olinger, L., and Stephens, R. S., 2001, Expression of *Chlamydia pneumoniae* polymorphic membrane protein family genes, *Infect. Immun.* **69:**2383–2389.
38. Vandahl, B. B., Birkelund, S., Demol, H., Hoorelbeke, B., Christiansen, G., Vandekerckhove, J., and Gevaert, K., 2001, Proteome analysis of the *Chlamydia pneumoniae* elementary body, *Electrophoresis* **22:**1204–1223.
39. Montigiani, S., Falugi, F., Scarselli, M., Finco, O., Petracca, R., Galli, G., Mariani, M., Manetti, R., Agnusdei, M., Cevenini, R., Donati, M., Nogarotto, R., Norais, N., Garaguso, I., Nuti, S., Saletti, G., Rosa, D., Ratti, G., and Grandi, G., 2002, Genomic approach for analysis of surface proteins in *Chlamydia pneumoniae*, *Infect. Immun.* **70:**368–379.
40. Rocha, E. P. C., Pradillon, O., Bui, H., Sayada, C., and Denamur, E., 2002, A new family of highly variable proteins in the *Chlamydophila pneumoniae* genome, *Nucleic Acids Res.* **30:**4351–4360.

11

Animal Models of *Chlamydia pneumoniae* Infection and Atherosclerosis

IGNATIUS W. FONG

1. INTRODUCTION

Chlamydia pneumoniae has been associated with cardiovascular disease and stroke in humans by numerous cross-sectional and case–control studies,[1,2] but prospective, longitudinal studies have produced mixed results and most have not confirmed this association.[3,4] However, pathological studies have established the association of *C. pneumoniae* antigen or DNA with atherosclerosis from surgical and autopsy vascular specimens, with odds ratio of around 20.[4,5] Although *in vitro* and cell culture studies have supported biological plausible mechanisms for *C. pneumoniae* to induce or accelerate atherosclerosis,[4] they lack the complexity of a real disease, thus limiting the scope of testing the hypothesis of causality.

Animal models are more likely to mimic real disease in humans, and are critical in many conditions in the understanding of the pathogenesis, cause-and-effect relationships and subsequent approaches to treatment and prevention of diseases. Thus, animal experimentation in medical research has played a major role in our understanding of diseases. The earliest recorded use of animal models extends as far back as 500 BC, when Alcmaeon of Croton defined the function of the optic nerve by transection in a living animal.[6] The Hippocratic treatise on the heart (circa 350 BC) discussed cutting the throat of a pig that drank colored water to study the act of swallowing.[6]

IGNATIUS W. FONG • St. Michael's Hospital, Division of Infectious Diseases, Department of Medicine, University of Toronto, Toronto, Canada.

1.1. Animal Models Used in Atherosclerotic Research

It is almost a hundred years now since the first evidence of experimental atherosclerosis was reported. Ignatowski in 1908 first reported thickening of the intima with the formation of large, clear cells in the aorta of rabbits fed a diet rich in animal proteins (meat, milk, and eggs).[7] Anitschkow in 1913 was responsible for establishing the cholesterol-fed rabbit model for atherosclerosis research.[8] He demonstrated that cholesterol caused these atherosclerotic changes in the rabbit arterial intima in a dose response fashion, which was very similar to human atherosclerosis. Since then, induction of atherosclerotic lesions in several animal models besides rabbits have been reported,[9] including rodents (mice, rats, hamsters, guinea pigs), avians (pigeons, chickens, quail), swine, carnivores (dogs, cats), and nonhuman primates. More recently transgenic/knockout mice have gained popularity in experimental models of vascular disease. Traditionally atherosclerosis in these models have been produced by inducing hypercholesterolemia by dietary manipulation, or by mechanical injury of the arterial intima and by genetic manipulation to induce spontaneous hypercholesterolemia.

Because atherosclerosis is a silent and asymptomatic disease, until complications arise late in the course with thrombosis-producing clinical symptoms, it is necessary to have models that reproduce human disease in its early stages. Unfortunately, not all experimental models of vascular disease have human resemblance and validity. There is no perfect or ideal animal model that completely mimics human atherosclerosis and the complications. The models that most closely resemble human disease are the nonhuman primates and pigs, but size of animals and cost have limited the use of these models, especially for studies requiring a large number of animals. Moreover, nonhuman primates are not available to many investigators and moral issues have been raised by animal activists of their use in experimentation. Experimental models of vascular disease have enhanced our understanding of the molecular and cellular mechanisms in the process of atherogenesis, pathophysiological processes leading to spontaneous and accelerated atherosclerosis and thrombosis.

Animal models have provided insight into the role of various components of lipids and lipoproteins, platelets, the renin–angiotensin system, cytokines, and growth factors in the evolution and progression of atherosclerosis. Knowledge has been gained from these models on the complex interactions of adhesion molecules, cytokines, and growth factors with endothelial cells, macrophages, smooth muscle cells, and collagen—all key components of the atherosclerotic plaque.

1.1.1. Rabbits

The rabbit was the first animal model used in atherosclerosis research. The rabbit shares several aspects of lipoprotein metabolism—except for deficiency in hepatic lipase—with humans. This animal model also develops advanced atherosclerotic plaques, but extremely high plasma levels of cholesterol are needed for this to occur.[9] Atherosclerotic lesions with high-fat/high-cholesterol diets are most prominent in the aortic arch and thoracic aorta rather than the abdominal aorta, which is most often affected in humans. The normocholesterolemic rabbit on a standard chow does not develop atherosclerosis. However, the Watanabe Heritable hyperlipidemic (WHHL) rabbits with natural LDL-receptor deficiency,[10] and the St. Thomas' Hospital strain, with combined hypertriglyceridemia and hyperlipidemia,[11] develop advanced spontaneous atherosclerosis.

1.1.2. *Rodents*

Mice: Rodents in general (mice and rats) are resistant to atherosclerosis even with a high-fat/high-cholesterol diet. Atherogenic diets cause an increase in total cholesterol only 1.1 to 4.3 times the average, and plasma triglyceride increase in only some stains of inbred mice.[9] The resistance to atherosclerosis in wild-type mice is attributable to elevated high density lipoprotein (HDL) level and low cholesterol absorption rate. However, early stages of atherosclerotic lesions (fatty streak) can be induced with atherogenic diets in C57BL/6J mice,[12] particularly in the aortic sinus. Genetic modification of mice has resulted in a number of gene "knockout" or transgenic mouse models that have become popular in atherosclerosis and lipid research. These animals, unlike their wild-type counterparts, can develop advanced atherosclerotic lesions rapidly. The apolipoprotein E knockout (apoE-KO) mice even on regular mouse chow will develop advanced atherosclerotic lesions,[13] and the addition of a cholesterol-rich chow will speed up the process and shorten the induction period.[9] The LDL-receptor knockout or human apoB100 transgenic mice will also develop advanced atherosclerotic lesions when fed a high cholesterol diet.[14,15] Other murine models susceptible to advanced atherosclerosis include the cholesterol ester transfer protein[16] and apoE* Leiden[17] transgenic mice, but these are less popular among researchers.

Rats: Similar to mice the rat model is also resistant to atherosclerosis. Rats do not have plasma cholesteryl ester transfer protein, and HDL is the major carrier of plasma cholesterol. Thus they are hyporesponsive to high dietary cholesterol. Strains of rats with heritable hyperlipidemia prone to atherosclerosis are also available.[9]

Guinea Pigs and Hamsters: Guinea pigs, like humans, have lipoprotein (a) in their plasma, and accumulation of this atherogenic lipoprotein can be found in atherosclerotic lesions.[9] Atherosclerotic changes have been seen with cholesterol-enriched diet varying from 0.25 to 2.5% by weight of the chow.

Hamsters may develop hypercholesterolemia and early atherosclerosis on atherogenic diets, such as fatty streaks in the ascending aorta in male Golden Syrian hamsters. With atherogenic diet very low density lipoprotein (VLDL) content increases over 200%, with a 20% increase in LDL and a 40% decrease in HDL cholesterol concentrations.[9] Whereas male hamsters develop hyperlipidemia and atherosclerotic changes of the aorta, female Syrian hamsters do not.

1.1.3. *Other Animals*

Avian models of atherosclerosis include chickens, pigeons, and quail. Chickens fed high cholesterol diet will develop atherosclerotic lesion of the thoracic and abdominal aorta within 2 weeks,[9] but the lesions are not advanced. A White Carneau (WC) strain of pigeons absorbs cholesterol efficiently and develops spontaneous atheromas of the thoracic and abdominal aorta and peripheral arteries, including the coronary and carotid arteries. Advanced atherosclerotic lesions and myocardial infarction with high dietary cholesterol also develop in genetic atherosclerosis susceptible Japanese quail.

Dogs and cats are rarely used in atherosclerotic research and are relatively resistant to development of atherosclerosis. However, high-fat/high-cholesterol diet can induce early atherosclerotic lesions in both animals.[9]

Spontaneous atherosclerosis can occur in pigs but very high cholesterol is needed to induce atherosclerotic lesions in coronary arteries of miniature pigs. Banding of the coronary artery with a copper band seems to enhance the development of atherosclerosis. Naturally defective pigs with Lpb5 and Lpul mutations develops hypercholesterolemia and atherosclerosis in the coronary, iliac, and femoral arteries even with low-fat/cholesterol-free diet,[9] and the severity and complexity of the lesions is related to the degree and duration of hypercholesterolemia (as in humans).

Nonhuman primates are the closest animal species to man and are attractive models. Development of atherosclerosis and myocardial infarction can be induced in monkeys with diet-induced hypercholesterolemia. High-fat/high-cholesterol diet results in dyslipidemia, with extensive atherosclerosis of the aorta and its major branches, the coronary and cerebral arteries, similar to lesions in humans.[9] However, the atherosclerotic changes are not exactly the same in nonhuman and human primates. The location of atheromas also varies among different species of nonhuman primates. A familial LDL receptor–deficient rhesus monkey with spontaneous development of atherosclerosis is also available.[9]

1.2. *C. pneumoniae* Lung Infection Model

Pulmonary infection with *C. pneumoniae* has been induced by intranasal or posterior nasopharyngeal inoculation in mice, rabbits, and nonhuman primates. Intranasal inoculation of 3×10^7 inclusion-forming units (IFUs) of *C. pneumoniae* in Swiss Webster mice result in a prolonged course of lung infection, with reisolation of the organisms from lungs at 42 days and persistence of lung pathology for >60 days.[18] Pathologically the pneumonitis was characterized by patchy interstitial neutrophilic infiltration in the early stages and mononuclear cell infiltration in the later stages. Further studies in this model showed dissemination to spleen and peritoneal macrophages,[19] indicating the organism has potential to spread to other organs (and the blood vessel wall) by infected macrophages. Other mouse strains tested, ICR, BALB/CAnN, C57BL/6N, C3H/HeN, and B6C3F1, were also shown to be susceptible to *C. pneumoniae* infection.[18]

Interestingly, even after 30 days when *C. pneumoniae* cannot be recovered from the lungs in mice, there is evidence of prolonged latent infection or persistence. Mice treated with corticosteroids after primary infection was cleared, as indicated by negative cultures, reactivate the infection with the subsequent recovery of viable organisms.[20,21]

Lung infection with *C. pneumoniae* in New Zealand White (NZW) rabbits is subclinical with no obvious ill effects, and is characterized by a brief course of patchy bronchiolitis and pneumonitis.[22,23] The infiltration is predominantly of mononuclear cells, with some plasma cells and histrocytic giant cells but no neutrophils, and involve the interstitium and alveolae. The pathological changes in the lung is most severe by day 7 after inoculation and normalize by days 21 to 28.[22] However, unlike the murine model, recovery of viable *C. pneumoniae* from lungs was not possible,[23] and the organism could only be cultured from the nasopharynx at days 2 to 3 in a few animals. Bronchial lavage or grounded lung tissue from uninfected rabbits when spiked with viable *C. pneumoniae* demonstrate inhibition of growth (unpublished data). The substance(s) in rabbit tissue inhibiting cultivation of *C. pneumoniae* is not known.

Evidence of widespread dissemination from the lungs to the spleen and liver is present in the rabbit model by demonstration of the organism by immunohistochemical stain.[22] Furthermore, we have been able to detect *C. pneumoniae* DNA by PCR in peripheral blood mononuclear cells (PBMC) in 23.7% of rabbits 3 days after intranasal inoculation (unpublished data).

C. pneumoniae has also been demonstrated to produce a mild subclinical respiratory infection in nonhuman primates.[24] Two Cynomolgus monkeys were inoculated in the nose, nasopharynx, and conjunctiva and *C. pneumoniae* could be isolated from these sites and rectum up to 5 weeks postinoculation. Clinical and histopathological ocular responses to *C. pneumoniae* were very mild compared with *Chlamydia trachomatis*. No lung pathology was performed.

2. MURINE MODELS OF *C. pneumoniae*–RELATED ATHEROSCLEROSIS

Studies addressing the issue of induction or acceleration of atherosclerosis in murine models have been performed in C57BL/6J, apolipoprotein E knockout (apoE-KO), and LDL-receptor knockout (LDLR-KO) mice. In initial studies using apoE-KO mice, which develop spontaneous atherosclerosis, *C. pneumoniae* could be recovered from 10 to 25% of abdominal atherosclerotic aortae within 1–2 weeks after intranasal inoculation, but PCR detected the DNA in 35–100% of aortae up to 8–16 weeks after inoculation.[25] In C57BL/6J mice, which develop very mild early atherosclerotic lesions on an atherogenic diet, *C. pneumoniae* was detected in only 8% of mice 2 weeks after inoculation.[25]

In male apoE-KO mice on C57BL/6 background intranasal inoculation with 3×10^7 IFU of *C. pneumoniae* (AR-39) at 8, 9, and 10 weeks of age, resulted in 2.4- and 1.6-fold increased area of atherosclerosis of the aortic arch at 16 and 20 weeks of age, respectively.[26] No difference in gross histologic findings were observed between infected and uninfected mice, and lesions varied from fatty streaks to raised atheromas. Using the same apoE-KO male mice two inoculations of *C. pneumoniae* intranasally at 6 and 8 weeks of age resulted in 70% increase in lesion size of the aortic sinus at 16 weeks of age, compared with uninfected controls in another study.[27] However, combined infection with cytomegalovirus did not result in additive or synergistic effect. In apoE-KO female mice inoculated at 10 and 12 weeks and killed at 20 weeks there was no significant increase of aortic lesions in *C. pneumoniae*–infected versus noninfected animals; but when inoculated at 12 and 14 weeks and sacrificed at 26 weeks the atherosclerotic index increased by almost 2-fold in the infected group, $p < 0.01$.[28]

In addition, endothelial dysfunction (which is believed to occur in the initial stages of atherogenesis) can be induced after repeated infection with *C. pneumoniae* in apoE-KO mice.[29] However, the results of exacerbation or acceleration of atherosclerosis in apoE-KO mice with *C. pneumoniae* infection have not been consistent. Two other groups failed to demonstrate an increase in the extent of atherosclerosis with *C. pneumoniae*–infected mice compared with uninfected controls.[30,31] In the Finnish study,[30] two strains of apoE-KO mice were used (FUB and C57BL/6J backgrounds), males and females, and were fed either regular, low-fat chow or a high-fat diet. Two inoculation schedules were examined: one set of male mice were inoculated intranasally with 3×10^6 IFU of *C. pneumoniae* at weeks 8, 9, and 10 and sacrificed 10 weeks later; 10 control and 10 infected mice were fed regular chow, and 6 control and 4 infected mice received a high-fat diet. Another group of mice (males and females) were inoculated four times at 3–4 week intervals with initially 1×10^6 IFU and subsequently 1×10^5 IFU of *C. pneumoniae* and sacrificed 18 weeks later. Of interest in this study, all infected animals demonstrated antibodies with titers ranging from 32 to 128 but none of the 20 infected mice had any detectable *C. pneumoniae* DNA (analyzed by PCR) of the aortic samples. Moreover, there were no significant changes in the lipid levels with infection.

In the other negative study,[31] female apoE-KO (on C57BL/6J background) and wild-type (C57BL/6J) 6- to 8-week-old mice were used. Thirteen apoE-KO and 8 wild-type mice were inoculated with 10^6 IFU of *C. pneumoniae* once and sacrificed at 22 weeks; a second group of 14 apoE-KO mice were reinfected at 18 weeks. All mice were fed standard mouse chow. No evidence of induction of atherosclerosis was evident in wild-type mice nor acceleration of atherosclerosis in the apoE-KO mice. This study differed significantly from previous studies demonstrating acceleration of atherosclerosis in apoE-KO mice in two main respects, the inoculation schedule was different and less frequent and female instead of male mice were used. There is evidence that female animals, like humans, are less susceptible to development of atherosclerosis; however, others have used female apoE-KO mice with success,[28] though sacrificed at a later time.

Wild-type C57BL/6J mice, resistant to atherosclerosis, has been shown by others not to develop atherosclerosis with or without *C. pneumoniae* infection on regular chow.[32] However, when fed a high-fat/high-cholesterol diet these mice show early changes of atherosclerotic lesions in the aortic sinus, and

C. pneumoniae infection increases these lesions by 2.5-fold at 18 weeks, after triple inoculation between weeks 8 and 11, and 3.3-fold by 24 weeks.[32] A further study by the same group demonstrated that *C. pneumoniae* did not accelerate the aortic lesions in wild-type mice if the atherogenic diet was started after the infection rather than at the same time.[33] *C. pneumoniae*, however, can induce mild inflammatory changes in the heart and aorta of normocholesterolemic wild-type mice.[34]

Another hypercholesterolemic murine model demonstrating accelerated atherosclerosis with *C. pneumoniae* infection is the LDLR-KO mice. In a study using female B6, 129 mice (4–5 weeks old) with LDLR gene deficiency fed normal chow did not result in atherosclerotic changes of the aorta with or without *C. pneumoniae* infection.[35] However, with a 2% cholesterol diet and monthly intranasal inoculations of 0.5–1 $\times$ 10^7 IFU of AR39 for 9 months resulted in an increase in the aortic lesion area, from 20% in controls to more than 30% in *C. pneumoniae*–infected but not *C. trachomatis*–infected mice. The lesions in the AR39 infected group were also more severe and advanced than the control group.[35] The cholesterol-enriched diet increased total cholesterol levels (primarily LDL) over standard chow by 2–3-fold, but chlamydial infection did not alter total cholesterol or LDL serum levels. A summary of the murine model results on atherosclerosis is shown in Table I.

3. RABBIT MODELS OF *C. pneumoniae* INFECTION AND ATHEROSCLEROSIS

C. pneumoniae was first demonstrated to produce early lesions of atherosclerosis (fatty streak) in the normocholesterolemic NZW rabbits.[22] Further studies in this model by our group with much larger number of animals demonstrated the *de novo* induction of early atherosclerotic lesions with regular chow in up to 26% of animals after a single inoculation intranasally of 1.0–2.6 $\times$ 10^7 IFU of *C. pneumoniae*.[36] These lesions were predominately microscope grade I lesion (fatty streak) or grade II lesion (advanced fatty streak with mixture of foamy macrophages and smooth muscle cells) (see Fig. 1). However, multiple inoculations on three occasions on regular chow resulted in more advanced lesions (fibromuscular lesion with and without calcification), grade III–IV (Fig. 1) in up to almost 35% of rabbits. Some of these lesions could now be seen macroscopically on the surface of the aorta as small, focal, raised patches.[36] The typical mature atheromas, as seen in humans, with central lipid core and fibrous capsule was not seen in these infected normocholesterolemic rabbits. Although the lesions of the infected normocholesterolemic rabbits were somewhat dissimilar to rabbits fed a high cholesterol (0.25–0.5%) diet, they were quite similar (histologically) to lesions in animals fed a lower cholesterol diet (0.15%) (See Fig. 2), which resulted in a total serum cholesterol level (4.0 mmol/l) considered normal for humans.

Of interest although, we could not demonstrate increased proinflammatory cytokines in these lesions compared with controls; 5–7 of 17 infected animals

TABLE I
Murine Models of *C. pneumoniae* Infection and Atherosclerosis

Ref.	Animals	Diet	Inoculation	Serum lipids	Aortic lesions	P value
1. Moazed *et al.*[26]	apoE-KO, 8-week males	Regular chow	3×10^7 IFU × 3, 1-week apart	Total cholesterol, no change	140% ↑ at 16 weeks 61% ↑ at 20 weeks	0.05 0.05
2. Burnett *et al.*[27]	apoE-KO, 6-week males	Regular chow	5×10^6 IFU × 2, 2 weeks apart	No data	70% ↑ at 16 weeks	<0.0001
3. Rothstein *et al.*[28]	apoE-KO, 10-week females, 12-week females	Regular chow	5×10^6 IFU × 2 1. 10 and 12 weeks 2. 14 and 15 weeks	No change in total cholesterol, LDL or HDL, ↑triglycerides ($P = 0.05$)	No ↑ at 20 weeks, △ 42% ↑ at 26 weeks	NS <0.01
4. Aalto-Setölä *et al.*[30]	apoE-KO, 8-week males and females	Regular and high-fat chow	3×10^6 IFU × 3 1. 1 week apart, killed 10 weeks later 2. 4 × at 3–4 weeks apart, killed 18 weeks later	No change in lipids except ↑ triglycerides at 6 weeks with regular chow ($P = 0.02$)	No ↑ lesions at 10 weeks or 18 weeks	NS
5. Caliguri *et al.*[31]	1. apoE-KO, 8-week females 2. C57BL/6J, 8-week females	Regular chow	10^6 IFU × at 8 and 18 weeks, killed at 22 weeks	No data	No ↑ in size nor components of lesions	NS
6. Blessing *et al.*[32]	C57BL/6J, 8-week males	1. Regular chow 2. High-fat/high-cholesterol chow	3×10^7 IFU × 3 at 8, 10, 12 weeks	Total cholesterol, no change with infection	1. No lesions with regular chow 2. High-fat/high-cholesterol chow ↑ lesions by 2.5-fold at 18 weeks ↑ by 3.3-fold at 24 weeks	 <0.15 <0.02
7. Blessing *et al.*[33]	C57BL/6J, 8-week males	High-fat/high-cholesterol chow 5 or 7 weeks after inoculation	3×10^7 IFU × 3 at 8, 10, 12 weeks	Total cholesterol and triglycerides no change with infection	1. atherogenic diet at 5 weeks ↑ 1.7-fold 2. atherogenic diet at 7 weeks ↑ 1.4-fold	0.15 0.55
8. Hu *et al.*[35]	LDLR-KO, 5-week females	Regular chow and 2% cholesterol chow	0.5–1.0 × 10^7 IFU × 9 monthly	Total cholesterol, LDL HDL, triglycerides no change with infection	1. Regular chow – no lesions 2. 2% cholesterol chow ↑ lesion area from 20% to 30%	NS <0.001

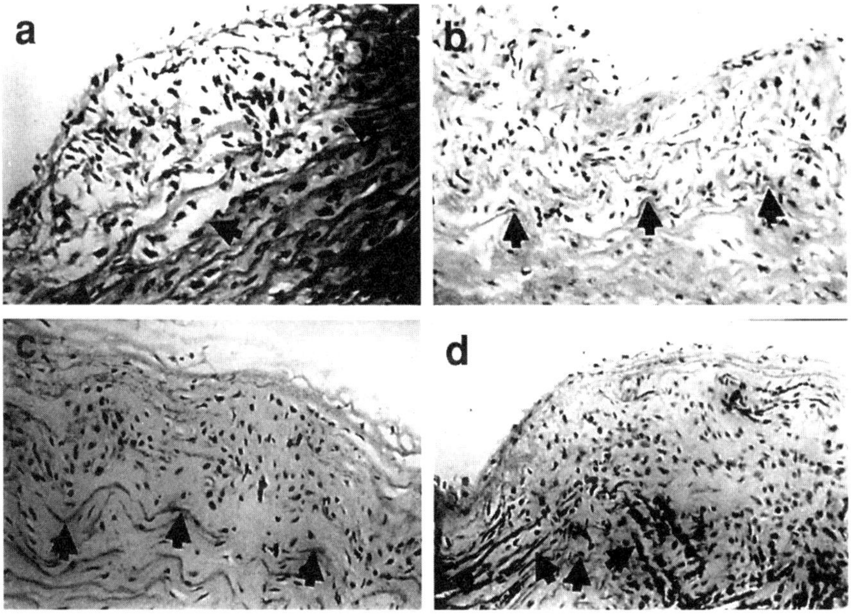

FIGURE 1. The histological changes of atherosclerosis of the aorta are shown in the normocholesterolemic (de novo) rabbit model. (a) Grade I lesion; foamy macrophages in intima (with clear, vacuolated cytoplasm). The arrow heads indicates internal elastic lamina. H & E stain, original magnification ×250. (b) Grade II lesion (advanced fatty streak); mixed foamy macrophages and spindle smooth-muscle cells predominating in the intima. Arrow heads indicate internal elastic lamina. H & E stain, original magnification ×250. (c) Grade III lesion (fibromuscular); predominantly smooth cells in the intima. Arrow heads indicate internal elastic lamina. H & E stain, original magnification ×250. (d) Grade IV lesion; advanced fibromusclar lesion with smooth muscle cell proliferation and clarifications (indicated by arrow heads). This lesion lacks foam cells, lipid core and fibrous cap as typical for Stary's type IV lesion or seen with high cholesterol diet in the rabbit. H & E stain, original magnification ×250.

demonstrated transforming growth factor (TGF)-β and platelet derived growth factor (PDGF) (tissue growth factors important in atherogenesis) in aortic lesions versus 0 of 25 in normal aorta of controls; $p = 0.0007$ to 0.0073, respectively (Fisher's exact test).[36] Furthermore, using reverse-transcriptase PCR it was demonstrated that *C. pneumoniae* infection of endothelial cells *in vitro* induced PDGF-β messenger RNA expression; and in the infected normocholesterolemic rabbits, increased maximal intimal thickness of the aorta was independently correlated with the presence of PDGF-β ($p = 0.009$) and *C. pneumoniae* antigen ($p = 0.043$) in aortic tissues.[37] *C. pneumoniae* could also be detected in aortic lesions by immunohistochemical stain in 41% of the infected animals with atherosclerotic changes. In addition, none of the sham-infected or *Mycoplasma pneumoniae* triple-infected rabbits had any changes representative of early atherosclerosis.[36]

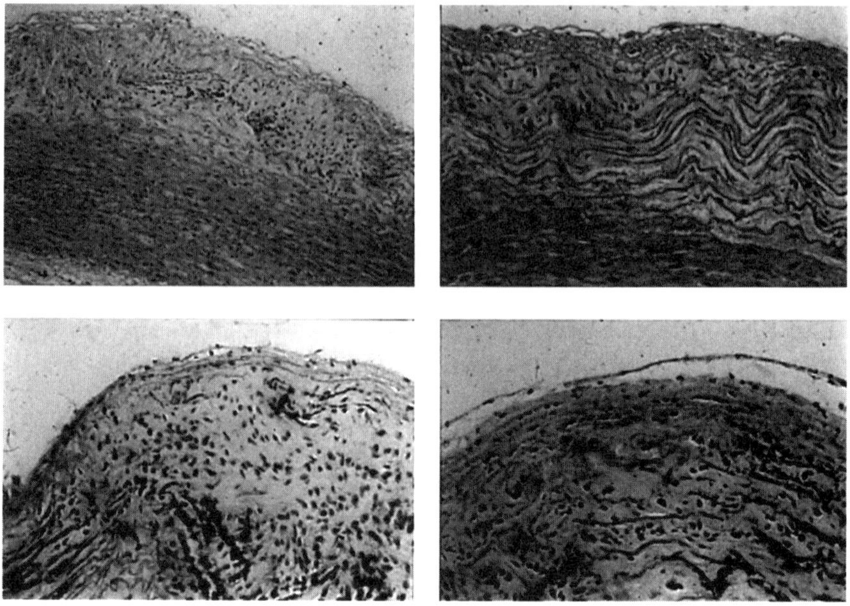

FIGURE 2. Comparison of the early lesions of atherosclerosis with *C. pneumoniae* infection in normocholesterolemic rabbits (left frames), compared to the changes in uninfected rabbits fed 0.15% cholesterol diet (cholesterol 4.0 mmol/L in blood) (right frames), note the similarity between the sections on the right and left. Top: Grade II lesion (advanced fatty streak), with mixture of foamy macrophages and smooth-muscle cells. Bottom: Grade IV advanced fibromuscular lesion with smooth-muscle cell proliferation and calcification (dark red, linear strains).

Finnish investigators using NZW rabbits fed standard chow, with two intranasal inoculation of *C. pneumoniae,* were able to detect inflammatory changes of the aorta, consisting of intimal thickening or fibroid plaques in 6 of 9 animals after 2 to 4 weeks following reinfection.[38]

NZW rabbits fed cholesterol-enriched chow develop changes of atherosclerosis, the severity and extent depending on the concentration of lipids in the diet or the level of hypercholesterolemia. It was first shown by Muhlestein *et al.*[39] that triple intranasal inoculations (3 weeks apart) with *C. pneumoniae* in rabbits fed a 0.25% cholesterol diet would increase the maximal intimal thickness (3.4-fold increase) versus uninfected controls ($p = 0.009$), and similarly the plaque area index of atherosclerosis was 8-fold greater ($p < 0.05$) in the aorta.[39] In this study, however, lipid levels were not analyzed.

Theoretically, *C. pneumoniae* infection could enhance or induce atherosclerosis in the rabbit model by direct or indirect mechanisms. Direct infection of the arterial wall could lead to endothelial dysfunction, increase inflammation and foam cell formation through the effect of lipoplysaccharide,[40] or enhance LDL oxidation through chlamydial heat shock protein.[41] Indirectly, infection could aggravate or accelerate atherosclerosis by alteration of serum lipid profile to a proatherogenic environment. In a study to address the latter

TABLE II
Effect of *C. pneumoniae* Infection on Atherosclerosis, Fibrinogen and Lipid Profile in Cholesterol fed (0.25%) Rabbits

	Controls	Infected	% Change	*p* value
Total cholesterol (mmol/l)	31.0 ± 15.7	34.2 ± 12.8	10.4	NS
LDL	4.4 ± 5.5	4.1 ± 6.0	6.8	NS
HDL	0.57 ± 0.3	0.49 ± 0.4	14	NS
VLDL	26.1 ± 10.4	31.6 ± 9.1	21	NS
Triglycerides	2.4 ± 1.2	4.01 ± 3.1	70	0.044
Fibrinogen (g/l)	1.0 ± 0.98	2.27 ± 0.6	127	0.001
Mean area of atherosclerosis (cm^2)	9.72 ± 3.33	15.27 ± 2.80	57	<0.001
Mean % of total area involved with atherosclerosis	51.6 ± 16.4%	80.9 ± 1.17%	56.8	<0.001

issue we infected (triple inoculation with *C. pneumoniae*) and sham-inoculated control rabbits fed a 0.25% cholesterol diet. The serum lipid profile and fibrinogen and atherosclerotic changes are shown in Table II.[42] Essentially, infection with *C. pneumoniae* was associated with significant increase in area of atherosclerosis (by 57%) of the aorta with a corresponding increase in serum fibrinogen and triglycerides but no changes in HDL, LDL, or very low density lipoprotein (VLDL). However, previous studies in rabbits demonstrated that the increase in triglycerides was associated with protection against hypercholesterolemia-induced atherosclerosis.[43] Although we could not demonstrate quantitative changes in the lipids to explain enhanced atherosclerosis, it is possible that infection could have induced functional changes in LDL and HDL, resulting in proatherogenic effects, which were not measured in this study. A summary of the rabbit model studies is shown in Table III.

4. EFFECT OF ANTIBIOTICS ON *C. pneumoniae* ANIMAL-INDUCED ATHEROSCLEROSIS

There are a few studies in animal models addressing the issue of antichlamydial antimicrobials' effect on preventing or ameliorating atherosclerosis accelerated or induced by *C. pneumoniae*. Muhlestein *et al.*[39] first demonstrated that a new macrolide–azalide, azithromycin 30 mg/kg given intramuscularly daily for 7 days, then twice weekly for 6 weeks, largely prevented the enhancing effect of *C. pneumoniae* infection on the intima–media thickness of NZW rabbit aortae, fed a 0.25% cholesterol diet. In this study, the antibiotic was started immediately after the final of a triple inoculation with *C. pneumoniae*.

However, in the apoE-KO mice, which demonstrated accelerated aortic atherosclerosis with *C. pneumoniae* infection after two inoculations, two oral doses of azithromycin 24 mg/kg administered 2 and 3 weeks after the second inoculation failed to the modify the extent or severity of the lesions.[28]

TABLE III

Rabbit Models of *C. pneumoniae* Infection and Atherosclerosis

Ref.	Animals	Diet	Inoculation	Serum lipids	Antibiotics	Aortic lesions	*P* value
1. Fong *et al.*[22]	NZW, 4 weeks male, $N = 12$ infected, 5 controls.	Regular chow	$1.0–5.0 \times 10^7$ IFU intranasally $\times 1$	N/A	None	1 fatty streak at day 7, 1 fibrous plaque (grade III) at 3 weeks, controls—all negative	–
2. Laitenan, *et al.*[38]	NZW, 5 months male, $N = 9$ reinfected, 4 controls	Regular chow	2×10^7 IFU intranasally $\times 2$, 3 weeks apart	N/A	None	6/9 reinfected rabbits had fibroid plaques and intimal thickening, controls negative	–
3. Fong, *et al.*[36]	NZW, 4 weeks male, $N = 46$ infected, 36 controls, 32 mycoplasma infected, killed after 3 months	Regular chow	$1.0–2.6 \times 10^7$ IFU $\times 1 - N = 23$ $\times 3 - N = 23$	No changes in total cholesterol with infection	None	1. Single inoculation 26.1% atherosclerotic lesions 2. Multiple inoculation 34.8% with lesions No lesions in controls or mycoplamsa infected	0.03 0.009
4. Muhlestein, *et al.*[39]	NZW, 2–4 month female, $N = 20$ infected (10 treated), $N = 10$ controls	0.25% cholesterol	$1–5 \times 10^6$ IFU intranasally $\times 3$, 3 weeks apart, killed 3 months later	N/A	Azithromycin 30 mg/kg/day/IM $\times 7$ days, then twice weekly $\times$ 6 weeks	1. MIT↑ 3.45 fold infected vs. controls 2. Rx reduced MIT almost to control level	0.009

5. Fong, *et al.*[42]	NZW, 1 month male $N = 36$, half-infected	0.25% cholesterol	$1–2 \times 10^7$ IFU $\times$ 3, 3 week apart, killed 4 months later	No change in total cholesterol, HDL, LDL, or VLDL; ↑ triglycerides by 70%	None	Mean area of atherosclerosis ↑ by 57%	<0.001
6. Fong, *et al.*[44]	NZW, 1-month male, $N = 24$ per group $\times$3.	Regular chow	$1.5–2.6 \times 10^7$ IFU $\times$ 3, 3 weeks apart, killed 10–11 weeks later	N/A	1. Azithromycin 30 mg/kg daily $\times$3 days, then every 6 days, 5 days after, 1st inoculation	Controls 34.8% with lesions, early Rx 4.2%	0.02
					2. Azithromycin same dose—2 weeks after 3rd inoculation until killed	Delayed Rx 33.3% with lesions	NS
7. Fong, *et al.*[45]	NZW, 1-month, male, $N = 24$ per group $\times$3.	Regular chow	$1.5–2.6 \times 10^7$ IFU $\times$ 3, 2 weeks apart, killed 12 weeks later	N/A	1. Clarithromycin 20 mg/kg for 8 days, 5days after each inoculation.	1. Early Rx reduced lesions by 75%	0.036
					2. Clarithromycin 20 mg/kg daily for 6 weeks, started 2 weeks after last inoculation	2. Delayed Rx reduced aortic lesions by 62.5%	0.07

MIT = Maximal intimal thickness (of aorta), Rx = Treatment, N/A = Notavailable, IM = Intramuscular.

In the *de novo* rabbit model without hypercholesterolemia, we demonstrated that azithromycin, when started 5 days after the first of three inoculations was highly effective in preventing atherosclerosis, but delayed treatment 2 weeks after the final inoculation was largely ineffective in preventing early changes of atherosclerosis.[44] Azithromycin 30 mg/kg daily for 3 days then every 5 days for 4–6 weeks was administered by oral gavage, and this dosing was used to mimic human dosing in the WIZARD trial.

Further studies in the *de novo* model was performed with another macrolide, clarithromycin, which has greater *in vitro* activity than azithromycin but has a shorter intracellular and serum half-life. Acute intermittent treatment with oral clarithromycin 20 mg/kg/day for 8 days starting 5 days after each of three inoculations (14 days apart) was highly effective in modifying or preventing early atherosclerotic lesions; but delayed treatment 2 weeks after the final inoculation and administered daily for 6 weeks was less effective, with a trend ($p = 0.07$) to a significant difference compared with untreated controls.[45] The dosage of clarithromycin used in this study provided serum levels that were just below that achieved with a 250-mg dose in humans. Thus, it is possible that a higher dosage of clarithromycin could be more effective. It should be noted that the effect of clarithromycin was mainly an antimicrobial effect and not a nonspecific antiinflammatory effect, as clarithromycin had minimal effect on the extent of atherosclerosis in the cholesterol-fed rabbits.[45]

5. SUMMARY AND FUTURE DIRECTIONS

In general, most of the studies in mice have shown that *C. pneumoniae* infection can enhance or accelerate hypercholesterolemia-induced atherosclerosis, either in genetically susceptible or wild-type mice fed an atherogenic diet. However, it does not appear that *C. pneumoniae* infection by itself can induce atherosclerosis in the murine model without hypercholesterolemia, but is capable of inducing early endothelial dysfunction or minor arterial inflammatory changes alone in this model.

In the rabbit model, *C. pneumoniae* infection is capable of inducing *de novo* early atherosclerotic lesions of the aorta in normocholesterolemic animals. These lesions vary from fatty streaks to fibromusclar lesions or calcified fibrotic lesions, but lack the typical appearance of a human mature atheroma with a lipid core and a fibrous capsule. In this model, *C. pneumoniae* may directly affect the arterial wall by stimulating foam cell accumulation and smooth muscle cell proliferation, partly by upregulation of PDGF-β and TGF-β. Furthermore, infection with *C. pneumoniae* can enhance or accelerate atherosclerosis in the presence of hypercholesterolemia. Although *C. pneumoniae* infection does not alter the concentration in serum of total cholesterol, LDL, HDL, or VLDL, it increases triglyceride concentration in the presence of an atherogenic diet, but this alone is not associated with worsening atherosclerosis in the rabbit model. To date no studies have been done on functional changes of LDL or HDL, or oxidation of LDL in the murine or rabbit model with *C. pneumoniae* infection.

Future studies need to address this issue on the mechanisms of enhancement of atherosclerosis with *C. pneumoniae* infection.

Current antimicrobial therapy in animals indicates that early treatment soon after infection can influence the development of early lesions, or largely prevent enhancement of atherosclerosis with an atherogenic diet. However, delayed treatment does not appear to be very effective. The most effective regimen or duration of treatment for optimal response is unknown and needs to be investigated further. In the murine pneumonitis model the combination of azithromycin plus rifampin showed the strongest activity and produced higher rates of eradication of *C. pneumoniae* from lung tissues than did azithromycin alone.[46] Thus, combination of macrolides with rifampin need to be studied in the atherosclerotic animal models several weeks after infection.

The effect of *C. pneumoniae* infection on atherosclerosis has only been reported in the murine and rabbit models. To establish causality it would be important to study this effect in other animals as well, such as avian models, mini-pigs, guinea pigs, and ultimately nonhuman primates. In a recent study of pigs with acute respiratory infection with *C. pneumoniae* endothelial dysfunction of both resistance and epicardial coronary arteries was found within two weeks, and favours a pro-coagulant status.[47] In another study in the pig model intra-coronary and intrapulmonary *C. pneumoniae* inoculation were associated with moderate intimal proliferation in the absence of a lipid-rich diet.[48] Hence further studies in this model would be of great value. It is of interest that *C. pneumoniae* (separate biovar) has been isolated from the koala and is associated with respiratory infection, and this maybe a suitable natural animal model to study.

Most important for future studies would be the development and investigation of a vaccine to try and prevent *de novo* lesions and acceleration of atherosclerosis.

5.1. Future Directions

The current animal models of atherosclerosis mainly address the two aspects of vascular disease: induction of the initial early lesions and acceleration of more advanced lesions. However, most of the current clinical studies are investigating interventions on the precipitating ischemic events that have not been addressed in the animal models. The pathogenesis of the acute ischemic attack involves different, although overlapping, mechanisms from initiation and growth of the plaque. Prevention of heart attack and stroke depends on detection of vulnerable plaques and development of plaque-stabilizing therapies. Infections such as *C. pneumoniae* may play a role in cardiovascular disease by destablizing plaques, through upregulation of metalloproteinases, which may cause thinning of the fibrous cap, or by precipitating acute thrombus on a plaque by procoagulant mechanisms.

Reproducing vulnerable plaque is one of the most difficult tasks in animal model design. Plaque mechanical properties are mostly determined by the extracellular matrix, fibrillar collagen, and extracellular lipids. Plaques with thin fibrous cap and large lipid/necrotic core are considered vulnerable. Hence the cap/core ratio is one of the most basic vulnerability end point.[49] The two approaches to designing animal models of plaque rupture are either searching for spontaneous rupture/thrombosis or actively triggering these mechanisms.

Spontaneous hemorrhage and rupture of plaques can be found in 39- to 54-month-old pigs with inherited hyper–LDL cholesterolemia in the coronary arteries.[50] High frequency of intraplaque hemorrhage associated with thinning of the fibrous cap can be found in the innominate artery of 42- to 54-week-old apoE-KO mice.[51] Similarly, 37- to 59-week-old apoE-KO mice fed a diet with 21% lard and 0.15% cholesterol develops luminal thrombi with ruptured plaques in the brachiocephalic artery.[52] Recently it has been reported that Dahl salt-sensitive hypertensive rats transgenic for human cholesteryl ester transfer protein develop occlusive thrombosis of left ventricular intramyocardial artery (without plaque rupture or erosion) with decreased survival.[53]

Induced plaque rupture in animal models by mechanical means such as balloon injury, local pressure, or injection of Russel's viper venom followed by histamine or combined with serotonin or angiotensin II can be accomplished in the rabbit model,[49] but are unphysiologic. A more appropriate model is that of double knockout mice (apoE-KO/LDR-KO), which develop acute ischemia and myocardial infarction following mental stress or hypoxia.[54] Although this model resembles the human scenario, no plaque rupture or associated plaque thrombosis were described.

These intriguing animal models could be incorporated into future studies with *C. pneumoniae* infection, to determine the effect of infection in increasing spontaneous plaque rupture or acute myocardial infarction. Furthermore, interventions with antimicrobial agents or a combination in these models may help guide future clinical studies on the value of the most optimal regimen.

REFERENCES

1. Saikku, P., 2000, Epidemiological association of *Chlamydia pneumoniae* and atherosclerosis: The initial serologic observation and more, *J. Infect. Dis.* **181** (Suppl. 3):S411–S413.
2. Fong, I. W., 2000, Emerging relations between infectious diseases and coronary artery disease and atherosclerosis, *CMAJ* **163:**49–56.
3. Danesh, J., Whincup, P., Walker, M., Lennon, L., Thomson, A., Appleby, P., Wong, Y., Bernardes-Silva, M., and Ward, M., 2000, *Chlamydia pneumoniae* IgG titers and coronary heart disease: Prospective study and metaanalysis, *Br. Med. J.* **321:**208–213.
4. Kolia, M., and Fong, I. W., 2002, *Chlamydia pneumoniae* and cardiovascular disease, *Curr. Infect. Dis. Rep.* **4:**35–43.
5. Kuo, C. C., and Campbell, L. A., 2000, Detection of *Chlamydia pneumoniae* in arterial tissues, *J. Infect. Dis.* **181** (Suppl. 3):S432–S436.
6. Maehle, A. H., and Tröhler, U., 1987, Animal experimentation from antiquity to the end of the 18th century: Attitudes and arguments, in: *Vivisection in Historical Perspective* (N. A. Rupke, ed.), Croom Helm, London, pp. 14–47.
7. Ignatowski, A. C., 1908, Influence of animal food on the organism of rabbits, *St. Peterberg, Izviest. Imp. Vogenno-Med. Akad.* **16:**154–173.
8. Finking, G., and Hanke, H., 1997, Nikolai Nikolajewitsch Anitschkow (1885–1964) established the cholesterol-fed rabbit as a model for atherosclerosis research, *Atherosclerosis* **135:**1–7.
9. Moghadasian, M. H., 2002, Experimental atherosclerosis: A historical overview (mini-review), *Life Sci.* **70:**855–865.

10. Aliev, G., and Burnstock, G., 1998, Watanabe rabbits with heritable hyperlipidemia: A model of atherosclerosis, *Histol. Histopathol.* **13:**797–817.
11. Beaty, T. H., Prenger, V. L., Virgil, D. G., Lewis, B., Kwiterovich, P. O., and Bachorik, P. S., 1992, A genetic model for control of hypertriglyceridemia and apolipoprotein B levels in Johns Hopkins colony of St. Thomas Hospital rabbits, *Genetics* **132:**1095–1104.
12. Nishina, P. M., Verstuyft, J., and Paigen, B., 1990, Synthetic low and high fat diets for the study of atherosclerosis in the mouse, *J. Lipid Res.* **31:**859–869.
13. Moghadasian, M. H., McManus, B. M., Godin, D. V., Rodriques, B., and Frohlich, J. J., 1999, Proatherogenic and antiatherogenic effect of probucol and phytosterols in apoE-deficient mice: Possible mechanisms of action, *Circulation* **99:**1733–1739.
14. Purcell-Huynh, D. A., Farese, R. V., Jr., Johnson, D. F., Flynn, L. M., Pierotti, V., Newland, D. L., Linton, M. F., Sanan, D. A., and Young, S. G., 1995, Transgenic mice expressing high levels of human apolipoprotein B develop severe atherosclerotic lesions in response to high fat diet, *J. Clin. Invest.* **95:**2246–2257.
15. Veniant, M. M., Zlot, C. H., Walzem, R. L., Pierotti, V., Driscoll, R., Dichek, D., Herz, J., and Young, S. G., 1998, Lipoprotein clearance mechanisms in LDL receptor–deficient "ApoB48-only" and "Apo-B100-only" mice, *J. Clin. Invest.* **102:**1559–1568.
16. Marotti, K., Castle, C. K., Boyle, T. P., Lin, A. H., Murray, R. W., and Melchior, G. W., 1993, Severe atherosclerosis in transgenic mice expressing simian cholesteryl ester transfer protein, *Nature* **364:**73–75.
17. Lutgens, E., Daemen, M., Kockx, M., Doevendans, P., Hofker, M., Havekes, L., Wellens, H., and De Muinck, E. D., 1999, Atherosclerosis in apoE*3-Leiden transgenic mice: From proliferative to atheromatous stage, *Circulation* **99:**276–283.
18. Yang, Z. P., Kuo, C. C., and Grayston, J. T., 1993, A mouse model of *Chlamydia pneumoniae* strain TWAR pneumonitis, *Infect. Immun.* **61:**2037–2040.
19. Yang, Z., Kuo, C. C., and Grayston, J. T., 1995, Systemic dissemination of *Chlamydia pneumoniae* following intranasal inoculation of mice, *J. Infect. Dis.* **171:**736–738.
20. Malinverni, R., Kuo, C. C., Campbell, L. A., and Graytson, J. T., 1995, Reactivation of *Chlamydia pneumoniae* lung infection in mice by cortisone, *J. Infect. Dis.* **172:**593–594.
21. Laitinen, K., Laurila, A. L., Leinonen, M., and Saikku, P., 1996, Reactivation of *Chlamydia pneumoniae* infection in mice by cortisone treatment, *Infect. Immun.* **64:**1488–1490.
22. Fong, I. W., Chiu, B., Viira, E., Fong, M. W., Jang, D., and Mahony, J., 1997, Rabbit model for *Chlamydia pneumoniae* infection, *Clin. Microbiol.* **35:**48–52.
23. Moazed, T. C., Kuo, C. C., Patton, D. L., Grayston, J. T., and Campbell, L. A., 1996, Experimental rabbit models of *Chlamydia pneumoniae* infection, *Am. J. Pathol.* **148:**667–676.
24. Holland, S. M., Taylor, H. R., Gaydos, C. A., Kappus, E. W., and Quinn, T. C., 1990, Experimental infection with *Chlamydia pneumoniae* in nonhuman primates, *Infect. Immun.* **58:**593–597.
25. Moazed, T. C., Kuo, C. C., Grayston, J. T., and Campbell, L. A., 1997, Murine models of *Chlamydia pneumoniae* infection and atherosclerosis, *J. Infect. Dis.* **175:**833–890.
26. Moazed, T. C., Campbell, L. A., Rosenfeld, M. E., Grayston, J. T., and Kuo, C. C., 1999, *Chlamydia pneumoniae* infection accelerates the progression of atherosclerosis in Apolipoprotein E–deficient mice, *J. Infect. Dis.* **180:**238–241.
27. Burnett, M. S., Gaydos, C. A., Madico, G. E., Glad, S. M., Paigen, B., Quinn, T. C., and Epstein, S. E., 2001, Atherosclerosis in apoE knockout mice infected with multiple pathogens, *J. Infect. Dis.* **183:**226–231.
28. Rothstein, N. M., Quinn, T. C., Madico, G., Gaydos, C. A., and Lowenstein, C. J., 2001, Effect of azithromycin on murine arteriosclerosis exacerbated by *C. pneumoniae*, *J. Infect. Dis.* **183:**232–238.
29. Liuba, P., Karnani, P., Pesonen, E., Paakkari, I., Forslid, A., Johansson, L., Persson, K., Wadstrom, T., and Laurini, R., 2000, Endothelial dysfunction after repeated *Chlamydia pneumoniae* infection in apolipoprotein E–knockout mice, *Circulation* **102:**1039–1044.
30. Aalto-Setälä, K., Laitinen, K., Erkkilä, L., Leinonen, M., Janhiainen, M., Ehnholm, C., Tamminen, M., Puolakkainen, M., Penttilä, I., and Saikku, P., 2001, *Chlamydia pneumoniae*

does not increase atherosclerosis in the aortic root of apolipoprotein E–deficient mice, *Arterioscler. Thromb. Vasc. Biol.* **21:**578–584.

31. Caligiuri, G., Rottenberg, M., Nicoletti, A., Wigzell, H., and Hansson, G. K., 2001, *Chlamydia pneumoniae* infection does not induce or modify atherosclerosis in mice, *Circulation* **103:**2834–2838.
32. Blessing, E., Campbell, L. A., Rosenfeld, M. E., Chough, N., and Kuo, C. C., 2001, *Chlamydia pneumoniae* infection accelerates hyperlipidemia induced atherosclerotic lesion development in C57BL/6J mice, *Atherosclerosis* **158:**13–17.
33. Blessing, E., Campbell, L. A., Rosenfeld, M. E., and Kuo, C. C., 2002, *Chlamydia pneumoniae* and hyperlipidemia are co-risk factors for atherosclerosis: Infection prior to induction of hyperlipidemia does not accelerate development of atherosclerotic lesions in C57BL/6J mice, *Infect. Immun.* **70:**5332–5334.
34. Blessing, E., Lin, T. M., Campbell, L. A., Rosenfeld, M. E., Lloyd, D., and Kuo, C. C., 2000, *Chlamydia pneumoniae* induces inflammatory changes in the heart and aorta of normocholesterolemic C57BL/6J mice, *Infect. Immun.* **68:**4765–4768.
35. Hu, H., Pierce, G. N., and Zhong, G., 1999, The atherogenic effects of chlamydia are dependent on serum cholesterol and specific to *Chlamydia pneumoniae, J. Clin. Invest.* **103:**747–753.
36. Fong, I. W., Chiu, B., Viira, E., Jang, D., and Mahony, J. B., 1999, *De novo* induction of atherosclerosis by *Chlamydia pneumoniae* in a rabbit model, *Infect. Immun.* **67:**6048–6055.
37. Coombes, B. K., Chiu, B., Fong, I. W., and Mahony, J. B., 2002, *Chlamydia pneumoniae* infection of endothelial cells induces transcriptional activation of platelet-dervied growth factor-β: A potential link to intimal thickening in a rabbit model of atherosclerosis, *J. Infect. Dis.* **185:**1621–1630.
38. Laitinen, K., Laurila, A., Pyhälä, L., Leinonen, M., and Saikku, P., 1997, *Chlamydia pneumoniae* infection induces inflammatory changes in the aortas of rabbits, *Infect. Immun.* **65:**4832–4835.
39. Muhlestein, J. B., Anderson, J. L., Hammond, E. H., Zhao, L., Trehan, S., Schwobe, E. P., and Carlquist, J. F., 1998, Infection with *C. pneumoniae* accelerates the development of atherosclerosis and treatment with azithromycin prevents it in a rabbit model, *Circulation* **97:**633–636.
40. Kalayoglu, M. V., and Byrne, G. I., 1998, A *Chlamydia pneumoniae* component that induces macrophage foam cell formation is chlamydial lipopolysaccharide, *Infect. Immun.* **66:**5067–5072.
41. Kalayoglu, M. V., Hoerneman, B., La Verda, D., Morrison, S. G., Morrison, R. P., and Byrne, G. I., 1999, Cellular oxidation of low density lipoprotein by *Chlamydia pneumoniae, J. Infect. Dis.* **180:**780–790.
42. Fong, I. W., Chiu, B., Mahony, J. B., Jang, D., Coombes, B., Dunn, P., Caterini, J., and Murdin, A., 2002, *C. pneumoniae* enhancing effect on atherosclerosis and relationships with lipid profile and fibrinogen in a rabbit model, in: *Chlamydial Infections* (J. Schachter, G. Christiansen, I. N. Clarke, M. R. Hammerschlag, B. Kaltenboeck, C. C. Kuo, R. G. Rank, G. L. Ridgway, P. Saikku, W. E. Stamm, R. S. Stephens, J. T. Summersgill, P. Timms, P. B. Wyrick, eds), Basum Yeri: Grafmat Basin ve Reklam Sanayi Tic. Ltd. Sti., Istanbul, pp. 321–324.
43. Van Heek, M., and Silversmit, D. B., 1988, Evidence for an inverse relation between plasma triglyceride and aortic cholesterol in the coconut oil, cholesterol-fed rabbit, *Atherosclerosis* **71:**185–192.
44. Fong, I. W., Chiu, B., Viira, E., Jang, D., Fong, M. W., and Mahony, J. B., 1999, Can an antibiotic (macrolide) prevent *Chlamydia pneumoniae* induced atherosclerosis in a rabbit model? *Clin. Diagn. Lab. Immunol.* **6:**891–894.
45. Fong, I. W., Chiu, B., Viira, E., Jang, D., and Mahony, J. B., 2002, Influence of clarithromycin on early atherosclerotic lesions after *Chlamydia pneumoniae* infection in a rabbit model, *Antimicrob. Agents Chemother.* **46:**2321–2326.
46. Wolf, K., and Malinverni, R., 1999, Effect of azithromycin plus rifampin versus that of azithromycin alone on eradication of *Chlamydia pneumoniae* from lung tissue in experi-

mental pneumonitis, *Antimicrob. Agents Chemother.* **43:**1491–1493.
47. Liuba, P., Pesonen, E., Paakkari, I., Batra, S., Forslid, A., Kovanen, P., Pentikainen, M., Persson, K., Sandstrom, S., 2003, Acute *Chlamydia pneumoniae* infection causes endothelial dysfunction in pigs, *Atherosclerosis.* **167:** 215–222.
48. Pislaru, S. V., Van Ranst, M., Pislaru, C., Szelid, Z., Theilmeier, G., Ossewaarde, J. M., Holvoet, P., Janssens, S., Verbeken, E., Van de Werf, F. J., 2003, *Chlamydia pneumoniae* induces neointima formation in coronary arteries of normal pigs, *Cardiovasc. Res.* **57:** 834–842.
49. Rekhter, M. D., 2002, How to evaluate plaque vulnerability in animal models of atherosclerosis? *Cardiovasc. Res.* **54:**36–41.
50. Prescott, M. F., McBride, C. H., Hasler, R. J., Von-Linden, J., and Rapacz, J., 1991, Development of complex atherosclerotic lesions in pigs with inherited hyper–LDL cholesterolemia bearing mutant alleles for apolipoprotein B, *Am. J. Pathol.* **139:**139–147.
51. Rosenfeld, M. E., Polinsky, P., Virmani, R., Kauser, K., Rubanyi, G., and Schwartz, S. M., 2000, Advanced atherosclerotic lesions in the innominate artery of apoE-knockout mice, *Arterioscler. Thromb. Vasc. Biol.* **20:**2587–2592.
52. Johnson, J. L., and Jackson, C. L., 2001, Atherosclerotic plaque rupture in the apolipoprotein E knockout mouse, *Atherosclerosis* **154:**399–406.
53. Herrera, V. L., Makrides, S. C., Xie, H. S., Adari, H., Krauss, R. M., Ryan, U. S., and Ruiz-Aparzo, N., 1999, Spontaneous combined hyperlipidemia, coronary heart disease and decreased survival in Dahl salt-sensitive hypertensive rats transgenic for human cholesteryl ester transfer protein, *Nat. Med.* **5:**1383–1389.
54. Caligiuri, G., Levy, B., Pernow, J., Thoren, P., and Hansson, G. K., 1999, Myocardial infarction mediated by endothelin receptor signaling in hypercholesterolemic mice, *Proc. Natl. Acad. Sci. U.S.A.* **96:**6920–6924.

12

Antiinfective Trials for the Treatment of *Chlamydia pneumoniae* in Coronary Artery Disease

JOSEPH B. MUHLESTEIN

1. INTRODUCTION

Because of the serologic, pathologic and experimental evidence linking *C. pneumoniae* to atherosclerotic cardiovascular disease, the hypothesis has been proposed that anti-infective therapy targeted against *C. pneumoniae* might result in clinical benefit. Consequently, a variety of approaches have been undertaken to test this hypothesis.

As noted in other chapters, *C. pneumoniae* has been found to be responsible for a variety of respiratory illnesses including 10% of cases of community-acquired pneumonia. Like the more familiar *C. trachomatis, C. pneumoniae* is an obligate intracellular pathogen with a unique life cycle (See Fig. 1).[1] Generally *C. pneumoniae* enters the body through a respiratory route and exists outside of cell's in a spore form called the elementary body. Once inside the host cell it makes use of the cell's own metabolic machinery and develops into a metabolically active but noninfectious form called the reticulate body. In this form, the bacterium has the ability to divide and differentiate into new elementary bodies which can then invade other host cells.

However, *C. pneumoniae* may also, unpredictably convert within the cell to a metabolically inactive form called the persistent body.[2] In this state, it may remain within the cell for extended periods essentially undetectable by the immune system and unresponsive antibiotics that interfere with bacterial metabolism. This conversion to a persistent body may have important

JOSEPH B. MUHLESTEIN • University of Utah, LDS Hospital, Salt Lake City, Utah 84143.

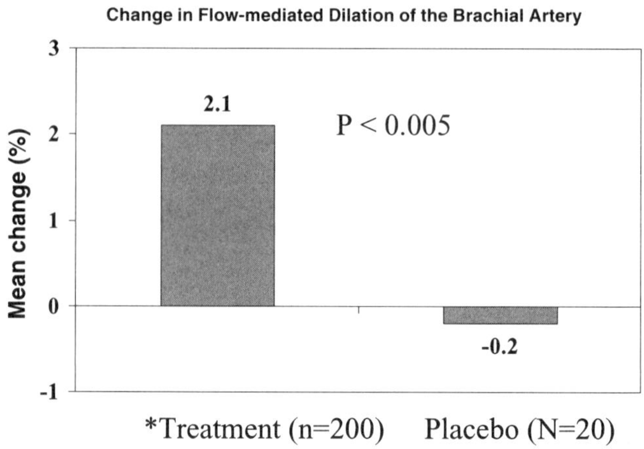

FIGURE 1. Effect of azithromycin on the incidence of death, myocardial infarction, and unstable angina in post–myocardial infarction patients with elevated *C. pneumoniae* titers at the 18-month follow-up.

implications regarding attempts to eradicate the organism. The major studies investigating the potential of antiinfective therapy have involved both animal and human subjects. Likewise, both observational and randomized studies have been performed.

2. ANIMAL STUDIES OF *Chlamydia pneumoniae*, ATHEROSCLEROSIS AND ANTIINFECTIVE THERAPY

In an initial attempt to evaluate the effect of antiinfective agents on atherosclerosis progression, our group looked at the rabbit model.[3] Rabbits fed 1–2%-cholesterol diets are a familiar laboratory model for accelerated atherosclerosis because they develop significant amounts of plaque.[4] Understanding the importance of lipid metabolism in the development of human atherosclerosis, it was hypothesized that combining *C. pneumoniae* infection with small amounts of cholesterol supplementation might significantly accelerate the development of atherosclerosis.

Thirty rabbits were fed 0.25%-cholesterol diets. Sham intranasal inoculations of saline were administered to 10 rabbits, and three separate intranasal inoculations of *C. pneumoniae* were administered to the others. Thereafter, infected rabbits were additionally randomized to receive azithromycin antiinfective therapy versus placebo.

Other studies, such as that performed by Dr. Fong *et al.*,[5] looked at the effect of the timing of antiinfective therapy. Three groups of New Zealand White rabbits (24 per group) were infected via the nasopharynx with *C. pneumoniae* on three separate occasions (2 weeks apart). Group I was untreated and sacrificed at 12 weeks; group II received clarithromycin at 20 mg/kg/day for 8 days, beginning 5 days after each inoculation (early treatment); and group III received a similar dose of clarithromycin starting 2 weeks after the third inoculation and continued for 6 weeks thereafter (delayed treatment). To test for a possible antiinflammatory effect of clarithromycin, two other groups of uninfected rabbits (12 animals in each) were fed 0.5% cholesterol-enriched chow, and one of these groups was treated with clarithromycin at 30 mg/kg/day for 6 weeks. Of 23 untreated infected rabbits, 8 developed early lesions of atherosclerosis, whereas 2 of the 24 early-treated (group II) had similar changes ($P = 0.036$ [75% efficacy]). However, in the delayed-treatment group, (group III) 3 of 24 rabbits developed early lesions of atherosclerosis, thus demonstrating 62.5% reduction compared with the untreated controls ($P = 0.07$ [trend to statistical significance]). *C. pneumoniae* antigen was detected in 8 of 23 group I (untreated) rabbits versus 1 of the 24 group II rabbits and 4 of 24 animals in group III ($P = 0.009$ and 0.138, respectively). All of the untreated, cholesterol-fed rabbits had moderate to advanced atherosclerosis (grade III or IV); clarithromycin had no effect on reducing the prevalence but did reduce the extent of atherosclerosis in the cholesterol-fed rabbits by 17% compared to untreated controls. Thus, clarithromycin administration modified *C. pneumoniae*–induced atherosclerotic lesions and reduced the ability to detect organisms in tissue. Early treatment was more effective than delayed treatment.

Rothstein *et al.*[6] tested the hypothesis that antibiotic therapy inhibits the atherogenic effects of *C. pneumoniae* infection, in 10-week-old apolipoprotein E (ApoE)-null mice who were infected with *C. pneumoniae* or placebo. Mice were treated for 2 weeks after infection with azithromycin or placebo, and were killed at 20 weeks of age. Infection did not affect the size of the aortic lesion, and antibiotic treatment had no effect. Another group of mice, 12-week-old ApoE mice, were infected with *C. pneumoniae* or placebo, were treated for 2 weeks after infection with azithromycin or placebo, and were killed at 26 weeks of age. *C. pneumoniae* infection increased the size of the lesion in infected mice, but azithromycin did not reduce the size of the aortic lesion in infected mice. They concluded that, therefore, immediate therapy of acute infection may be necessary to prevent the proatherogenic effects of *C. pneumoniae* infection.

These animals models, therefore, provide some hope that antiinfective therapy might be helpful in the treatment of *C. pneumoniae*–induced atherosclerotic complications. However, there remains a significant number of questions regarding the necessary timing of treatment. Likewise, animal models are not exact replicas of the human experience and, therefore, results from these studies must be interpreted with caution. Ultimately, studies must be performed directly in patients.

3. OBSERVATIONAL STUDIES OF THE EFFECT OF ANTIMICROBIALS ON THE COURSE OF CORONARY ARTERY DISEASE

One method of evaluating the effect of antiinfective therapy on coronary artery disease has been to retrospectively analyze patient registries to determine an association between the incidence of myocardial infarction and prior use of antibiotic therapy. Results of this endeavor have been mixed. In one of the first published studies, Meir, *et al.*,[7] in order to determine whether previous use of antibiotics decreases the risk of developing a first-time acute myocardial infarction, evaluated a total of 3,315 case patients aged 75 years or younger with a diagnosis of first-time acute myocardial infarction between 1992 and 1997 and 13,139 controls without myocardial infarction matched to cases for age, sex, general practice attended, and calendar time. Cases were significantly less likely to have used tetracycline antibiotics (adjusted odds ratio [OR], 0.70; 95% confidence interval [CI], 0.55–0.90) or quinolones (adjusted OR, 0.45; 95% CI, 0.21–0.95). No effect was found for previous use of macrolides (primarily erythromycin), sulfonamides, penicillins, or cephalosporins.

However, Jackson, *et al.*,[8] who performed a very similar study, found little or no association between the use of erythromycin, tetracycline, or doxycycline during the previous 5 years and the risk for first myocardial infarction in the primary prevention setting. More recently, Herings, *et al.*[9] reported the results of a nested case–control study (628 cases with myocardial infarction; 1,615 age-, sex-, exposure window- and pharmacy-matched controls) of patients enrolled in large Dutch health maintenance organization in which prior use of antibiotics were evaluated in relationship to future incidence of myocardial infarction. Prior use of fluoroquinolones, quinolones, tetracyclines, macrolides and other antibiotics were evaluated. Only high doses of fluoroquinolones were associated with a lower risk of acute MI. For those who took more than 1 course of fluoroquinolones, the odds ratio was 0.12 (95% CI, 0.02–0.94). For all other antibiotics, no significant association was observed.

Because of potential confounding variables, observational studies such as above, while being potentially informative, cannot be definitive regarding the usefulness of antibiotics to prevent complications of coronary artery disease. Prospective randomized clinical trials are necessary.

4. RANDOMIZED SECONDARY PREVENTION TRIALS FOR CLINICAL PERIPHERAL VASCULAR DISEASE

C. pneumoniae has not only been discovered within atherosclerotic plaques of coronary arteries, but also it is commonly discovered with plaques of virtually any arterial bed.[10] Because some peripheral arteries, such as carotid arteries or the aorta, are more easily nonivasively visualized than coronary arteries, a variety of randomized placebo-controlled studies testing the effect of antibiotic therapy on atherosclerotic progression have been performed targeting them. Sanders, *et al.*[11] evaluated the effect of roxithromycin therapy (150 mg twice daily for 30 days) on the progression of the intima-to-media thickness (IMT) of the common carotid artery using duplex ultrasonography in a prospective and randomized trial with a follow-up of 2 years in 272 consecutive patients with ischemic stroke aged over 55 years in whom the first IMT measurement and *C. pneumoniae* testing (IgG and IgA) were performed at least 3 years before the roxithromycin treatment. *C. pneumoniae* IgG antibodies ($\geq$1:64) were initially found in 123 (45%) patients and IgA antibodies ($\geq$1:16) in 112 (41%) patients. During the 3 years before antibiotic therapy, *C. pneumoniae*–positive patients showed an enhanced IMT progression, even after adjustment for other cardiovascular risk factors (0.12 [95% CI, 0.11 to 0.14] vs. 0.07 [0.05 to 0.09] mm/year; $P<0.005$). The 62 *C. pneumoniae*–positive patients given roxithromycin showed a significantly decreased IMT progression after 2 years compared with the *C. pneumoniae*–positive patients without therapy (0.07 [0.045 to 0.095] vs. 0.11 [0.088 to 0.132] mm/year; $P<0.01$). No significant difference in the occurrence of future cardiovascular events was found between both groups during follow-up. No change of IMT was observed in *C. pneumoniae*–negative patients given roxithromycin ($n=74$) compared with those without therapy (0.06 [0.03 to 0.09] vs. 0.07 [0.05 to 0.09] mm/year). They concluded that these findings suggest a positive impact of antibiotic therapy on early atherosclerosis progression in *C. pneumoniae*–seropositive patients with cerebrovascular disease.

In a similar study looking at occlusive peripheral vascular disease Wiesli, *et al.*,[12] Forty *C. pneumoniae*–seropositive men suffering from peripheral arterial occlusive disease were randomly assigned to receive either roxithromycin (300 mg daily) or placebo for 28 days. During the 2.7-year follow-up, the number of invasive revascularizations per patient, the walking distance before intervention (in patients without intervention at study end), and the change of carotid plaque size were assessed. Five interventions were performed on 4 patients (20%) in the roxithromycin group, and 29 interventions were performed on 9 patients (45%) in the placebo group. Limitation of walking distance to 200 m or less was observed in 4 patients (20%) in the roxithromycin group and in 13 patients (65%) in the placebo group. The effect of macrolide treatment on the number of interventions per patient and on preinterventional walking distance was significant. Possible confounding variables such as classical vascular risk factors were excluded by multiple regression analyses. Carotid plaque areas monitored over 6 months decreased in the roxithromycin group (mean relative value, 94.4%) but remained constant in the placebo group (100.2%). Regression of carotid plaque size observed in roxithromycin-treated patients was significant for soft plaques.

The hypothesis that antibiotic therapy may be helpful in the prevention of future expansion of abdominal aortic aneurysms (AAA) was tested by Mosorin, *et al.*[13] Owing to evidence of its efficacy against *C. pneumoniae* infection and inhibition of elastolytic matrix metalloproteinases (MMP), doxycycline was chosen. The study group consisted of 32 of 34 initially eligible patients who had an AAA diameter perpendicular to the aortic axis of 30 mm or more in size or a ratio of infrarenal to suprarenal aortic diameter of 1.2 or more and a diameter less than 55 mm. Patients were randomly assigned to receive either doxycycline (150 mg daily) or placebo during a 3-month period and underwent ultrasound surveillance during an 18-month period. Outcome measures included aneurysm expansion rates, the number of patients who had AAA rupture or repair, *C. pneumoniae* antibody titers, and serum concentrations of C-reactive protein. The aneurysm expansion rate in the doxycycline group was significantly lower than that in the placebo group during the 6- to 12-month ($P=0.01$) and the 12- to 18-month periods ($P=0.01$). Five patients (41%) in the placebo group and 1 patient (7%) in the doxycycline group had an overall expansion of the aneurysm of 5 mm or more during the 18-month follow-up. Among the placebo group patients, a higher expansion rate was observed in those with enhanced *C. pneumoniae* immunoglobulin G antibody titers (>128) than in those with lower titers ($P = 0.03$). Doxycycline treatment had no clear effect on antibody titers. However, at 6-month follow-up, C-reactive protein levels in the doxycycline group were significantly lower than the baseline levels ($P = 0.01$). This study could not ascertain whether the clinical benefit came from doxycycline's antimicrobial effect or its inhibition of harmful MMPs.

Finally, the effect of diseased peripheral vessels on endothelial function was tested by Parchure, *et al.*'s[14] randomized, prospective, double-blind, placebo-controlled trial in 40 male patients (mean age, 55 $\pm$ 9 years) with documented coronary artery disease and positive *C. pneumoniae*–IgG antibody titers was performed. After baseline evaluation, patients were randomized to receive either azithromycin or placebo for 5 weeks. Flow-mediated dilation (FMD) of the brachial artery and E-selectin, von Willebrand factor, and C-reactive protein (CRP) levels were assessed at study entry and at the end of the treatment period. Patients who received azithromycin had a significant improvement in FMD (mean change, 2.1 $\pm$ 1.1%; $P < 0.005$). In contrast, FMD was not significantly changed in the placebo group (mean change, $-0.02 \pm 0.2\%$, $P = 0.64$) (see Fig. 1). Azithromycin therapy also resulted in a significant decrease of E-selectin and von Willebrand factor levels. CRP levels were not significantly altered by treatment with either azithromycin or placebo. Beneficial effects of azithromycin treatment were independent from the presence of low (<1:32) or high ($\geq$1:32) *C. pneumoniae* antibody titers. The authors concluded that treatment with azithromycin has a favorable effect on endothelial function in patients with documented coronary artery disease and evidence of *C. pneumoniae* infection irrespective of antibody titer levels.

Although these randomized clinical trials have demonstrated significant promise for benefit from antibiotic therapy in the management of atherosclerotic cardiovascular disease, most of the endpoints evaluated were secondary or surrogate ones. Before an entirely new therapeutic approach can be accepted, it must be demonstrated that it has positive effects on primary outcomes such as death, myocardial infarction, stroke, etc. Therefore further studies directed at hard clinical outcomes, especially among patients with coronary artery disease, are also necessary.

5. RANDOMIZED SECONDARY PREVENTION TRIALS FOR CLINICAL CORONARY ARTERY DISEASE

Several small pilot randomized clinical trials of antibiotic therapy for the secondary prevention of coronary artery disease have been reported. Their results are summarized below.

In a small study from London,[15] 60 stable post–myocardial infarction male patients who were seropositive to *Chlamydia pneumoniae* were randomized to receive azithromycin (500 mg/day for 3 days [$n = 28$] or 500 mg/day for 6 days [$n = 12$]) or placebo and followed for 18 months, looking for the endpoints

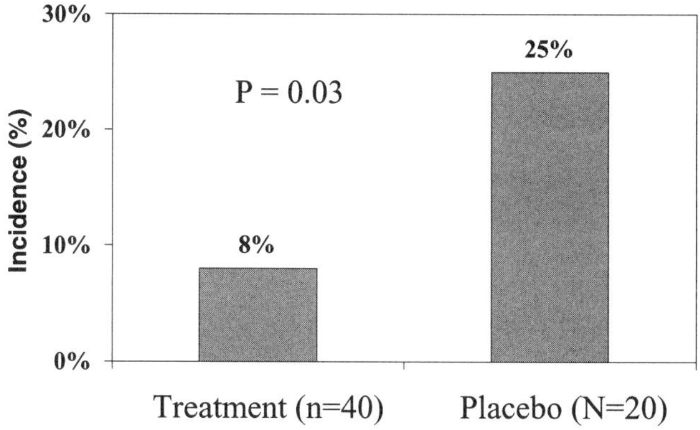

FIGURE 2. Effect of roxithromycin on the incidence of death, myocardial infarction, and unstable angina in patients presenting with acute coronary syndrome after 30 days and 6 months.

of death, myocardial infarction, or the need for coronary revascularization. The results are shown in Fig. 2. Although the numbers were small, there was a statistically significant reduction in the number of events in the group receiving antibiotic treatment (25 vs. 8%, $p = 0.03$).

In another study from Argentina,[16] the effect of roxithromycin was assessed in a double-blind, randomized, prospective, multicenter, parallel-group, placebo-controlled pilot study of 202 patients with unstable angina or non–Q-wave myocardial infarction. No serologic test was required for inclusion in this study. Patients were randomly assigned either roxithromycin 150 mg orally twice a day ($n = 102$) or placebo orally twice a day ($n = 100$). The treatment was for 30 days. Patients were followed up for 6 months. The primary clinical endpoints were cardiac ischaemic death, myocardial infarction, and severe recurrent ischaemia assessed at day 31. Figure 3 shows the results. A statistically significant reduction in the primary composite triple-endpoint rates was

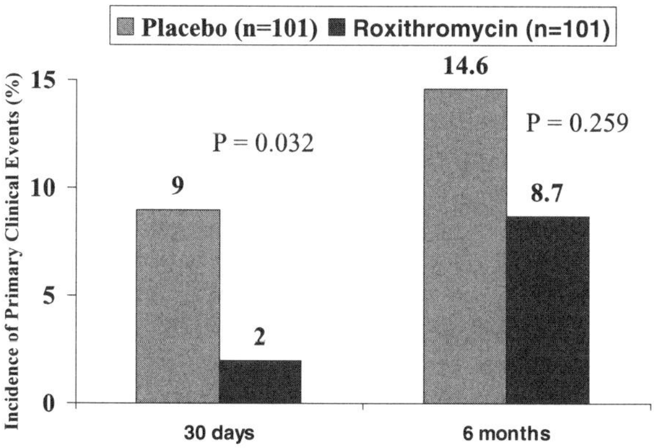

FIGURE 3. Effect of 3 months' therapy with azithromycin on C-reactive protein levels in patients with stable coronary artery disease after 3 and 6 months.

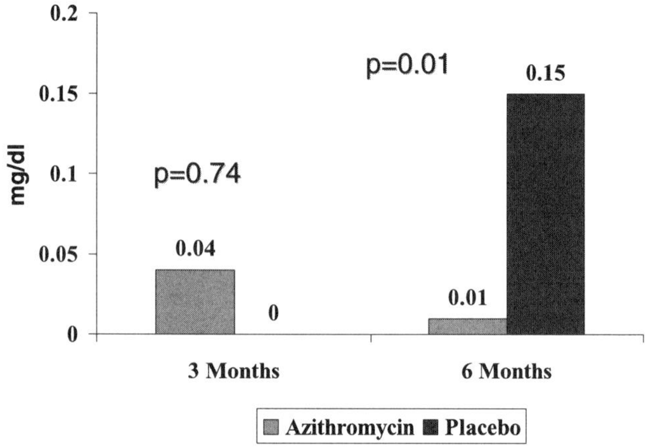

FIGURE 4. Two-year event-free survival curve of patients in the ACADEMIC study randomized to 3 months of azithromycin or placebo.

observed in the roxithromycin group: $p = 0.032$. The rate of severe recurrent ischemia, myocardial infarction, and ischaemic death were 5.4, 2.2, and 2.2% in the placebo group and 1.1, 0, and 0%, respectively, in the roxithromycin group. After follow-up to six months was accomplished,[17] there remained a difference between the treated and control groups in regards to adverse events, but the difference had narrowed such that statistical significance was no longer present.

A third, intermediate-size study of 302 patients was reported by our group.[18] In this study, patients with known coronary artery disease who were seropositive to *C. pneumoniae* were randomized to receive either azithromycin 500 mg per day for 3 days followed by 500 mg per week for 3 months or placebo. By 6 months, 3 months after discontinuation of the antibiotic, a statistically significant reduction in levels of C-reactive protein and interleukin-6 were noted in the treatment group (see Fig. 4). However, after 2 years of clinical follow-up[19] there was no significant difference in the cardiovascular endpoints between the two groups (hazard ratio [HR] for azithromycin = 0.89; 95% CI = 0.51–1.61, $p = 0.74$) (see Fig. 5), although a trend toward a reduction in events in the azithromycin arm was noted during the second year of the study (HR = 0.59, 95% CI = 0.23–1.50, $p = 0.26$) (see Fig. 6).

The ISAR-3 Trial[20] tested the hypothesis that antibiotic therapy may be helpful in the prevention of restenosis after coronary stent deployment. A total of 1,010 patients undergoing percutaneous coronary intervention were randomized to receive roxithromycin 300 mg daily for 4 weeks versus placebo and followed for 6 months. No significant differences in angiographic restenosis rate at 6 months and target vessel revascularization or major cardiac events at 30 days in the group were noted overall. There was, however, a differential effect dependent on *C. pneumoniae* titers. In patients with high titers, roxithromycin reduced the rate of restenosis.

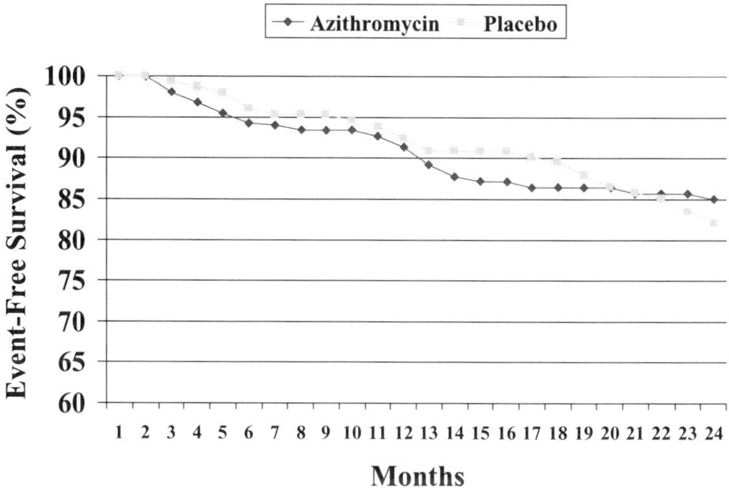

FIGURE 5. Total clinical events of patients enrolled into the ACADEMIC study stratified by year after enrollment.

Recently, a randomized clinical trial entitled STAMINA[21] was reported in which 325 patients presenting with acute coronary syndrome were randomized to one of three treatment regimens, each lasting 1 week: (1) azithromycin 500 mg/day, omeprazole 20 mg bid, and metronidazole 400 mg bid (designed to be an anti-chlamydial regimen); (2) amoxycillin 500 mg bid, omeprazole 20 mg bid, and metronidazole 400 mg bid (designed to be an antihelicobacter regimen); or (3) placebo. Follow-up period extended to 12 months. All patients received standard treatment for CHD. There was no statistically significant difference in frequency or timing of major adverse cardiac events for either the azithromycin- or the amoxycillin-treated groups compared to placebo. However,

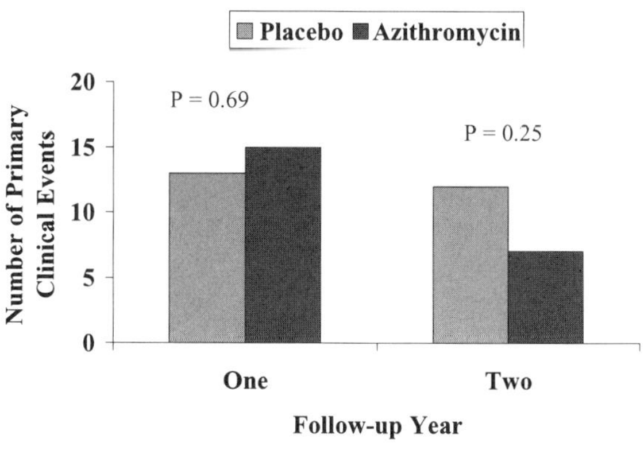

FIGURE 6. Effect of 5 weeks' azithromycin treatment on endothelial function in *C. pneumoniae*–seropositive male patients with coronary artery disease.

when combined, subjects receiving either one or other active antibiotic treatment regimens had a 40% reduction in MACE compared to placebo ($p = 0.034$). The full benefit was observed by 12 weeks and persisted to 1 year. No differences in inflammatory markers were noted. Seropositivity to *C. pneumoniae* or *Helicobacter pylori* also had no effect. The authors made the point that this study could not distinguish whether the benefits were related to the antimicrobial or antiinflammatory properties of the antibiotics used. Larger clinical trials were recommended.

Sinisalo *et al.*(22) also tested the effect of antibiotic therapy on the secondary prevention of acute coronary syndrome in the CLARIFY (Clarithromycin in Acute Coronary Syndrome Patients in Finland) trial. Altogether, 148 patients with acute non–Q-wave infarction or unstable angina were randomly assigned to receive double-blind treatment with either clarithromycin or placebo for 3 months. The primary endpoint was a composite of death, myocardial infarction, or unstable angina during treatment; the secondary endpoint was occurrence of any cardiovascular event during the entire follow-up period (average 555 days, range 138–924 days) (see Fig. 7). There was a trend toward fewer patients meeting primary endpoint criteria in the clarithromycin group than in the placebo group (11 vs. 19 patients, respectively; risk ratio 0.54, 95% CI 0.25–1.14; $P = 0.10$). By the end of the entire follow-up, 16 patients in the clarithromycin group and 27 in the placebo group had experienced a cardiovascular event (risk ratio 0.49, 95% CI 0.26–0.92; $P = 0.03$). It was therefore concluded that clarithromycin appears to reduce the risk of ischemic cardiovascular events in patients presenting with acute non–Q-wave infarction or unstable angina.

The first large trial, the WIZARD (Weekly Intervention with Zithromax for Atherosclerosis and its Related Disorders) trial,(23–24) was completed in early 2002. A total of 7,724 stable patients with a history of myocardial infarction and

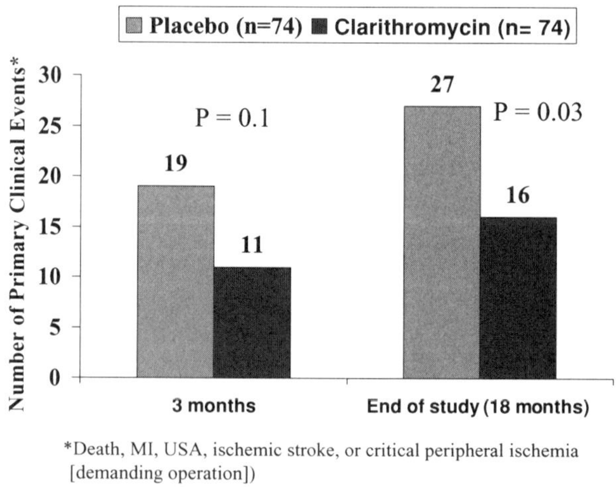

FIGURE 7. Effect of 3 months' therapy with clarithromycin in patients presenting with acute coronary syndrome: Results of the CLARIFY study.

documented presence of seropositivity to *C. pneumoniae* were randomized to receive either placebo or 3-month treatment of azithromycin (600 mg/week). The primary clinical endpoint was death, myocardial infarction, admission for unstable angina, and need for repeat revascularization at the end of the study. Overall, short-term (3-month) azithromycin therapy was safe and well tolerated. However, it only resulted in a 7% overall nonsignificant reduction in the incidence of recurrent cardiovascular disease in this population of stable *C. pneumoniae*–positive, post–myocardial infarction patients. No evidence of a treatment effect by baseline *C. pneumoniae* titer was observed. Post hoc analyses did suggest a possible early treatment benefit [33% reduction in death/myocardial infarction at 6 months ($p = 0.03$)] that was not sustained over the observation period. This raised the question as to whether prolonged antimicrobial therapy might produce a more sustained clinical benefit.

Finally, the Azithromycin in Acute Coronary Syndrome (AZACS)[(25)] has recently been completed. In this study, patients with acute coronary syndromes (unstable angina or myocardial infarction) were randomized in a double-blind, placebo-controlled fashion to either azithromycin 500 mg/day followed by 250 mg/day for 4 days or matching placebo. The primary endpoints were death from any cause, nonfatal myocardial infarction, and recurrent ischemia at 6 months. Secondary endpoints included worsening of ischemia and new congestive heart failure requiring hospitalization. A total of 1439 patients were randomized, 723 to the placebo arm and 716 to the antibiotic. Short-term (4-day) treatment with azithromycin did not have any effect on the recurrence of the primary or secondary ischemic events during a 6-month follow-up period. Likewise, there was no difference between patients who tested positive for the presence of *C. pneumoniae* antibodies and those who did not. Interestingly, again similar to the WIZARD Study, there was a trend toward early benefit from antibiotic therapy during the first 2 months after antibiotic therapy, but this was not sustained over the entire 6 months of the study.

6. ONGOING CLINICAL ANTIINFECTIVE TRIALS

The ongoing ACES (Azithromycin and Coronary Events) study[(26)] is an NIH-sponsored randomized, double-blind, placebo-controlled trial of azithromycin among adults with stable coronary artery disease. The study is based on the hypothesis that infection with *C. pneumoniae* may be causally associated with cardiovascular disease and that therefore treatment directed against this organism might reduce the risk of subsequent coronary events. Seropositivity to *C. pneumoniae* is not required in this trial although serologic levels are being followed. Participants randomized to treatment receive 600 mg of azithromycin orally once a week for 1 year and are being followed a mean of 4 years for the composite primary outcome of coronary heart disease death, nonfatal myocardial infarction, hospitalization for unstable angina, and coronary revascularization. Secondary objectives include those related to a better understanding of the relationship between antibody titer and inflammatory markers with treatment status and outcome; therefore, all participants are having additional specimens collected periodically during follow-up. More than 4,000 patients have been enrolled in the study and are into the follow-up period.

Finally, another large randomized trial, entitled PROVE-IT, is also underway. In this study, >3,000 patients presenting with acute coronary syndrome are randomized in a 2×2 factorial fashion to one of two statins, pravastatin or atorvastatin, in different relative doses, and then to an intermittent course of gatifloxacin or placebo. The primary endpoint is major cardiovascular clinical endpoints after at least 18 months' follow-up. The study has combined the evaluation of two different statins, each with very different lipid-lowering potencies but potentially similar antiinflammatory effects with the testing of a quinolone with potential bactericidal antimicrobial activity. This study will potentially answer a number of questions regarding the importance of inflammation and infection in the pathogenesis of acute coronary syndromes in relationship to lipid-related risk. It is the first major clinical trial testing an agent other than azithromycin.

7. POTENTIAL CONCERNS

While we await for the results of the presently ongoing large clinical trials, several concerns still exist. First, can the antibiotic effectively eradicate the organism? Many of the proposed antibiotics are generally felt to be bacteriostatic rather than bactericidal. Chronic intracellular organisms such as *C. pneumoniae* may be able to survive even large doses of such therapeutic agents. Evidence of this potential was recently reported by Gieffers, *et al.*,[27] who demonstrated that *C. pneumoniae* may be able to survive and even thrive within monocytes treated with high-dose, but short-duration, azithromycin. It may therefore be necessary to use multidrug therapy in a manner similar to that required to eradicate tuberculosis.[28] Second, can we ascertain organism eradication in a specific patient? Serologic studies have not correlated well with the presence of intraplaque organisms. Evaluation of peripheral blood mononuclear cells is not standardized, and throat cultures also have not been found to be useful. Consequently, all presently existing clinical antibiotic secondary prevention trials have resorted to the use of major adverse cardiovascular events for primary clinical endpoints. Because of these limitations, there is still no clear understanding regarding which, how much, and how long antibiotics should be given. Further animal studies are presently going on that will hopefully improve the situation. Third, what will be the effect on antibiotic resistance of the generalized long-term use of antibiotics? The present level of antibiotic use throughout the world has already made many experts uneasy about the future of antibiotic resistance. There certainly is a potential of significantly worsening this problem by the implementation of a generalized secondary-prevention antibiotic program. This risk must be taken into account as the potential benefit of such a strategy is considered. Fourth, the hypothesis that infectious agents play a causative role in the development and progression of atherosclerosis remains just that, a hypothesis. Although positive secondary-prevention clinical benefit from the use of antibiotic therapy will strengthen the hypothesis, it is still not likely to prove it. Many antibiotics, especially macrolides, have potentially beneficial antiinflammatory effects independent of any antibiotic effect. Consequently, any positive clinical benefit cannot be necessarily ascribed to the removal of a causative infectious organism.

For these reasons we feel strongly that it is not yet appropriate to treat patients with coronary artery disease with antibiotic therapy as a secondary prevention measure. This significant change in our standard paradigm demands solid evidence.

REFERENCES

1. Grayston, I. T., Campbell, L. A., Kuo, C.-C., Mordhorst C. H., Saíkku p., Thom D. H., Wang S. P., 1990, A new respiratory tract pathogen: *Chlamydia pneumoniae* strain TWAR, *J. Infect. Dis.* **161**:618–625.
2. Beatty, W. L., Morrison, R. P., and Byrne, G. I., 1994, Persistent chlamydiae: From cell culture to a paradigm for chlamydial pathogenesis, *Microbiol. Rev.* **58**:686–699.
3. Muhlestein, J. B., Anderson, J. L., Hammond, E. F., Zhao, L., Trehan, S., Schwobe, E. P., and Carlquist, J. F., 1998, Infection with *Chlamydia pneumoniae* accelerates the development of atherosclerosis and treatment with azithromycin prevents it in a rabbit model, *Circulation* **97**:633–636.
4. Constantanides, P., Booth, J., and Carlson, G., 1960, Production of advanced cholesterol atherosclerosis in the rabbit, *Arch. Pathol.* **70**:80–92.
5. Fong, I. W., Chiu, B., Viira, E., Jang, D., and Mahony, J. B., 2002, Influence of clarithromycin on early atherosclerotic lesions after Chlamydia pneumoniae infection in a rabbit model, *Antimicrob. Agents Chemother.* **46**:2321–2326.
6. Rothstein, N. M., Quinn, T. C., Madico, G., Gaydos, C. A., and Lowenstein, C. J., 2001, Effect of azithromycin on murine arteriosclerosis exacerbated by *Chlamydia pneumoniae, J. Infect. Dis.* **183**:232–238.
7. Meier, C. R., Derby, L. E., Jick, S. S., Vasilakis, C., and Jick, H., 1999, Antibiotics and risk of subsequent first-time acute myocardial infarction, *JAMA* **281**:427–431.
8. Jackson, L. A., Smith, N. L., Heckbert, S. R., Grayston, J. T., Siscovick, D. S., and Psaty, B. M., 1999, Lack of association between first myocardial infarction and past use of erythromycin, tetracycline, or doxycycline, *Emerg. Infect. Dis.* **5**:281–284.
9. Herings, R. M. C., Leufkens, H. G. M., and Vandenbroucke, J. P., 2000, Acute myocardial infarction and prior antibiotic use, *JAMA* **284**:2990–2999.
10. Muhlestein, J. B., 2000, Chronic infection and coronary artery disease, *Med. Clin. N. Am.* **84**:123–148.
11. Sander, D., Winbeck, K., Klingelhofer, J., Etgen, T., and Conrad, B., 2002, Reduced progression of early carotid atherosclerosis after antibiotic treatment and *Chlamydia pneumoniae* seropositivity, *Circulation* **106**:2428–2433.
12. Wiesli, P., Czerwenka, W., Meniconi, A., Maly, F. E., Hoffmann, U., Vetter, W., and Schulthess, G., 2002, Roxithromycin treatment prevents progression of peripheral arterial occlusive disease in *Chlamydia pneumoniae* seropositive men: A randomized, double-blind, placebo-controlled trial, *Circulation* **105**:2646–2652.
13. Mosorin, M., Juvonen, J., Biancari, F., Satta, J., Surcel, H. M., Leinonen, M., Saikku, P., and Juvonen, T., 2001, Use of doxycycline to decrease the growth rate of abdominal aortic aneurysms: A randomized, double-blind, placebo-controlled pilot study, *J. Vasc. Surg.* **34**:606–610.
14. Parchure, N., Zouridakis, E. G., and Kaski, J. C., 2002, Effect of azithromycin treatment on endothelial function in patients with coronary artery disease and evidence of *Chlamydia pneumoniae* infection, *Circulation* **105**:1298–1303.
15. Gupta, S., Leatham, E. W., Carrington, D., Mendall, M. A., and Kaski, J. C., 1997,

Elevated *Chlamydia pneumoniae* antibodies, cardiovascular events, and azithromycin in male survivors of myocardial infarction, *Circulation* **96:**404–407.
16. Gurfinkel, E., Bozovich, G., Daroca, A., Beck, E., and Mautner, B., 1997, Randomized trial of roxithromycin in non–Q-wave coronary syndromes: ROXIS Pilot study. ROXIS Study Group, *Lancet* **350:**404–407.
17. Gurfinkel, E., Bozovich, G., Beck, E., Testa, E., Livellara, B., and Mautner, B., 1999, Treatment with the antibiotic roxithromycin in patients with acute non–Q-wave coronary syndromes. The final report of the ROXIS Study, *Eur. Heart J.* **20:**121–127.
18. Anderson, J. L., Muhlestein, J. B., Carlquist, J., Allen, A., Trehan, S., Nielson, C., Hall, S., Brady, J., Egger, M., Horne, B., and Lim, T., 1999, Randomized secondary prevention trial of azithromycin in patients with coronary artery disease and serological evidence for *Chlamydia pneumoniae* infection. The azithromycin in coronary artery disease: Elimination of myocardial infection with Chlamydia (ACADEMIC) study, *Circulation* **99:**1540–1547.
19. Muhlestein, J. B., Anderson, J. L., Carlquist, J. F., Salunkhe, K., Horne, B. D., Pearson, R. R., Bunch, J. T., Allen, A., Trehan, S., and Nielson, C., 2000, Randomized secondary prevention trial of azithromycin in patients with coronary artery disease: Primary clinical results of the ACADEMIC Study, *Circulation* **102:**1755–1760.
20. Neumann, F., Kastrati, A., Miethke, T., Pogatsa-Murray, G., Mehilli, J., Valina, C., Jogethaei, N., da Costa, C. P., Wagner, H., and Schomig, A., 2001, Treatment of Chlamydia pneumoniae infection with roxithromycin and effect on neointima proliferation after coronary stent placement (ISAR-3): A randomised, double-blind, placebo-controlled trial, *Lancet* **357:**2085–2089.
21. Stone, A. F. M., Mendall, M., Kaski, J.-C., Gupta, S., Camm, J., and Northfield, T., 2001, Antibiotics against *Chlamydia pneumoniae* and *Helicobacter pylori* reduce further cardiovascular events in patients with acute coronary syndromes, *J. Am. Coll. Cardiol.* **37**(2A)**:**1A–648A.
22. Sinisalo, J., Mattila, K., Valtonen, V., Anttonen, O., Juvonen, J., Melin, J., Vuorinen-Markkola, H., and Nieminen, M. S., 2002, The Clarithromycin in Acute Coronary Syndrome Patients in Finland (CLARIFY) Study Group. Effect of 3 months of antimicrobial treatment with clarithromycin in acute non–Q-wave coronary syndrome, *Circulation* **105:**1555–1560.
23. Dunne, M. W., 2000, Rationale and design of a secondary prevention trial of antibiotic use in patients after myocardial infarction: The WIZARD (weekly intervention with Zithromax [azithromycin] for atherosclerosis and its related disorders) trial, *J. Infect. Dis.* **181** (Suppl. 3)**:**S572–S578.
24. Coletta, A., Thackray, S., Nikitin, N., and Cleland, J. G., 2002, Clinical trials update: Highlights of the scientific sessions of The American College of Cardiology 2002: LIFE, DANAMI 2, MADIT-2, MIRACLE-ICD, OVERTURE, OCTAVE, ENABLE 1 & 2, CHRISTMAS, AFFIRM, RACE, WIZARD, AZACS, REMATCH, BNP trial and HARDBALL, *Eur. J. Heart Failure* **4:**381–388.
25. Cercek, B., Shah, P. K., Noc, M., Zahger, D., Zeymer, U., Matetzky, S., Maurer, G., Mahrer, P., and AZACS Investigators, 2003, Effect of short-term treatment with azithromycin on recurrent ischaemic events in patients with acute coronary syndrome in the Azithromycin in Acute Coronary Syndrome (AZACS) trial: A randomised controlled trial, *Lancet* **361:**809–813.
26. Jackson, L. A., 2000, Description and status of the azithromycin and coronary events study (ACES), *J. Infect. Dis.* **181** (Suppl. 3)**:**S579–S581.
27. Gieffers, J., Füüllgraf, H., Jahn, J., Klinger, M., Dalhoff, K., Katus, H. A., Solbach, W., and Maass, M., *Chlamydia pneumoniae* infection in circulating human monocytes is refractory to antibiotic treatment, *Circulation* **103:**351–356.
28. Bin, X. X., Wolf, K., Schaffner, T., and Malinverni, R., 2000, Effect of azithromycin plus rifampin versus amoxicillin alone on eradication and inflammation in the chronic course of *Chlamydia pneumoniae* pneumonitis in mice, *Antimicrob. Agents Chemother.* **44:**1761–1764.

13

Chlamydia pneumoniae and Myocarditis

HÅKAN G. GNARPE and JUDY A. GNARPE

1. INTRODUCTION

Myocarditis is an inflammation of the heart that often also involves the pericardium. It often appears in younger age groups <45 years of age. It is most often caused by viral agents, but bacterial and noninfectious causes are well known. Myocarditis or perimyocarditis is often mild, and a majority of cases appear as sequelae to other infections, e.g. pneumonia.

The clinical diagnosis of myocarditis or perimyocarditis can be made with reasonable certainty by electrocardiography and by measuring serum markers for myocyte damage, but the etiology is often difficult to establish. On rare occasions, a diagnosis is made from the histopathological analysis of endomyocardial biopsies. A minority of cases of myocarditis or perimyocarditis may develop into dilated cardiomyopathy, and the patient may then present with cardiac insufficiency. Viral agents, like the Coxsackie B_{1-5}, viruses are well-documented causes of myocarditis. There have been several published reports of *Chlamydia psittaci* as a cause of myocarditis, perimyocarditis, and endocarditis.[1,2] These reports were published long before it was found that the complement fixation test (CF) measured antibodies to the lipopolysaccharide antigen, which is shared by all three Chlamydiae clinically important for human infection: *C. psittaci, Chlamydia trachomatis*, and *Chlamydia pneumoniae*. Several of these reports were published before *C. pneumoniae* was defined as the third species.[3] A few of these early cases of myocarditis that have been attributed to *C. psittaci* may have been caused by *C. pneumoniae*, and this organism may well be the etiological agent for some of the patients reported by Sutton *et al.* in *Circulation*[1] and in *American Heart Journal*, 1971.[2]

HÅKAN G. GNARPE • Institute of Medical Sciences, University of Uppsala, Uppsala, Sweden. JUDY A. GNARPE • Institute of Medical Microbiology and Immunology, University of Alberta, Edmonton, Canada.

C. pneumoniae is by far the most common chlamydial species affecting man. Earlier reports that *C. pneumoniae* is uncommon in early childhood are based on the serological findings of microimmunofluorescence analysis (MIF), and MIF is only occasionally positive in children less than 6 years of age.(4) *C. pneumoniae* diagnosed by PCR is common in early childhood,(5) as shown in day care settings by Normann *et al.*(6) and in general practice.(7) Asymptomatic carriers have been reported.(8) More than 50% of adults have specific IgG antibodies to *C. pneumoniae* as a marker of exposure, and reinfection is common.(9,10) Most individuals will be exposed several times during their lives.(10) Specific IgM antibodies are often absent in reinfection; IgA antibodies seem to be an important immunoglobulin class to monitor in repeat infections.(11–15) IgA antibodies are shortlived, with a half-life in the circulation of about 5 to 7 days.(16) Therefore, increasing amounts of specific IgA antibodies have been suggested as markers of persistent or chronic *C. pneumoniae* infections in cases with cardiovascular(11) and respiratory tract diseases.(14,15) *C. pneumoniae* is a well-established cause not only of pneumonia(9,17,18) but also of acute bronchitis(13) and other infections of the respiratory tract.(14,15)

2. MYOCARDITIS

Very few clinical reports have been published about *C. pneumoniae* as a cause of myocarditis/perimyocarditis since *C. pneumoniae* was established as a distinct *Chlamydial* species(3) (see Table I). Most publications are case studies, where patients often have concomitant pneumonia or other complicating factors. The diagnosis has been made by serology in most cases; cultures for *C. pneumoniae* have been uniformly negative in all cases where culture was attempted. *C. pneumoniae* has been diagnosed by PCR in four individual cases, of which three had fatal outcomes. In two of these cases,(19,20) and in the fourth case,(21) the results of serology were suggestive of *C. pneumoniae* infection.

TABLE I
Reports of *C. pneumoniae* in Cases With Myocarditis, Perimyocarditis, and/or Endocarditis

Author(s)	Year of publication	Number of cases	Disease[a]	Concomitant disease(s)	Diagnosis made by		
					Serology	PCR	IHC
Wesslen *et al.*	1992	1	M		+	+	
Gran *et al.*	1993	1	M	Pneumonia, arthritis	+		
Tong *et al.*	1993	1	M		+		
Gnarpe *et al.*	1997	6/20	M		+		
Gnarpe *et al.*	1998	1	M, PM	Earlier pneumonia			+
Marrie *et al.*	1991	1	E		+		
Etienne *et al.*	1992	10	E		+		
Norton *et al.*	1995	1	E	Pneumonia	+		
Gdoura *et al.*	2002	1	E			+	
Song *et al.*	2001	26	DCM		+		

[a]M = myocarditis, PM = perimyocarditis, E = endocarditis, DCM = dilated cardiomyopathy.

Wesslen *et al.*[19] reported a series of six cases of myocarditis with fatal outcome among Swedish elite orienteers aged 17–29 between 1989 and 1992. All cases had degenerative changes in the myocardium in addition to lymphocytic infiltrates and patchy fibrosis but no obvious other pathological changes at autopsy. The last of the six cases, a 26-year-old man, was described in more detail. He had pneumonia a couple of months earlier with a cough that lasted almost 2 months, and did not train during this period. He died suddenly after a skiing tour with teammates a few months later.

The heart showed small foci of lymphocyte infiltration, degenerative changes within myocytes, and some focal fibrosis in the left ventricle and in the septum at autopsy. He had both specific *C. pneumoniae* IgM (1:64) and IgG (1:512) antibodies demonstrated by MIF and a low titer of IgM antibodies (1:16) to *C. psittaci*, which was judged as cross-reactive. PCR was positive for *C. pneumoniae* in specimens obtained from the septum and lung. Analyses of antibodies to a number of other possible viral and bacterial agents were negative.

Gran *et al.*[22] reported a case of pneumonia, myocarditis, and reactive arthritis in a 37-year-old male patient in 1993. The patient's illness started with a sore throat, malaise, and fever followed by erythema nodosum in the lower extremities and reactive arthritis of both ankles and the right wrist. He was admitted to a hospital and diagnosed with pneumonia and myocarditis in addition to arthritis. Echocardiography showed a dilated heart with decreased contractility. He had an increased ESR (78 mm/h) and his CRP was >200. Serology for *Yersinia enterocolitica*, Widal's test, and AST were negative, as was a test for HLA B27. Routine bacteriological cultures were negative for pathogenic bacteria. A complement fixation test showed a titer of 1:256 to Chlamydia and a MIF test for *C. pneumoniae* showed a specific IgG titer of 1:2048 and IgM of <1:8. Specific IgA antibodies were not investigated. He was given treatment with high dosages of i.v. penicillin and netilmicin. His clinical condition improved, and no clinical signs remained after 1 month, but he complained of fatigue when exposed to physical activity. The authors conclude that his condition was caused by *C. pneumoniae*. Although not mentioned, it is plausible that his myocarditis was secondary to reactive arthritis and pneumonia.

Tong and coworkers[23] reported a case of *C. pneumoniae* myocarditis, possibly in combination with influenza. The patient was an 18-year-old male who was admitted with vomiting, myalgia, and fever after participating in a 400-m race. His condition worsened with generalized convulsions; chest X-rays showed widespread unilateral shadowing, indicating consolidation or alveolar hemorrhage, but bronchoscopy did not reveal any hemorrhage. He required mechanical ventilation for several days and treatment for renal failure with hemofiltration. The management of the patient was complicated by repeated episodes of pulmonary edema. ECGs were abnormal. A wide variety of serological tests for viral and bacterial agents were negative, except a rising CFT titer to influenza A and to *Chlamydia* spp. Specific Chlamydia serology (MIF) showed a rising titer of specific IgG and positive IgM antibodies to *C. pneumoniae*.

The patient improved after being treated with i.v. erythromycin and cefotaxime. He was given a course of doxycycline to prevent relapse. Cardiac and renal functions were normalized at follow-up. The authors conclude that

C. pneumoniae was the likely cause of his myocarditis. They speculate that intense physical exercise may have been a triggering factor.

We have investigated a material of 20 male patients, 13–43 years of age (median 28, mean 27.5) with a clinical diagnosis of myocarditis ($n = 7$), perimyocarditis ($n = 9$), and pericarditis ($n = 4$).[24] All patients had ECGs suggestive of myocarditis. Sera from these patients were compared to sera from new, healthy male blood donors aged 18–43 (median 30, mean 31) ($n = 111$). All sera were analyzed using MIF for specific antibodies to *C. trachomatis* D-K, *C. psittaci*, and *C. pneumoniae.* All sera initially positive for IgM or IgA antibodies were absorbed with protein A–Sepharose CL-4B to remove possible interference from IgG and retested.[25] Thirty percent of the patients (6/20) had specific *C. pneumoniae* IgG antibodies in titers of 1:512 or more compared with 4.5% of blood donors (5/111). Sixty percent of the patients (13/20) had increased specific *C. pneumoniae* IgA antibodies compared with 8.1% (9/111) of the blood donors. The patients with myocarditis, perimyocarditis, or pericarditis had also much higher geometric mean titers (GMT) of specific antibodies to *C. pneumoniae* than the age- and sex-matched blood donors (363 vs. 107 for IgG; 65 vs. 13 for IgA). We concluded that *C. pneumoniae* was the likely cause for some of the cases.

We have also reported a case of perimyocarditis with multiorgan failure and a fatal outcome in a 51-year-old male teacher with a presumed year-long *C. pneumoniae* infection.[26] He was admitted to the hospital suffering from pronounced fatigue, hypotension, and a CRP value of >200 mg/l. He had bilateral pneumonia on admittance and an enlarged heart compared with chest X-ray 1 month earlier. He was given i.v. penicillin and fluid substitution because of low diuresis. He suddenly collapsed the next morning. During attempts at resuscitation, 500 ml of pericardial fluid was removed. Serum electrophoresis showed low C_3 and C_4, suggesting complement activation. Chlamydia serology (MIF) showed a specific *C. pneumoniae* IgG titer of 1:256 and IgA of 1:64 but no IgM; no specific antibodies to *C. trachomatis* or *C. psittaci* were seen. Serology for *Legionella* spp., Mycoplasma, CMV, EBV, and enteroviruses were all negative. Antinuclear factors, Goodpasture's antigen, c-ANCA, and p-ANCA were negative. Rheumatoid factor was not found.

The autopsy revealed a fibropurulent pericarditis with mild myocardial hypertrophy and mild to moderate coronary atherosclerosis. There was mild chronic portal inflammation of the liver. Immunohistochemical staining of specimens taken from the heart and lung revealed the presence of *C. pneumoniae* and a massive inflammatory infiltrate both in the heart and in the lung.

3. ENDOCARDITIS CASES

Marrie and coworkers reported a 59-year-old male patient with culture-negative endocarditis and aortic stenosis.[27] The patient had chills and a productive cough for 1 month preceding admission to the hospital. Serological investigations were negative or showed no rise over 3 weeks to Hepatitis B surface Ag and Ab, CMV, *Coxiella burnetii*, EBV, *Brucella* sp., *Mycoplasma pneumoniae* and *Borrelia burgdorferi*. When serum was tested with complement fixation test to Chlamydia group antigen, the titer was 1:1280. When the sera were investigated using MIF over a period of 121 days, the *C. pneumoniae*–specific IgG antibody titer had decreased from 1:256 to 1:128, and the specific *C. pneumoniae* IgM antibody titer had declined from 1:128 at 60 days to 0 at 121 days. The sera had no specific antibodies to *C. trachomatis* or to nine different strains of *C. psittaci*.

The patient was given tetracycline treatment and was afebrile in 1 week. He was discharged and given doxycycline, 100 mg twice daily to prevent a relapse. The patients' aortic valve was replaced after 3 months. The valvular material was thickened and fibrotic. There was marked destruction and distortion of the valve when examined histologically. A mixed infiltrate of lymphocytes, plasma cells, eosinophils, and a few granulocytes were seen together with large macrophages and active fibroblasts. The authors suggest that *C. pneumoniae* was the causative agent, although no organism could be detected or cultured.

Etienne and coworkers reported a series of 10 cases of Chlamydial endocarditis collected between 1983 and 1990 in 1992.[28] The cases were all men with a mean age of 42 (26–59). None had a history of bird contacts. They had all been symptomatic for at least 2 months with weight loss, anorexia, and fever (8/10). Hemodynamic failure was found in seven patients and neurological signs in four. Repeated blood cultures were negative in all. Complement fixation tests for *Chlamydia* spp. were positive in 6 of 10 cases and MIF showed cross-reacting antibodies in all nine cases studied, with transient IgM antibodies in six cases. IgG titers varied from 16 to 2048 for *C. trachomatis* and *C. psittaci*, and from 64 to 4096 for *C. pneumoniae*. IgA was found in 5 of 10 cases but no titer values were reported. Serology for *C. burnetii*, Candida, Aspergillus, *B. burgdorferi*, *M. pneumoniae*, and Brucella were all negative. Six of the 10 patients had preexisting cardiac anomalies. Infection involved the aortic valve in seven patients, the mitral valve in one, and both in two. Nine of the 10 patients had vegetations identified by echocardiography. All patients required valve replacement for cardiac insufficiency. Three patients died within 8 months of surgery.

The replaced heart valves were fibrotic, with focal heavy infiltrations of macrophages, plasmocytes, lymphocytes, and a few granulocytes. There were necrotic areas in seven cases. With the Machiavello stain, red granulae could be seen within the macrophages. These granulae stained positive with a monoclonal antibody to *Chlamydia* spp.

The authors conclude that *C. pneumoniae* was the most likely etiologic agent as none of the patients had been exposed to a sick bird, but state that direct culture from the valves may be the only way to establish a true etiology.

In another case report, Norton *et al.* describe a patient with pneumonia complicated by endocarditis, disseminated intravascular coagulation, and multiorgan failure terminating in a fatal outcome.[29] A tracheal aspirate was negative for seven respiratory viruses and for *M. pneumoniae*; serology was negative for Q-fever, cytomegalovirus, *Legionella pneumophila*, and *Legionella longbeachae*. Chlamydia serology showed an increase in specific IgG antibodies to *C. pneumoniae* from 1:64 to 1:1024 over a 5-day period. Chlamydia antigen was found in the lung at postmortem examination, but no other bacterial or viral agents were detected. Fibrinous vegetations on the tricuspid, mitral, and aortic valves were present but immunofluorescent staining of valvular material for *C. pneumoniae* with a species-specific monoclonal antibody as well as with a genus-specific monoclonal antibody were negative. No other bacterial or viral agents were detected, and the authors suggested that *C. pneumoniae* could have been the causative agent for the pneumonia and the endocarditis.

Gdoura *et al.*[21] report a case of culture-negative endocarditis in a 54-year-old woman who was clinically diagnosed with endocarditis resulting in a need for replacement of both the mitral and aortic valves. Sera were obtained at admission and 2 and 4 weeks later. The titer of specific *C. pneumoniae* IgG antibodies were high in all three sera (>1:4096 with MIF); all three sera as well were positive for specific *C. pneumoniae* IgM antibodies. The titer of specific IgM antibodies declined later during the observation period. Sera were tested for antibodies to *L. pneumophila*, *M. pneumoniae*, and *C. burnetii*, but were negative. The mitral and aortic valves were surgically replaced and showed marked destruction with calcification, fibrosis, and an inflammatory infiltrate. Both the mitral and aortic valves were positive for *C. pneumoniae* by PCR but negative for *C. trachomatis* and *C. psittaci*. *In situ* hybridization was positive for *C. pneumoniae*. The authors conclude that the patients' endocarditis with resulting need for valve replacement was caused by *C. pneumoniae* as indicated by the serological findings and the demonstration of *C. pneumoniae* DNA in the resected mitral and aortic valves.

4. DILATED CARDIOMYOPATHY

Myocarditis has occasionally been shown to be associated with the development of dilated cardiomyopathy (DCM) with dysfunction of the heart.

In a short report Song *et al.* describe a patient material of 26 consecutive patients with DCM and compared these with 28 age- and sex-matched controls regarding markers for chronic *C. pneumoniae* infection.[30] They found that there was a significant difference both in the amount and prevalence of specific *C. pneumoniae* IgA antibodies but not of specific IgG antibodies between patients and controls. They also showed that the serum concentrations of CRP and of fibrinogen were significantly higher in DCM patients indicating ongoing inflammatory activity. As persistence of the short-lived specific IgA antibodies may be taken as an indication of a persistent infection, they conclude that a chronic *C. pneumoniae* infection was the etiology for some of the DCM patients.

5. ANIMAL MODELS OF MYOCARDITIS

There is very little information available about *C. pneumoniae* and myocarditis in animal experimental settings. Most animal models are focused on the possible role of Chlamydia in the etiology of atherosclerosis[31–33].

However, Bachmeier and his group reported in *Science* (1999) that there is an association between *C. pneumoniae* and induced autoimmune murine myocarditis.[34] A 60-kDa cysteine-rich outer membrane protein in *C. pneumoniae* has a sequence homology to a peptide from the murine heart-muscle-specific α-myosin heavy chain. When the Chlamydia 60-kDa outer-membrane-derived peptides were injected into mice, an autoimmune myocarditis developed with perivascular inflammation, fibrotic changes, and blood vessel occlusion in the heart. T and B cells were also induced to react to the homologous endogenous heart-muscle-specific peptide. The authors also demonstrated that Chlamydia DNA functioned as an adjuvant to trigger the autoimmune, peptide-induced inflammatory heart disease. They conclude that bacteria of the genus Chlamydia could mediate inflammatory heart disease through antigen mimicry.

In experimental autoimmune myocarditis, the production of monocyte chemoattractant protein 1 is enhanced,(35) which in turn increases the migration of monocytes and T cells to an inflammatory focus.

6. POSSIBLE MECHANISMS

C. pneumoniae causes infections of the respiratory tract. It is likely that some of the infectious elementary bodies are transported down to the alveoli and are there phagocytosed by alveolar macrophages. *C. pneumoniae*–infected monocytes may differentiate into macrophages as reported by Yamaguchi *et al.*,(36) and these cells may enter the circulation. In animal models, *C. pneumoniae* has been shown to be disseminated through the bloodstream after nasal inoculation,(37) and there are numerous reports about the detection of and even isolation of *C. pneumoniae* in atherosclerotic blood vessels.(38–40)

C. pneumoniae has, like other species of Chlamydiae, a tendency to cause chronic or persistent infection. *C. trachomatis* has been shown inhibiting apoptosis in Chlamydia-infected cells by inhibition of activation downstream of caspase 3 and cleavage of poly(ADP-ribose) polymerase as well as preventing release of mitochondrial cytochrome c.(41) It was recently shown that *C. pneumoniae* has a similar effect.(42) This inhibition of apoptosis is in part due to induction of IL-10.(43) Rajalingam *et al.* have also reported that epithelial cells infected with *C. pneumoniae* are resistant to apoptosis.(44)

C. pneumoniae–infected monocytes have been shown to exhibit increased adherence to human aortic endothelial cells(45) and to human coronary artery endothelial cells.(46) *C. pneumoniae* can infect human macrophages, endothelial cells, and aortic smooth muscle cells, and many strains are capable of replication in these cells.(47) It was recently shown by Molestina *et al.* that endothelial cells infected with *C. pneumoniae* stimulated transendothelial migration of neutrophil granulocytes and monocytes.(48) Vehmaan-Kreula *et al.* reported that adherent *C. pneumoniae* can stimulate the production of a 92-kDa gelatinase from macrophages, which promotes the destruction of extracellular matrix and facilitates tissue invasion by inflammatory cells.(49)

Macrophages infected with *C. pneumoniae* in the presence of LDL undergo foam cell degeneration(50) and the component causing this was found to be chlamydial lipopolysaccharide.(51) It has recently been shown that *C. pneumoniae* and chlamydial Hsp60 can induce cellular oxidation of LDL to a highly atherogenic form.(52) Chlamydial Hsp60 seems to be an important stimulus for inflammatory reactivity in infections with *C. pneumoniae*.(53,54)

C. pneumoniae is a common finding in arteries with atherosclerotic changes but not in normal arteries[(39)] and has been cultured from coronary arteries,[(38)] aortic aneurysms,[(40)] and an explanted heart.[(55)] Although *C. pneumoniae* is present in the coronary arteries of many patients, very few individuals develop myocarditis, perimyocarditis, or endocarditis. One possible explanation may be that only specific strains of *C. pneumoniae* have an antigenic profile that may induce the development of autoimmune myocarditis similar to the myocarditis demonstrated in a murine model by Bachmeier *et al.* It is known that there is a certain antigenic variation in different isolates of *C. pneumoniae,*[(56,57)] but it is not known whether these differences have any clinical importance.

Another possible mechanism to explain why so few infected patients develop myocarditis may be the absence or presence of specific host factor(s) that may render individual patients more susceptible for the development of myocarditis. Individual differences in susceptibility to *C. pneumoniae* infection has been shown by Haranga *et al.,*[(58)] who found that *C. pneumoniae* can infect and multiply *in vitro* in human lymphocytes but only 4 of 11 donor's lymphocytes supported infection. Thus there seems to be individual host variability in the susceptibility for *C. pneumoniae* infection. Recently, Rugonfalvi-Kiss *et al.*[59] reported that severe cardiovascular disease may develop in individuals with a variation in the mannose-binding gene.

We suggest that the reason why there are so few reported cases with *C. pneumoniae*–induced myocarditis could possibly be that there is a strain specificity for development of disease in individuals lacking specific receptors or with a putative genetic disposition for development of myocardial disease.

7. SUMMARY

Myocarditis, perimyocarditis, or endocarditis caused by *C. pneumoniae* seems to be unusual in clinical practice. Only a few cases have been reported to date. *C. pneumoniae* is a ubiquitous microbe: over 50% of the adult population has been exposed. Most of the mechanisms necessary to provide access to the myocardium are present in cases with low-grade chronic infection with *C. pneumoniae.* There are antigenic differences between different strains of *C. pneumoniae,* but it has still to be shown in experimental models if some strains are more prone to invade the myocardium. In all probability, individual host factor(s) might exist, making certain individuals more prone to develop myocarditis, perimyocarditis, or endocarditis when infected by *C. pneumoniae.* There seems to be no direct association between the development of myocarditis and atherosclerosis even if most of the necessary risk factors are present.

REFERENCES

1. Sutton, G. C., Morrissey, R. A., Tobin, J. R., *et al.*, 1967, Pericardial and myocardial disease associated with evidence of infection by agents of the Psittacosis-Lymphogranuloma venereum group (Chlamydiaceae), *Circulation* **36:**830–838.

2. Sutton, G. C., Demakis, J. A., Anderson, T. O., and Morrissey, R. A., 1971, Serologic evidence of a sporadic outbreak in Illinois of infection by Chlamydia (psittacosis-LGV agent) in patients with primary myocardial disease and respiratory disease, *Am. Heart J.* **81:**597–607.
3. Grayston, J. T., Kuo, C.-C., Campbell, L. A., and Wang, S. O. P., 1989, *Chlamydia pneumoniae* sp. Nov. for *Chlamydia* sp. Strain TWAR, *Int. J. Syst. Bacteriol.* **39:**88–90.
4. Lund-Olsen, I., Lundbäck, A., Gnarpe, J., and Gnarpe, H., 1994, Prevalence of specific antibodies to *Chlamydia pneumoniae* in children with acute respiratory infections, *Acta Pediatr.* **83:**1143–1145.
5. Normann, E., Gnarpe, J., Gnarpe, H., and Wettergren, B., 1998, *Chlamydia pneumoniae* in children with acute respiratory infections, *Acta Paediatr.* **87:**23–27.
6. Normann, E., Gnarpe, J., Gnarpe, H., and Wettergren, B., 1988, *Chlamydia pneumoniae* in children attending day-care centers in Gävle, Sweden, *Pediatr. Infect. Dis. J.* **17:**474–478.
7. Falck, G., Gnarpe, J., and Gnarpe, H., 1997, Prevalence of *Chlamydia pneumoniae* in healthy children and in children with respiratory tract infection, *Pediatr. Infect. Dis. J.* **16:**549–554.
8. Gnarpe, J., Gnarpe, H., and Sundelöf, B., 1991, Endemic prevalence of *Chlamydia pneumoniae* in subjectively healthy persons, *Scand. J. Infect. Dis.* **23:**387–388.
9. Grayston, J. T., Kuo, C.-C., Wang, S.-P., and Altman, J., 1986, A new *Chlamydia psittaci* strain, TWAR, isolated in acute respiratory tract infections, *N. Engl. J. Med.* **315:**161–168.
10. Grayston, J. T., 1992, *Chlamydia pneumoniae* strain TWAR pneumonia, *Ann. Intern. Med.* **43:**317–323.
11. Saikku, P., Leinonen, M., Tenkanen, L., *et al.*, 1992, Chronic *Chlamydia pneumoniae* infection as a risk factor for coronary heart disease, *Ann. Intern. Med.* **116:**273–278.
12. Elkind, M. C. V., Liu, I.-F., Grayston, J. T., and Sacco, R. L., 2000, *Chlamydia pneumoniae* and the risk of first ischemic stroke, *Stroke* **31:**1521–1525.
13. Falck, G., Heyman, L., Gnarpe, J., and Gnarpe, H., 1994, *Chlamydia pneumoniae* (TWAR): A common agent in acute bronchitis, *Scand. J. Infect. Dis.* **26:**179–186.
14. Falck, G., Gnarpe, J., and Gnarpe, H., 1996, Persistent *Chlamydia pneumoniae* infection in a Swedish family, *Scand. J. Infect. Dis.* **28:**271–273.
15. Falck, G., Engstrand, I., Gad, A., Gnarpe, J., *et al.*, 1997, Demonstration of *Chlamydia pneumoniae* in patients with chronic pharyngitis, *Scand. J. Infect. Dis.* **29:**585–589.
16. Tomasi, T. B. and Grey, H. M., 1972, Structure and function of immunoglobulin A, *Progr. Allergy* **16:**181–213.
17. Marrie, T. J., Grayston, J. T., Wang, S.-P., and Kuo, C.-C., 1987, Pneumonia associated with the TWAR strain of Chlamydia, *Ann. Int. Med.* **106:**507–511.
18. Sundelöf, B., Gnarpe, J., Gnarpe, H., *et al.*, 1993, *Chlamydia pneumoniae* in Swedish patients, *Scand. J. Infect. Dis.* **25:**429–433.
19. Wesslen, L., Påhlson, C., Friman, G., *et al.*, 1992, Myocarditis caused by *Chlamydia pneumoniae* (TWAR) and sudden unexpected death in a Swedish elite orienteer, *Lancet* **340:** 427–428.
20. Schaad, H. J., Malinverni, R., Campbell, L. A., and Matter, L., 1999, Myocardial infarction, culture-negative endocarditis and *Chlamydia pneumoniae* infection: A dilemma? *Clin. Infect. Dis.* **28:**162–163.
21. Gdoura, R., *et al.*, 2002, Culture-negative endocarditis due to *Chlamydia pneumoniae*, *J. Clin. Microb.* **40:**718–720.
22. Gran, J. T., Hjetland, R., and Andreassen, A. H., 1993, Pneumonia, myocarditis and reactive arthritis due to *Chlamydia pneumoniae*, *Scand. J. Rheumatol.* **22:**43–44.
23. Tong, C. Y. W., Potter, F., Worthington, E., and Mullins, P., 1995, *Chlamydia pneumoniae* myocarditis, *Lancet* **346:**710–711.
24. Gnarpe, H., Gnarpe, J., Gästrin, B., and Hallander, H., 1997, *Chlamydia pneumoniae* and myocarditis, *Scand. J. Infect. Dis.* **104**(Suppl.)**:**50–52.

25. Jaukiainen, T., Tuomi, T., Leinonen, M., *et al.*, 1994, Interference of immunoglobulin G antibodies in IgA antibody determinations for *Chlamydia pneumoniae* by immunofluorescence test, *J. Clin. Microb.* **32:**839–840.
26. Gnarpe, J., Gnarpe, H., Nissen, K., *et al.*, 1998, *Chlamydia pneumoniae* infection associated with multi-organ failure and fatal outcome in a previously healthy patient, *Scand. J. Infect. Dis.* **30:**523–524.
27. Marrie, T. J., Harczy, M., Mann, O. E., *et al.*, 1990, Culture-negative endocarditis probably due to *Chlamydia pneumoniae, J. Infect. Dis.* **161:**127–129.
28. Etienne, J., Ory, D., Thouvenot, D., *et al.*, 1992, Chlamydial endocarditis: A report of ten cases, *Eur. Heart J.* **13:**1422–1426.
29. Norton, R., *et al.*, 1995, *Chlamydia pneumoniae* with endocarditis, *Lancet* **345:**1376–1377.
30. Song, H., *et al.*, 2001, Dilated cardiomyopathy and *Chlamydia pneumoniae* infection, *Heart* **86:**456–458.
31. Laitinen, K., Laurila, A., Pyhälä, L., *et al.*, 1997, *Chlamydia pneumoniae* infection induces inflammatory changes in the aortas of rabbits, *Infect. Immun.* **65:**4832–4835.
32. Moazed, T. C., Kuo, C.-C., Grayston, J. T., and Campbell, L. A., 1997, Murine models of *Chlamydia pneumoniae* infection and atherosclerosis, *J. Infect. Dis.* **175:**883–890.
33. Fong, I. W., Chiu, B., Viira, E., Fong, M. W., *et al.*, 1999, *De novo* induction of atherosclerosis by *Chlamydia pneumoniae* in a rabbit model, *Infect. Immun.* **67:**6048–6055.
34. Bachmeier, K., Neu, N., de la Maza, L. M., *et al.*, 1999, Chlamydia infections and heart disease linked through antigenic mimicry, *Science* **283:**1335–1339.
35. Fuse, K., Kodama, M., Hanawa, H., *et al.*, 2001, Enhanced expression and production of monocyte chemoattractant protein-1 in myocarditis, *Clin. Exp. Immunol.* **124:**346–352.
36. Yamaguchi, H., Haranga, S., Widen, R., *et al.*, 2002, *Chlamydia pneumoniae* infection induces differentiation of monocytes into macrophages, *Infect. Immun.* **70:**2392–2398.
37. Yang, Z., Kuo, C.-C., and Grayston, J. T., 1995, Systemic dissemination of *Chlamydia pneumoniae* following intranasal inoculation in mice, *J. Infect. Dis.* **171:**736–738.
38. Maass, M., Bartels, C., Engel, P. M., *et al.*, 1998, Endovascular presence of viable *Chlamydia pneumoniae* is a common phenomenon in coronary artery disease, *JAAC* **31:**827–832.
39. Taylor-Robinson, D., Thomas, B. J., Goldin, R., and Stanbridge, R., 2002, *Chlamydia pneumoniae* in infrequently examined blood vessels, *J. Clin. Pathol.* **55:**218–220.
40. Karlssson, L., Gnarpe, J., Nääs, J., *et al.*, 2000, Detection of viable *Chlamydia pneumoniae* in abdominal aortic aneurysms, *Eur. J. Vasc. Endovasc. Surg.* **19:**630–635.
41. Fan, T., Lu, H., Hu, H., Shi, L., *et al.*, 1998, Inhibition of apoptosis in Chlamydia-infected cells: Blockade of mitochondrial cytochrome c release and caspase activation, *J. Exp. Med.* **187:**487–496.
42. Fischer, S., Schwarz, C., Vier, J., and Häcker, G., 2001, Characterization of antiapoptotic activities of *Chlamydia pneumoniae* in human cells, *Infect. Immun.* **69:**7121–7129.
43. Geng, Y., Shane, R. B., Berensci, K., *et al.*, 2000, *Chlamydia pneumoniae* inhibits apoptosis in human peripheral blood mononuclear cells through induction of IL-10, *J. Immun.* **164:**5522–5529.
44. Rajalingam, K., Al-Younes, H., Muller, A., *et al.*, 2001, Epithelial cells infected with Chlamydophila pneumoniae (*Chlamydia pneumoniae*) are resistant to apoptosis, *Infect. Immun.* **69:**7880–7888.
45. Kalayoglu, M. V., Perkins, B. N., and Byrne, G. I., 2001, *Chlamydia pneumoniae*–infected monocytes exhibit increased adherence to human aortic endothelial cells, *Microb. Infect.* **3:**963–969.
46. Kaul, R., and Wenman, W., 2001, *Chlamydia pneumoniae* facilitates monocyte adhesion to endothelial and smooth muscle cells, *Microb. Pathogen.* **30:**149–155.
47. Gaydos, C. A., Summersgill, J. T., and Sahney, N. N., 1996, Replication of *Chlamydia pneumoniae in vitro* in human macrophages, endothelial cells and aortic artery smooth muscle cells, *Infect. Immun.* **64:**1614–1620.

48. Molestina, R. E., Miller, R. D., Ramirez, J. A., and Summersgill, J. T., 1999, Infection of human endothelial cells with *Chlamydia pneumoniae* stimulates transendothelial migration of neutrophils and monocytes, *Infect. Immun.* **67:**1323–1330.
49. Vehmaan-Kreula, P., Poulakkainen, M., Sarvas, M., *et al.*, 2001, *Chlamydia pneumoniae* proteins induce secretion of the 92-kDa gelatinase by human monocyte–derived macrophages, *Arterioscler. Thromb. Vasc. Biol.* **21:**e1–e8.
50. Kalayoglu, M. V., and Byrne, G. I., 1998, Induction of macrophage foam cell formation by *Chlamydia pneumoniae*, *J. Infect. Dis.* **177:**725–729.
51. Kalayoglu, M. V., and Byrne, G. I., 1998, A *Chlamydia pneumoniae* component that induces macrophage foam cell formation is chlamydial lipopolysaccharide, *Infect. Immun.* **66:**5067–5072.
52. Kalayoglu, M. V., Hoerneman, B., LaVerda, D., *et al.*, 1999, Cellular oxidation of low-density lipoprotein by *Chlamydia pneumoniae*, *J. Infect. Dis.* **180:**780–790.
53. Prazeres da Costa, C., Kirschning, J., Busch, D., *et al.*, 2002, Role of heat shock protein 60 in the stimulation of innate immune cells by *Chlamydia pneumoniae*, *Eur. J. Immunol.* **32:**2460–2470.
54. Mayr, M., Metzler, B., Kiechl, S., *et al.*, 1999, Endothelial cytotoxicity mediated by serum antibodies to heat shock proteins of *Escherichia coli* and *Chlamydia pneumoniae*, *Circulation* **99:**1560–1566.
55. Ramirez, J. A., 1996, Isolation of *Chlamydia pneumoniae* from the coronary artery of a patient with coronary atherosclerosis, *Ann. Intern. Med.* **125:**979–982.
56. Black, C. M., Johnson, J. E., Farshay, C. E., *et al.*, 1991, Antigenic variation among strains of *Chlamydia pneumoniae*, *J. Clin. Microb.* **29:**1312–1316.
57. Jantos, C. A., Heck, S., Roggendorf, R., *et al.*, 1997, Antigenic and molecular analyses of different *Chlamydia pneumoniae* strains, *J. Clin. Microb.* **35:**620–623.
58. Haranaga, S., Yamaguchi, H., Friedman, H., *et al.*, 2001, *Chlamydia pneumoniae* infects and multiplies in lymphocytes *in vitro*, *Infect. Immun.* **69:**7753–7759.
59. Rugonfalvi-Kiss, S., Endresz, V., Madsen, H. O., *et al.*, 2002, Association of *Chlamydia pneumoniae* with coronary artery disease and its progression is dependent on the modifying effect of mannose-binding lectin, *Circulation* **106:**1071–1076.

14

Chlamydia pneumoniae as a Candidate Pathogen in Multiple Sclerosis

CHARLES W. STRATTON and SUBRAMANIAM SRIRAM

1. INTRODUCTION

Multiple sclerosis (MS) is the most common demyelinating disease of the central nervous system (CNS).[1,2] It commonly affects adults between the ages of 18 and 50 years, with the mean onset of initial symptoms being 29 years of age. In the general population, the prevalence of MS is approximately 1/1000. Women are generally affected more commonly than men. Most MS patients initially present with a relapsing remitting course. This relapsing remitting course slowly evolves into a progressive disease. The exact pathogenesis of MS remains unknown, which adversely affects attempts at medical intervention. The following discussion reevaluates the current hypothesis concerning autoimmunity in MS and describes how an infectious process may influence the development of MS and proposes that *C. pneumoniae* is a likely candidate pathogen in the development of MS.[3–6]

2. AUTOIMMUNE BASIS FOR MS

Current opinion favors the hypothesis that MS is an autoimmune disease of the CNS.[2,7] Autoimmunity is presumed to result from self-reactivity to neural antigens, leading to an inflammatory response in the CNS with attendant tissue injury. This autoimmune hypothesis for MS is largely based on studies done in various experimental animal models. These studies began in 1933 when Rivers

CHARLES W. STRATTON • Department of Pathology, Vanderbilt School of Medicine, Nashville, Tennessee. SUBRAMANIAM SRIRAM • Department of Neurology, Vanderbilt School of Medicine, Nashville, Tennessee.

and colleagues induced an acute paralytic disease in monkeys following the injection of whole brain homogenate and thus introduced the term "experimental allergic encephalitis" (EAE).[8] "EAE" has since been modified to mean "experimental autoimmune encephalomyelitis" to reflect more clearly the underlying immune mechanism of disease produced in these animal models. The first encephalitogenic component of myelin to be isolated and purified was myelin basic protein (MBP). Subsequently, it was demonstrated that EAE is induced in animals following immunization with other myelin antigens such as myelin oligodendrocyte glycoprotein (MOG) and proteolipid protein (PLP).

3. MS AS AN AUTOIMMUNE DISEASE: UNRESOLVED ISSUES

Although current opinion favors MS to be an autoimmune disease in humans that is similar to EAE in animals, evidence of autoimmunity in MS is not without controversy.[9,10] If MS is truly an autoimmune disease, it should be possible to identify the neural antigens that are responsible for the autoimmune process. In EAE, the encephalitogenic neural antigens capable of inducing disease are well defined. In contrast, the encephalitogenic antigen(s) in MS remain unknown. Some MS patients, but not all, do show an increase in the number of circulating T cells that recognize different MBP peptides.[7,11,12] However, similar reactivity to MBP is seen in healthy volunteers. Attempts to show the reactivity of lymphocytes present in the cerebrospinal fluid (CSF) to MBP in MS patients has met with mixed results. Numerous attempts to show the presence of MBP-reactive T lymphocytes in brain tissue from MS patients have been equally unrewarding. Thus, the consistent identification in MS of reactive encephalitogenic neural antigens as well as autoimmune cells in peripheral blood and in CNS tissue remains an unresolved issue.

Another crucial element that discounts the role of myelin antigen as a principal encephalitogenic neural antigen in MS is the absence of antibodies to myelin antigen in the CSF of MS patients. Although MS patients show an increase in the levels of immunoglobulins in the CSF, none of these immunoglobulins in the CSF react to any known myelin or other neural antigen. This is in striking contrast to EAE, in which antibodies to myelin antigens are seen in the CSF, as well as to other CNS inflammatory infectious diseases in which oligoclonal bands show reactivity to antigens of the specific infectious pathogen.[13–16]

A comparison of EAE with MS must also include the examination of the "immune profile" of the inflammatory process in both the CSF and the CNS. Since EAE is a Th1 autoimmune disease mediated by CD4 cells, it would be predicted that a similar immune phenotype would be seen in MS. However, in distinct contrast to EAE, the CNS pathology in MS shows a marked activation of macrophages, with over 80% of the cellular phenotype consisting of activated macrophages and the presence of CD4 T cells not being absolute.[17,18]

Cytokines produced by an autoimmune process have been thought to be important in the pathogenesis of MS. Preliminary evidence supporting this notion came about when an increase in clinical relapses of MS was seen in patients who received gamma interferon. This was, however, a small study done prior to the advent of modern imaging technology, and it is not clear if there was indeed an increase in the number of inflammatory CNS lesions.[19] In spite of extensive analysis, there is no clear consensus for the presence of a well-defined cytokine/immune profile in either CNS tissues or in CSF from MS patients. Every conceivable cytokine that may play a role in inflammation has been found in the CNS/CSF of patients with MS. Another cytokine that has received attention is TNF, since levels of TNF appeared to correlate with disease severity and TNF is toxic to oligodendrocytes.[20,21] However, such guilt by association has not proven to be true for TNF, as *in vivo* treatment with anti-TNF antibodies actually worsened MS.[22,23]

Because current opinion suggests that MS is an autoimmune disease, specific and nonspecific immunological agents have been used therapeutically in MS. Broad-spectrum immunosuppressive agents continue to be used today; recently mitoxantrone has been approved for the treatment of secondary progressive MS. Unlike other immunological diseases such as myasthenia gravis (an autoimmune disease with a clearly defined autoantigen), the efficacy of virtually all immuosuppressive regimens in MS has either failed or has met with only limited success.[24,25] For example, if MS is a T cell–mediated disease caused by CD4 T cells, we would have expected to see an improvement in patients who received anti-CD4 antibody therapy. Clinical studies have failed to show benefit in patients receiving anti-CD4 antibodies in spite of significant reduction in the number of CD4-positive T cells.[22,26]

In summary, there are a number of issues that challenge the autoimmune hypothesis for MS. In particular, these observations raise questions as to the usefulness of EAE as a model of MS and whether successful therapies in MS should be based on the ability to "cure" EAE.

4. INFECTION AS AN ALTERNATE HYPOTHESIS FOR MS

Most chronic diseases are either caused by genetic defects or occur in response to environmental agents. There has been considerable evaluation of the genetic component of MS but as yet no definitive gene has been identified. Family studies have also not shown any clear pattern of inheritance. This suggests that MS is unlikely to be due primarily to a genetic defect. Hence, the role of environment and by implication an infectious agent is attractive.

One of the tenets of evolutionary biology is the role of fitness on disease. The essential feature of this theory is that all genetic diseases will have a negative impact on the fitness of the individual and will therefore eventually disappear. The higher the negative impact on fitness the earlier will be the deletion of the defective allele from the population. Hence, any disease that has been maintained for extended periods of time in a population is very likely to result from an environmental cause and is likely to be infectious.[27] A corollary to this view is that if a genetic defect is known to be present for many generations, then it has quite likely conferred a benefit on the fitness of the individual. In most cases, this involves protection against an infection (e.g., the protection of sickle cell disease from malaria). The association of infection with genetic factors is not surprising, as it is well accepted that host genes can regulate the expression of infectious disease(s). Development of tuberculosis, malaria, and HIV are all modified by host genetic factors.[28]

Historical descriptions of disease that have features of MS have been known for over 300 years, although MS was described as a clinical entity only 125 years ago. The tenets of evolutionary fitness would suggest that MS is caused by environmental agents and, therefore, most likely is infectious. MS has been found to have an uneven geographical distribution in a north–south gradient of declining incidence.[10,29] It has been thought that some of this variation may be due to underlying racial differences: the highly susceptible populations such as Scandinavians settled primarily in the northern United States. However, in Australia the racial makeup of their population is largely homogeneous south to north.[30]

Migration studies provide additional support that an environmental agent may play a role in the development of MS. These studies have demonstrated that immigrants tend to have the same risk of developing MS as the native population.[31] This has been seen for migration both to and away from high- and low-risk regions. These studies suggest that migration needs to occur at or before age 15 (around puberty) for migrants to have the prevalence risk of the indigenous population. In some cases, ethnicity does appear to confer a protective effect for developing MS. Native Japanese have a very low prevalence of MS (2/100,000) as compared to the Japanese in Hawaii (7/100,000). The rate for Caucasian Hawaiians was 10.5/100,000 and for white immigrants 34/100,000. In California, the rate for people of Japanese descent remained 7/100,000 despite the Caucasian rate of 30/100,000. Although the risk of developing MS for Japanese is higher after immigration to an area with greater endemic MS, there appears to be a limit to the proportion of individuals who can develop MS.[32]

Another line of epidemiological evidence supports the contention of MS being caused by an infectious agent. There have been several MS "epidemics" described. The best studied has been in the Faroe Islands population. Prior to World War II, there were no reported cases of MS in the Faroe Islands. From 1943 to 1982, 46 cases were identified, and additional 4 cases had onset of their illness in the late 1980s. These data suggested the existence of four successive waves of MS incidence thought to have started with the stationing of 8,000 British troops on the islands during World War II.[29] The authors suggest that the British troops were responsible for exposing the Faroese people to the primary MS affection (PMSA) that led to clinical/neurological MS (CNMS). PMSA was thought to be a common but specific infectious agent that rarely resulted in CNMS. The population was thought to be susceptible following 2 years of exposure and when between the ages of 11 and 45. There was thought to be a 6-year incubation period. The following diminishing waves of MS incidence were due to Faroe Islanders previously exposed to PMSA but not necessarily having CNMS.

More compelling evidence for the role of infection in MS is provided by the immunoglobulin patterns typically found in the CSF of MS patients. There is increased synthesis of immunoglobulins within the CSF. These intrathecal antibodies are found in the CSF and not in the serum, migrate to the cathodal regions on isoelectric focusing (IEF) gels, and are called oligoclonal bands (OCBs). OCBs occur in approximately 90% of MS patients, but are not specific for MS, as such bands are also found in the CSF of patients with a wide variety of CNS infections. In general, the presence of OCBs in CSF is thought to represent an intrathecal immune response to an infectious agent. A number of chronic infections of the CNS are characterized by the presence of OCBs in the CSF. These CNS infections include tuberculosis, syphilis, neuroborreliosis, cryptococcal meningitis, herpes simplex encephalitis, HTLV-1 myelitis, and subacute sclerosing panencephalitis (SSPE).[13,14,33]

In MS patients, the specificity of the OCBs for any particular antigen has been lacking. However, several studies have found that elevated IgG titers for a variety of infectious agents are found in the CSF of MS patients. This suggests that more immune activity in a general sense occurs in the CNS of MS patients.

These data support the idea that an environmental agent is responsible for at least triggering the development of MS. However, these data do not tell us anything about the nature of any putative infectious agent. Furthermore, these data do not specifically support the notion of a single infectious agent as being responsible for causing all MS. An alternative explanation would be that a naive or virgin susceptible population (i.e., Faroe islanders) could be vulnerable to a wide variety of common pathogens, which could trigger an epidemic. This notion may be more consistent with the idea that MS susceptibility is likely polygenic. Some individuals may have MS triggered by only one pathogen whereas, in other individuals, with a different underlying pattern or degree of genetic susceptibility, several infectious agents could be responsible for development of their MS.

5. CANDIDATE PATHOGENS

A wide variety of infectious agents have been proposed to be the cause of MS. Unfortunately, most of these observations have failed the test of reproducibility. Most efforts to identify a pathogen have focused on viruses. Viruses are attractive candidates because of several factors. Viruses are known to cause demyelinating disease in animal models and humans, often with long periods of latency that may present clinically with a relapsing and remitting course.[34,35] As mentioned earlier, antibody titers are often elevated to viruses in both the CSF and serum of MS patients compared with controls. Several laboratories found measles-specific antibodies within the OCB immunoglobulins. Subsequent work found that OCBs in MS patients reacted to a variety of viruses, but these responses were nonspecific.[36] A number of candidate pathogens, mostly viral, have been proposed to be causative for MS (Table I). Because of either lack of reproducibility or specificity, these candidate agents have been eliminated as MS-specific pathogens. This has created an atmosphere of marked incredulousness when new candidates are proposed. Nevertheless, in the last several years, candidate pathogens have emerged that may have a role in the etiology of MS.[37]

TABLE I
Candidate Pathogens in Ms

Viruses found in CSF/CNS of MS	Bacteria linked to MS
Rabies	*Treponema pallidum*
HSV-1 and 2	Berrelia
Parainfluenza virus	*M. pneumoniae*
Paramyxovirus	*C. pneumoniae*
Corona virus	
HIV	
HLTV-1	
Other retroviruses	
Measles virus	
Poliomyelitis	
Ebstein–Barr virus	
HHV-6	

5.1. Human Herpes Virus-6 (HHV-6)

HHV-6 is the cause of exanthema subitum (roseola infantum), which is a common infectious exanthem in humans. This virus has been found to have CNS tropism and can result in meningoencephalitis.[38] Studies by other investigators have found increased IgM and IgG responses to HHV-6 in MS patients as well as HHV-6 DNA in CSF and serum.[39,40] Other labs have failed to confirm these results or have not found differences in HHV-6 antibody titers in MS patients and controls.[41] Another study examined the frequency of detection of the HHV-6 genome in peripheral blood mononuclear cells from relapsing–remitting and progressive MS patients versus healthy and neurological disease controls.[42] They were unable to demonstrate a correlation between HHV-6 infection of peripheral blood mononuclear cells and the development of MS.

5.2. Retroviruses

Pathogenic retroviruses in humans (HTLVs and HIVs) have been considered candidate pathogens for MS because of their ability to impair the immune system and to promote CNS inflammation and demyelination.[43] For example, HTLV-1 can cause a chronic inflammatory demyelinating disease in the CNS. Extensive investigations have been made to find an HTLV-1-like retrovirus in MS patients. Despite initial claims, the presence of HTLV-1 or HTLV-like retroviruses has not been confirmed.

A retrovirus associated with MS was derived from the choroid plexus cells and from Epstein–Barr virus (EBV)-immortalized B cells of MS patients. The virus is designated as multiple sclerosis–associated retrovirus (MSRV), and it is an oncovirus. It forms extracellular virions; moreover, RNA has been amplified only from the CSF and serum of MS patients and controls, including cases of myelitis and encephalitis. It remains unclear if MSRV is an exogenous retrovirus or an endogenous provirus.

The role of MSRV in the etiology or pathogenesis of MS remains unknown. It has been proposed that EBV may have an important role as a cofactor triggering retroviral reactivation.[44,45] Therefore, a dual infection/reactivation may be required for the development of MS. EBV may be required to transactivate MSRV or other retroviruses. A recent case–control study from the United Kingdom found that patients with a history of infectious mononucleosis (IM) had more than 5 times the risk of subsequent MS.[46] The clinical significance of these findings remains unclear because of the small number of cases of infectious mononucleosis ($n = 11$, out of 225 cases and 900 controls).

5.3. *Chlamydia pneumoniae*

C. pneumoniae belongs to a genus of intracellular pathogens that are infectious to man and other vertebrates.[47] A case report of CNS infection with *C. pneumoniae* in a patient with rapidly progressive MS has been described.[48] Antimicrobial therapy directed against this organism resulted in marked neurological improvement that has been largely sustained for 3 years. Additional work found *C. pneumoniae* present in the CSF of patients with newly diagnosed relapsing–remitting MS and with progressive MS but not in other neurological disease controls.[49] CSF samples were examined for *C. pneumoniae* by culture, PCR assays, and CSF immunoglobulin (Ig) reactivity with *C. pneumoniae* elementary body (EB) antigens. *C. pneumoniae* was isolated by tissue culture from the CSF in 64% of MS patients versus 11% of the controls. PCR assays demonstrated the presence of *C. pneumoniae* major outer membrane protein (MOMP) gene in the CSF of 97% of MS patients and 18% of the controls. Finally, 86% of the MS patients had increased CSF antibodies to *C. pneumoniae* EB antigens, as demonstrated by enzyme-linked immunoadsorbent assay adsorbance values that were 3 standard deviations greater than those seen in the controls.

More recently, the specificity of the intrathecal immune response to *C. pneumoniae* antigens has been evaluated in relapsing–remitting and progressive MS patients. Affinity-driven immunoblot studies found that in 17 MS patients, CSF antibodies to *C. pneumoniae* EB antigens focused on the cathodic region of the IEF gel where OCBs seen in MS normally are found. Furthermore, in 14 of 17 patients, these OCBs were adsorbed either partially or completely from the CSF by EB antigens of *C. pneumoniae*, but not by control viral or bacterial antigens. Adsorption of OCBs with EB antigens did not occur with CSF from three control patients with SSPE, although the bands in these three patients were partially adsorbed by measles antigen.[50]

6. MECHANISM OF *C. pneumoniae*–MEDIATED ACCENTUATION OF AUTOIMMUNE DISEASE

Infections and MS are intertwined processes. MS is very likely triggered by an infectious agent and, furthermore, worsening of MS is often preceded by an upper respiratory infection. We have recently examined the role of infection in experimental allergic encephalitis (EAE), an animal model of MS. EAE was recently used to evaluate a possible association of EAE with *C. pneumoniae* infection. Intraperitoneal inoculation of mice with *C. pneumoniae*, following immunization with neural antigens, increased the severity of EAE. Accentuation of EAE required live infectious *C. pneumoniae*, and the severity of the disease could be attenuated with antiinfective therapy. Following immunization with neural antigens, systemic infection with *C. pneumoniae* led to the dissemination of the organism into the CNS in mice, which then caused accentuated EAE. In contrast, inoculation with *C. trachomatis* did not worsen EAE, and infectious organisms were not seen in the CNS. This study demonstrates the potential effect of a *C. pneumoniae* infection in the CNS on the amplification of autoimmune disease following immunization with three different neural antigens (MSCH, MBP, and MOG). A causal association between *C. pneumoniae* infection and accentuated EAE was inferred not only from the direct presence of a replicating organism in the CNS, but also from the attenuation of EAE following therapy with the antimicrobial agent fluorphenicol.[51] We have shown that infection of *C. pneumoniae* in the CNS can worsen EAE and believe it also may worsen MS. Systemic infection with *C. trachomatis* enhanced the *in vitro* proliferative response to MBP, which was higher than controls'. Unlike *C. pneumoniae*, *C. trachomatis* did not infect the CNS in mice and thus did not cause worsening of EAE.

One mechanism by which infections can potentially induce autoimmune disease is through molecular mimicry. Following immunization with chlamydial peptides that show homology with MBP, rats developed severe EAE.[52] Our study did not show evidence of molecular mimicry between *C. pneumoniae* and neural antigens in SJL mice. Animals showed worsening of EAE induced by three different classes of encephalitogenic antigens—MBP, MOG, and MSCH—suggesting that molecular mimicry is an unlikely explanation for the worsening of the disease.

Infectious agents are well known to rapidly expand the pool of immune cells that recognize the invading pathogen. An increase in population of T cells that recognize other antigens including those that react to self-proteins may occur consequent to the secretion of cytokines and may be sufficient to cause disease.[53] In mouse hepatitis model of CNS infection, prominent demyelination mediated by CD8 T cells was seen in mice in the absence of cognate antigen in the CNS, suggesting that demyelination did not require the presence of antigen-specific T cells.[54] In the mouse keratitis model induced by HSV-1, both antigenic mimicry and bystander activation are thought to be responsible for tissue injury.[55–57] In trying to reconcile the views of molecular mimicry and bystander activation in autoimmunity, it was proposed that these two processes may not necessarily be exclusive and may depend upon the circulating levels of autoreactive cells.[4]

Persistence of *C. pneumoniae* in the CNS is likely to provide an environment that can lead to the activation of autoreactive T cells and contribute to the pathogenesis of a chronic disease such as MS. Infection of the CNS is necessary for accentuation of EAE, which may be facilitated in the presence of an ongoing CNS inflammation. We suggest that a similar scenario may occur in MS, in which a ubiquitous pathogen may amplify an autoimmune response. We predict that if an infectious agent can persist and amplify an immune response, it can modify the expression of a T cell–mediated autoimmune disease in an organ-specific manner. A direct interplay between an infectious agent and autoimmunity is also likely to have immediate therapeutic implications. Although an animal model has suggested that infection with *C. pneumoniae* can worsen EAE, it does not necessarily prove that this may be the situation in human MS, but nonetheless offers a model system to test these hypotheses.

C. pneumoniae in some respects is a compelling candidate as an infectious cause for MS. One of the hallmarks of chlamydial infection is tissue persistence and development of chronic infection. Periodic exacerbations and remissions clinically characterize trachoma, with progressive inflammation to corneal scarring. Meningoencephalitis and other neurological complications have been described for patients infected with *C. pneumoniae.*[(58,59)]

Epidemiological strategies are useful in establishing causality. Case–control surveys involving infectious, genetic or other studies in conjunction with surveys evaluating risk factors and exposures could be carried out. If *C. pneumoniae* or any other agent were identified, a prospective cohort study of the exposed and nonexposed population could be done. Because of differential recall bias, case control surveys are problematic. Any prospective cohort study would require a large number of subjects as well as at least a decade of follow-up evaluation. An alternative method to a case control study is a historical cohort study. If two cohorts exist, one may be able to determine if an excess of common cases exists after cross-referencing one versus the other (e.g., trauma registry versus multiple sclerosis cohort). This approach minimizes recall bias; however, few retrospective cohorts exist; the number of subjects in such cohorts may be too small to detect a difference; and if the risk due to the exposure is small, it may go undetected. A therapeutic trial in MS patients with evidence of *C. pneumoniae* infection in the CSF would potentially establish a role for *C. pneumoniae* in the pathogenesis of MS. It would not necessarily establish a causal relationship between *C. pneumoniae* infection and the development of MS. Because of the breakdown of the blood brain barrier, *C. pneumoniae* may be present as a "bystander" in the CNS. It may not have a role in the pathogenesis of MS, but its presence in MS plaques may exacerbate, prolong, or worsen inflammation in the CNS such as that seen in the animal model of EAE. A well-designed and appropriately powered clinical trial would provide additional answers about the role *C. pneumoniae* has in inflammatory demyelinating disease as well as in the clinical course of MS.

7. CONCLUSIONS

Establishing the etiology of MS continues to be a difficult problem. The polygenic inheritance pattern, varying clinical features, heterogeneity of the clinical course, and environmental factors all must be considered when considering a putative infectious agent in MS. Nonetheless, the infectious nature of MS is inescapable. It may be that MS is a syndrome similar to pneumonia and more than one etiologic infectious agent may be involved. This would account for the heterogeneity of the disease. Host genes may further modulate disease expression as seen with infectious diseases such as tuberculosis and AIDS. The recent sequencing of the human genome may allow for more rapid identification of host genetic factors that could interact in some fashion with an infectious pathogen(s) to result in the cascade of events that we identify as inflammatory demyelinating disease pathologically and multiple sclerosis clinically.

REFERENCES

1. Noseworthy, J. H., 1999, Progress in determining the causes and treatment of multiple sclerosis, *Nature* **399:**40–47.
2. Noseworthy, J. H., Lucchinetti, C., Rodriguez, M., and Weinshenker, B. G., 2000, Multiple Sclerosis, *New. Eng. J. Med.* **343:**938–946.
3. Whitton, J. L., and Fujinami, R. S., 1999, Viruses as triggers of autoimmunity: Facts and fantasies, *Curr. Opin. Microbiol.* **2:**392–397.
4. Wucherpfennig, K. W., 2001, Mechanism for the induction of autoimmunity by infectious agents, *J. Clin. Invest.* **108:**1097–1104.
5. Fairweather, D., Kaya, Z., Shellam, G. R., Lawson, C. M., and Rose, N. R., 2001, From infection to autoimmunity, *J. Autoimmun.* **16:**175–186.
6. Yucesan, C., and Sriram, S., 2001, *Chlamydia pneumoniae* infection of the central nervous system, *Curr. Opin. Neurol.* **14:**355–359.
7. Antel, J. P., and Owens, T., 1999, Immune regulation and CNS autoimmune disease, *J. Neuroimmunol.* **100:**181–189.
8. Rivers, T. M., Sprunt, D. H., and Berry, G. P., 1933, Observations on the attempts to produce acute disseminated encephalomyelitis in monkeys, *J. Exp. Med.* **58:**39–53.
9. Steiner, I., Nisipianu, P., and Wirguin, I., 2001, Infection and the etiology and pathogenesis of multiple sclerosis, *Curr. Neurol. Neurosci. Rep.* **1:**271–276.
10. Kurtzke, J. F., 1993, Epidemiologic evidence for MS as an infection, *Clin. Microbiol. Rev.* **6:**382–427.
11. Wekerke, H., 2000, Immunology of MS, in: *McAlpines Multiple Sclerosis,* 3rd ed., Churchill Livingstone, Edinburgh, pp. 379–407.
12. Ota, K., Matsui, M., Milford, E. L., Mackin, G. A., Weiner, H. L., and Hafler, D. A., 1990, T cell recognition of an immunodominant myelin basic protein epitope in multiple sclerosis *Nature,* **346:**183–185.
13. Mehta, P. D., 1991, Diagnostic usefulness of cerebrospinal fluid in multiple sclerosis, *Crit. Rev. Clin. Lab. Sci.* **28:**233–251.
14. Andersson, M., Alvarez-Cermeno, J., Bernardi, G., Cogato, I., Fredman, P., Frederiksen, J., Fredrikson, S., Gallo, P., Grimaldi, L. M., Gronning, M., *et al.*, 1994, Cerebrospinal fluid in the diagnosis of multiple sclerosis: A consensus report, *J. Neurol. Neurosurg. Psychiatry* **57:**897–902.
15. Zeman, A., McLean, B., Keir, G., Luxton, R., Sharief, M., and Thompson, E., 1993, The significance of serum oligoclonal bands in neurological diseases, *J. Neurol. Neurosurg. Psychiatry* **56:**32–35.
16. Hershey, L. A., and Trotter, J. L., 1980, The use and abuse of the cerebrospinal fluid IgG profile in the adult: A practical evaluation, *Ann. Neurol.* **8:**426–434.
17. Lucchinetti, C. F., Bruck, W., Rodriguez, M., and Lassman, H., 1996, Distinct patterns of MS pathology indicated heterogeneity in pathogenesis, *Brain Pathol.* **6:**259–274.
18. Sriram, S., and Rodriguez, M., 1997, Indictment of the microglia as the villain in MS, *Neurology.* **48:** 464–470.
19. Panitch, H. S., and Bever, C. T., 1993, Clinical trials of interferons in MS. What have we learnt? *J. Neuroimmunol.* **46:**155–164.

20. Selmaj, K., Raine, C. S., Farooq, M., and Brosnan, C., 1991, Cytokine cytotoxicity against oligodendrocytes: Apoptosis induced by lymphotoxin, *J. Immunol.* **147:**1522–1530.
21. Sharief, M. K., and Hentges, R., 1991, Association between TNF alpha and disease progression in patients with MS, *N. Engl. J. Med.* **325:**467–472.
22. van Oosten, B. W., Barkhof, F., Truyen, L., Boringa, J. B., Bertelsmann, F. W., von Blomberg, B. M., Woody, J. N., Hartung, H. P., and Polman, C. H., 1996, Increased MRI activity and immune activation in two multiple sclerosis patients treated with the monoclonal anti-tumor necrosis factor antibody cA2, *Neurology* **47:**1531–1534.
23. Group, L. S., 1999, TNF neutralization in MS: Results of a randomized, placebo-controlled multicenter study. The Lenercept Multiple Sclerosis Study Group and The University of British Columbia MS/MRI Analysis Group, *Neurology* **53:**457–465.
24. Corboy, J. R., Goodin, D. S., and Frohman, E. M., 2003, Disease-modifying therapies for multiple sclerosis, *Curr. Treat. Options Neurol.* **5:**35–54.
25. Hartung, H. P., Gonsette, R., Konig, N., Kwiecinski, H., Guseo, A., Morrissey, S. P., Krapf, H., and Zwingers, T., 2002, Mitoxantrone in progressive multiple sclerosis: A placebo-controlled, double-blind, randomised, multicentre trial, *Lancet* **360:**2018–2025.
26. van Oosten, B. W., Lai, M., Hodgkinson, S., Barkhof, F., Miller, D. H., Moseley, I. F., Thompson, A. J., Rudge, P., McDougall, A., McLeod, J. G., Ader, H. J., and Polman, C. H., 1997, Treatment of multiple sclerosis with the monoclonal anti-CD4 antibody cM-T412: Results of a randomized, double-blind, placebo-controlled, MR-monitored phase II trial, *Neurology* **49:**351–357.
27. Cochran, G. M., Ewald, P. W., and Cochran, K. D., 2000, Infectious causation of disease: An evolutionary perspective, *Perspect. Biol. Med.* **43:**406–448.
28. Cooke, G. S., and Hill, A. V., 2001, Genetics of susceptibility to human infectious disease, *Nat. Rev. Genet.* **2:**967–977.
29. Kurtzke, J. F., 2000, Multiple sclerosis in time and space—geographic clues to cause, *J. Neurovirol.* **6:**S134–S140.
30. McLeod, J. G., Hammond, S. R., and Hallpike, J. F., 1994, Epidemiology of multiple sclerosis in Australia. With NSW and SA survey results, *Med. J. Aust.* **160:**117–122.
31. Alter, M., Leibowitz, U., and Speer, J., 1966, Risk of multiple sclerosis related to age at immigration to Israel, *Arch. Neurol.* **15:**234–237.
32. Moses, H., Jr., and Sriram, S., 2001, An infectious basis for multiple sclerosis: Perspectives on the role of *Chlamydia pneumoniae* and other agents, *BioDrugs* **15:**199–206.
33. Dorries, R., and Ter Meulen, V., 1984, Detection and identification of virus-specific, oligoclonal IgG in unconcentrated cerebrospinal fluid by immunoblot technique, *J. Neuroimmunol.* **7:**77–89.
34. Johnson, R. T., 1998, Viral Infections of the Nervous System, Lippincott-Raven, Philadelphia.
35. Gilden, D. H., Devlin, M. E., Burgoon, M. P., and Owens, G. P., 1996, The search for virus in multiple sclerosis brain, *Mult. Scler.* **2:**179–183.
36. Salmi, A. A., Hyypia, T., Ilonen, J., Reunanen, M., and Remes, M., 1989, Production of viral antibodies *in vitro* by CSF cells from mumps meningitis and multiple sclerosis patients, *J. Neurol. Sci.* **90:**315–324.
37. Swanborg, R. H., Whittum-Hudson, J. A., and Hudson, A. P., 2002, Human herpesvirus 6 and *Chlamydia pneumoniae* as etiologic agents in multiple sclerosis—a critical review, *Microb. Infect.* **4:**1327–1333.
38. Carrigan, D. R., and Knox, K. K., 1997, Human herpesvirus six and multiple sclerosis, *Mult. Scler.* **3:**390–394.
39. Soldan, S. S., Berti, R., Salem, N., Secchiero, P., Flamand, L., Calabresi, P. A., Brennan, M. B., Maloni, H. W., McFarland, H. F., Lin, H. C., Patnaik, M., and Jacobson, S., 1997, Association of human herpes virus 6 (HHV-6) with multiple sclerosis: Increased IgM response to HHV-6 early antigen and detection of serum HHV-6 DNA, *Nat. Med.* **3:**1394–1397.

40. Soldan, S. S., Fogdell-Hahn, A., Brennan, M. B., Mittleman, B. B., Ballerini, C., Massacesi, L., Seya, T., McFarland, H. F., and Jacobson, S., 2001, Elevated serum and cerebrospinal fluid levels of soluble human herpesvirus type 6 cellular receptor, membrane cofactor protein, in patients with multiple sclerosis, *Ann. Neurol.* **50:**486–493.
41. Enbom, M., 2001, Human herpesvirus 6 in the pathogenesis of multiple sclerosis, *Apmis* **109:**401–411.
42. Mayne, M., Krishnan, J., Metz, L., Nath, A., Auty, A., Sahai, B. M., and Power, C., 1998, Infrequent detection of human herpesvirus 6 DNA in peripheral blood mononuclear cells from multiple sclerosis patients, *Ann. Neurol.* **44:**391–394.
43. Soldan, S. S., and Jacobson, S., 2001, Role of viruses in etiology and pathogenesis of multiple sclerosis, *Adv. Virus Res.* **56:**517–555.
44. Haahr, S., and Munch, M., 2000, The association between multiple sclerosis and infection with Epstein–Barr virus and retrovirus, *J. Neurovirol.* **6**(Suppl. 2)**:**S76–S79.
45. Haahr, S., Sommerlund, M., Moller-Larsen, A., Mogensen, S., and Andersen, H. M., 1992, Is multiple sclerosis caused by a dual infection with retrovirus and Epstein–Barr virus? *Neuroepidemiology* **11:**299–303.
46. Marrie, R. A., Wolfson, C., Sturkenboom, M. C., Gout, O., Heinzlef, O., Roullet, E., and Abenhaim, L., 2000, Multiple sclerosis and antecedent infections: A case–control study, *Neurology* **54:**2307–2310.
47. Stevens, R. S., 1993, Challenge of Chlamydia research, *Infect. Agents Dis.* **1:**279–293.
48. Sriram, S., Mitchell, W., and Stratton, C., 1998, Multiple sclerosis associated with *Chlamydia pneumoniae* infection of the CNS, *Neurology* **50:**571–572.
49. Sriram, S., Stratton, C. W., Yao, S., Tharp, A., Ding, L., Bannan, J. D., and Mitchell, W. M., 1999, *C. pneumoniae* infection of the CNS in MS, *Ann. Neurol.* **46:**6–14.
50. Yao, S. Y., Stratton, C. W., Mitchell, W. M., and Sriram, S., 2001, CSF oligoclonal bands in MS include antibodies against Chlamydophila antigens, *Neurology* **56:**1168–1176.
51. Du, C., Yi-Yao, S., Rose, A., and Sriram, S., 2002, *Chlamydia pneumoniae* infection of the central nervous system worsens EAE, *J. Exp. Med.*, **196:** 1639–1644.
52. Lenz, D. C., Lu, L., Conant, S. B., Wolf, N. A., Gerard, H. C., Whittum-Hudson, J. A., Hudson, A. P., and Swanborg, R. H., 2001, A *Chlamydia pneumoniae*–specific peptide induces experimental autoimmune encephalomyelitis in rats, *J. Immunol.* **167:**1803–1808.
53. Murali-Krishna, K., 1998, Counting antigen specific CD8 T cell: A reevaluation of bystander activation during viral infection, *Immunity* **8:**177–187.
54. Haring, J. S., Pewe, L. L., and Perlman, S., 2002, Bystander CD8 T cell–mediated demyelination after viral infection of the central nervous system, *J. Immunol.* **169:**1550–1555.
55. Deshpande, S. P., Lee, S., Zheng, M., Song, B., Knipe, D., Kapp, J. A., and Rouse, B. T., 2001, Herpes simplex virus–induced keratitis: Evaluation of the role of molecular mimicry in lesion pathogenesis, *J. Virol.* **75:**3077–3088.
56. Gangappa, S., Deshpande, S. P., and Rouse, B. T., 2000, Bystander activation of CD4 T cells accounts for herpetic ocular lesions, *Invest. Ophthalmol. Vis. Sci.* **41:**453–459.
57. Zhao, Z. S., Granucci, F., Yeh, L., Schaffer, P. A., and Cantor, H., 1998, Molecular mimicry by HSV-1: Autoimmune disease after viral infections, *Science* **279:**1344–1347.
58. Koskiniemi, M., Genacy, M., Salonen, O., Puolakkainen, M., Farkkila, M., Saikku, P., and Vaheri, A., 1996, *C. Pneumoniae* associated with CNS infections, *Eur. J. Neurol.* **36:**160–163.
59. Korman, T. M., Turnidge, J. D., and Grayston, M. L., 1997, Neurologic complications of chlamydial infections: Case report and review of the literature, *Clin. Infect. Dis.* **25:**847–851.

15

Chlamydia pneumoniae in the Pathogenesis of Alzheimer's Disease

BRIAN J. BALIN, CHRISTINE J. HAMMOND, C. SCOTT LITTLE, ANGELA MacINTYRE, and DENAH M. APPELT

1. INTRODUCTION

Alzheimer's disease (AD) is a progressive neurodegenerative condition that accounts for the most common and severe form of dementia in the elderly.[1,2] Symptoms of AD progress from mild to severe, resulting in progressive memory loss, decreased cognitive function, emotional and behavioral deficits, and daily living impairment.[3,4] These symptoms are a direct result of nerve cell damage and loss reflected by accumulations of abnormal protein deposits in the brain. Great strides have been made in the past 20 years with regard to understanding the pathological entities that arise in the AD brain. The pathology observed in the brain includes neuritic senile plaques (NSPs), neurofibrillary tangles (NFTs), neuropil threads (NPs), and deposits of cerebrovascular amyloid. Genetic, biochemical, and immunological analyses have provided a greater knowledge of these entities, but our understanding of the "trigger" events leading to the many cascades resulting in this pathology and neurodegeneration is still quite limited. For this reason, the etiology of AD has remained elusive. However, a number of recent studies have implicated infection in the etiology and pathogenesis of AD. This chapter focuses specifically on infection with

BRIAN J. BALIN, CHRISTINE J. HAMMOND, and C. SCOTT LITTLE • Department of Pathology, Microbiology, and Immunology, Philadelphia College of Osteopathic Medicine, Philadelphia, Pennsylvania 19131. ANGELA MacINTYRE and DENAH M. APPELT • Department of Biomedical Sciences, Philadelphia College of Osteopathic Medicine, Philadelphia, Pennsylvania 19131.

Chlamydophila *Chlamydia pneumoniae* (*C. pneumoniae*) and Alzheimer's disease, and how this infection may be involved with the pathogenesis of sporadic AD.

2. CLASSIFICATION AND CHARACTERISTICS OF ALZHEIMER'S DISEASE

AD is classified typically as either Familial AD (FAD) or late-onset sporadic AD. FAD is the hereditary form of the disease, and represents a small percentage (~2%) of total AD cases worldwide.[5] In contrast, sporadic AD accounts for the vast majority of AD cases (estimates of 95% or more), and more typically occurs as a late-onset event [>60 years of age (yo)].

In FAD, which usually occurs with an earlier onset in life (<60 yo), a number of different genetic mutations have been identified that result in the increased deposition of β-amyloid and account for the majority of early-onset FAD cases. These mutated genes include those that encode for the β-amyloid precursor protein (βAPP),[6] and presenilin 1 and 2 proteins (PS1, PS2).[7–9] As a result of these mutations, increased proteolysis by secretases of the βAPP protein near its carboxyl terminus leads to the generation of 40–43 amino acid peptides that comprise β-amyloid.[10] These peptides undergo conformational change into β-sheets that deposit into plaques in regions of the brain more susceptible to this damage. This deposition appears to be critical in the neuronal degeneration observed in AD.[11] With regards to sporadic AD, β-amyloid deposition also occurs and is thought to predispose susceptible regions of the brain to neuronal degeneration. Unlike FAD, in this form of the disease, other influences must prevail as no specific genetic mutations have been recognized.

3. RISK FACTORS IN ALZHEIMER'S DISEASE

Other genetic risk factors, excluding mutations, have been identified for AD, including the expression of the apolipoprotein E (apoE) ε4 allele.[12] The apoE protein is found as one of three polymorphic isoforms: ε2, ε3, and ε4, with ε4 being the least expressed of the three in the general population.[13] The apoE-4 allele appears to act in a dose-dependent manner to increase risk for both FAD and sporadic AD, and is associated with earlier onset and more rapid progression in FAD.[13,14] However, more than one-third of late-onset AD cases do not have a single apoE-4 allele.[15] ApoE functions in the transport of cholesterol and triglycerides as it is a protein constituent of chylomicrons, VLDL, and HDL.[16] As has been postulated previously, there may be a relationship between AD, apoEε4 expression, and atherosclerosis,[17] another risk factor in AD.

Other, nongenetic risk factors have been identified for sporadic AD. Although age remains the number one risk factor, atherosclerosis, neurotrauma resulting from head injury, and chronic inflammation all have been implicated in the development of AD neurodegeneration. One of the most elusive risk factors may be the association of infection with the development of sporadic AD. Herpes Simplex Virus 1 (HSV-1) infection in the brain in individuals expressing the apoE-4 allele is considered to be a risk factor for development of AD,[18] although specifics as to how the viral infection leads to neurodegeneration are still in question. In a similar vein, intracellular bacteria have been considered, and although most have been excluded as causative agents in AD, spirochetes[19] and *C. pneumoniae*[20] remain in question.

4. ASSOCIATION OF *C. pneumoniae* AND ALZHEIMER'S DISEASE

The first report of an association of *C. pneumoniae* with AD demonstrated by polymerase chain reaction (PCR) that the DNA of the organism was present in 90% of postmortem brain samples examined from sporadic AD.[20] As compared to these results, only 5% of postmortem brain samples from age-matched, non-AD, control individuals contained DNA from *C. pneumoniae*. In this study, PCR was conducted using highly specific and sensitive probes for sequences of *C. pneumoniae* chromosomal DNA.[21] Areas of the brain demonstrating significant neuropathology (e.g., temporal cortices, hippocampus, parietal cortex, prefrontal cortex) as well as areas less often demonstrating AD pathology (e.g., cerebellum) were sampled by PCR. Positive samples were obtained from at least one area demonstrating neuropathology, and in four cases, from the cerebellum. Interestingly, in the latter four cases, severe neuropathology was observed throughout, whereas in the two AD brains that were PCR-negative, very mild pathology was observed.[20]

For confirmation of the PCR findings, other techniques were used to determine if *C. pneumoniae* antigens or viable organisms were present in the brain tissues. In tissues from which samples were taken for previous PCR testing, immunohistochemistry and electron microscopy were used to identify the organism. These analyses demonstrated in AD samples that antigens for *C. pneumoniae* were apparent within perivascular macrophages, microglia, and astroglial cells in areas of the temporal cortices, hippocampus, parietal cortex, and prefrontal cortex, but not in control samples. Electron microscopy revealed chlamydial inclusions that contained elementary (EB) and reticulate (RB) bodies. Immunoelectron microscopy verified *C. pneumoniae* in the samples following labeling of the organism with a monoclonal antibody to a *C. pneumoniae* outer membrane protein (OMP) and a gold-conjugated secondary antibody. Immunogold labeling was not evident in comparable control sections negative by PCR. Upon confirmation that *C. pneumoniae* was present in the tissues, frozen tissue samples were analyzed by RT-PCR to determine whether RNA transcripts could be identified. This analysis demonstrated that two transcripts could be identified, one for KDO transferase and the other for an $M_r = 76{,}000$ protein. Given that the RNA for these transcripts was recoverable from the frozen tissues, homogenates of representative PCR and RT-PCR positive samples were prepared and placed on THP-1 monocytes in culture to determine if *C. pneumoniae* could be cultured from the frozen brain tissues. Recovery of viable organisms was successful from two different AD brains and negative from two control brains.[20] Taken together, these data confirmed that *C. pneumoniae* was present in areas of AD neuropathology, was viable from AD brain tissues, and was capable of being cultured from those tissues.

In the analysis of the AD and control brains, 11 of the 17 samples positive for *C. pneumoniae* had at least one allele for the apoEε4 isoform (64%), which was consistent with the apoEε4's being a risk factor for AD.[(14)] In a separate study in patients with arthritis in which *C. pneumoniae* DNA was found in their synovia, 68% had at least one copy of ε4.[(22)] The high percentage of patients infected with *C. pneumoniae* and possessing apoEε4 suggests that this apolipoprotein may be involved as a co-risk factor with *C. pneumoniae* in the pathobiology of numerous diseases in which this organism is a primary suspect.[(20,22)]

The uniqueness of our early study and the importance of replicating these findings has led others to attempt to identify and associate the presence of *C. pneumoniae* with sporadic AD. There have been mixed results reported from these attempts. Two reports from independent laboratories identified *C. pneumoniae* in brain samples from late-onset sporadic AD.[(23–25)] Three early reports and one more recently found no association of *C. pneumoniae* with AD.[(26–29)] In review of the literature, there have been discrepancies in clearly identifying an association of *C. pneumoniae* with other diseases. One example is the association of *C. pneumoniae* with atherosclerosis and cardiovascular disease in which early reports were met with mixed results. Today, there is much greater acceptance of this organism having a role in these processes. At this time, multiple reasons for these discrepancies could be given, including sampling error, methodology, and absence of standardized techniques. This conundrum has led us to evaluate more specifically, *in vitro* and in an animal model, how infection with *C. pneumoniae* interrelates with the known pathological processes identified in sporadic AD. In this regard, we have embarked on determining how *C. pneumoniae* may enter the central nervous system (CNS), stimulate a neuroinflammatory process, and act as a "trigger" for amyloid processing and deposition into plaques, a hallmark of AD pathology.

5. ENTRY OF *C. pneumoniae* INTO THE NERVOUS SYSTEM

Infection of the oral and nasal mucosae of the respiratory tract by *C. pneumoniae* is considered to be the normal route of entry for this obligate, intracellular pathogen into the body.[(30)] However, the exact mechanisms by which the organism enters the systemic circulation are not well defined. Two routes by which *C. pneumoniae* may enter the CNS are through the intravascular and olfactory pathways (Fig. 1). Evidence for these routes has been obtained in our studies of the association of this organism with AD.[(20,31)] *C. pneumoniae*-infected glial cells, perivascular macrophages, and monocytes have been identified around blood vessels in the AD brain.[(20,32)] As the monocyte may be the principal peripheral blood cell in which *C. pneumoniae* is harbored[(33)] and through which the organism gains initial entry to the circulation,[(34)] the monocyte is likely to be the vehicle for trafficking *C. pneumoniae* across the blood–brain barrier (BBB).[(32)]

There is precedence for chronic persistent infection of monocytes with *C. pneumoniae*,[(33)] and this could facilitate systemic and CNS infection with this organism. Recent evidence implicates monocytes and human brain microvascular endothelial cells (HBMECs) in the entry of *C. pneumoniae* through an *in vitro* model of the blood–brain barrier.[(32)] *C. pneumoniae* infection of HBMECs

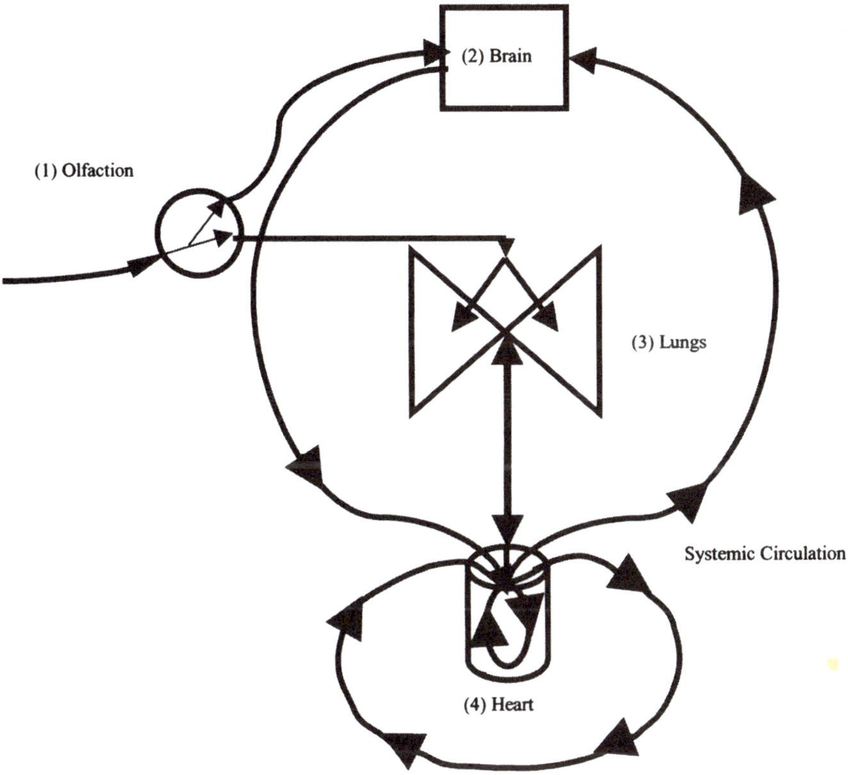

FIGURE 1. Hypothetical routes of infection with *C. pneumoniae.* (1) Olfaction, entry of bacteria into nasal and/or oral mucosae. From this site, *C. pneumoniae* trafficks to the (2) brain and/or to the (3) lungs. Following lung infection, *C. pneumoniae* gains access to the systemic circulation. From the circulation, distant sites such as the (4) heart and brain are susceptible to establishment of infection.

increased the expression of the surface adhesion molecules, intercellular adhesion molecule–1 (ICAM-1) and vascular cellular adhesion molecule–1 (VCAM-1). In a similar manner, *C. pneumoniae* infection of THP-1 monocytes resulted in increased surface expression of the β integrins LFA-1 and MAC-1, the ligands for ICAM-1, and the $\alpha 4\beta 1$ integrin VLA-4, the ligand for VCAM-1. With increased expression of the surface adhesins on the endothelial cells and the integrins on the monocytes, there was a 3-fold increase in transmigration of the monocytes through the *in vitro* barrier relative to the transmigration of uninfected monocytes through an uninfected endothelial barrier. Thus, brain microvascular endothelia and peripheral monocytes could play a major role in promoting the entry of *C. pneumoniae* into the CNS.

In conjunction with these studies, junctional protein expression was examined to determine how the zonula adherens and zonula occludens junctions were affected following endothelial and monocyte infection with *C. pneumoniae.*[35] Analysis of surface-to-junction cross-talk examined the role of cadherins, catenin, and occludin in maintaining the endothelial junctional integrity following infection. VE-cadherin, which is specific to the intercellular junctions of endothelial cells, communicates with the actin cytoskeleton through β-catenin. Perturbations of the cytoskeleton can directly affect the endothelial junctional complex assembly. Likewise, N-cadherin, important to the junctional assembly of the zonula adherens, is present on the entire brain endothelial cell surface. This study demonstrated that infection of HBMECs with *C. pneumoniae* led to upregulation of VE-cadheren, N-cadheren, and β-catenin. In contrast, infection of the HBMECs with *C. pneumoniae* resulted in the downregulation of the tight junctional protein, occludin, at 36–48 h postinfection, with recovery of occludin expression at 72 h postinfection. These data suggest that a compensatory response occurred at the level of the adherens junction to maintain barrier integrity during the downregulation of tight junctional proteins at the time when barrier permeability increased. Occludin expression returned to control levels at 72 h postinfection, which suggests that the permeability changes were transient. These transient changes increase the likelihood that transmigration of monocytes through the HBMEC barrier would occur.[(35)] The alteration in the blood–brain barrier transport mechanism could therefore lead to increased immune cell infiltration and pathogen entry into the brain. Thus, these *in vitro* studies suggest that infection with *C. pneumoniae* of monocytes and brain endothelial cells could trigger the entry of infection into the brain, thereby setting the stage for neuroinflammation and eventual neurodegeneration characteristic of AD.

Another route of entry for *C. pneumoniae* into the CNS is through the olfactory pathway. Since *C. pneumoniae* readily infects epithelial cells and has direct access to the olfactory neuroepithelium of the nasal olfactory system, this route of infection would be likely, given that *C. pneumoniae* is a respiratory pathogen. Examination of the olfactory bulbs obtained at autopsy from two AD cases revealed by PCR and RT-PCR that *C. pneumoniae* genetic material was present in these structures.[(31)] Some of the earliest pathology observed in AD occurs in the olfactory and entorhinal cortices, in particular layers II and III of the entorhinal cortex of the parahippocampal gyrus from which neural projections of the perforant pathway arise to innervate the hippocampal formation.[(36)] Our earlier studies found evidence of the organism in the entorhinal cortex, hippocampus, and temporal cortex.[(20)] These findings bring into question how infection, inflammation, and/or damage of the olfactory bulbs could lead to damage in deeper cortical and limbic structures in the AD brain. Whether infection of the olfactory system with *C. pneumoniae* can ultimately establish inroads for more extensive infection in the brain remains to be determined.

6. *C. pneumoniae* AND NEUROINFLAMMATION

Immunopathogenesis from inflammation is a hallmark of Chlamydia-induced disease. Chlamydial infections *in vivo* typically result in chronic inflammation characterized cellularly by the presence of activated monocytes and macrophages.[(37)] For example, in reactive arthritis associated with chlamydial infection, systemic chronic disease is manifested with activation of TH1/TH2 CD4+ cells and macrophages at sites of inflammation.[38] In addition, at sites of chlamydial infections, proinflammatory cytokines (IL-1β, IL-6, TNFα) and TH1-associated cytokines (IFNγ and IL-12) have been identified.[37] Promotion of any or all of these responses could be evoked by chronic or persistent infection with *C. pneumoniae* as well as by chlamydial products such as lipopolysaccharide (LPS), heat shock proteins, and outer membrane proteins. The expression of LPS alone by this organism could account for numerous aspects of AD pathology. Previous work by others demonstrated that *E. coli* LPS injected at low dose directly into the brains of rats resulted in inflammation characterized by increased cytokine production and microglial activation.[39] In addition, pathology comparable to that observed in AD was observed in the rat temporal lobe, as demonstrated by the induction of the amyloid precursor protein. Thus, these studies suggest that products of infection that are either produced by the infectant, or by the host in response to infection, may stimulate the inflammatory process in the brain, resulting in neurodegeneration characteristic of AD.

In the AD brain, inflammation is thought to arise as a result of β-amyloid deposition, and has been advanced as a major mechanism in the overall pathogenesis of AD.[40] Clinical trials investigating the effects of nonsteroidal antiinflammatory drugs in older populations also implicate inflammation as a factor in AD, as some of these trials have shown that use of these drugs can delay the onset of sporadic AD.[41] The resident cells in the brain responsible for inflammation are typically the microglia and to a lesser extent, the astroglia. Microglia and astroglia have been shown to be activated in the AD brain, and often are identified in and around amyloid plaques.[42] Microglia are the tissue macrophages in the brain and respond to insult with the production of proinflammatory cytokines, and the generation of reactive oxygen species, among numerous other products. Identification of *C. pneumoniae* in the CNS has led us to speculate on the role of this infection in the pathology observed in the AD brain.

A distinct superimposition of the inflammation induced by *C. pneumoniae* infections and that documented in AD is apparent. Data on the association of *C. pneumoniae* with AD demonstrated that microglia, astroglia, and perivascular macrophages were the cells principally infected in the AD brain with *C. pneumoniae*,[20] and these infected cells were observed often in areas of amyloid deposition. The influx of activated monocytes infected with *C. pneumoniae* through the blood–brain barrier could have dire consequences in the brain. In diseases associated with other infectants such as human immunodeficiency virus–1 (HIV-1), perivascular monocytic infiltration with subsequent symptoms of dementia has been documented.[43] Activation of microglia and astroglia[44] in response to the presence of infected, activated monocytes could promote increased production of a variety of cytokines and chemokines such as interleukins IL-1β, IL-6, TNFα, and IFNγ among others.[45–47] As perivascular macrophages, pericytes, microglia, and astroglia were shown to be infected with *C. pneumoniae*,[20] this infection could account for a significant proportion of the neuroinflammation and underlying pathology in the AD brain. In other neurological disorders in which neuroinflammation is a primary factor in pathogenesis,[48] *C. pneumoniae* has been implicated. Thus, infection with *C. pneumoniae* is likely to play a role in neuroinflammation.

7. CONSEQUENCES OF *C. pneumoniae* INFECTION *IN VITRO*

The role that *C. pneumoniae* plays in "triggering" events resulting in AD pathology has been analyzed following *in vitro* infection of the human cell lines, THP-1 monocytes, and human brain microvascular endothelial cells (HBMECs). Experiments have been designed to determine whether infection with *C. pneumoniae* influences the production and/or processing of cellular proteins found to be important in the pathology characteristic of sporadic AD. Data from these studies suggest that expression of the amyloid precursor protein is increased as well as the processing of this protein into fragments that are immunoreactive for Aβ 1–40 and Aβ 1–42. THP-1 monocytes acutely or chronically infected with *C. pneumoniae* appeared to increase both message and protein expression for the amyloid precursor protein, with increased breakdown of the precursor into fragments that contained Aβ 1–40 immunoreactive epitopes (Fig. 2a). The increase in immunoreactive fragments was determined following analysis of whole cell lysates. In this regard, the 4-kDa fragment of Aβ was not apparent on the Western blots as this processed peptide is typically secreted from the cell following cleavage events. The monocytes also produced IL-1β and TNFαs as determined by immunocytochemistry following infection with *C. pneumoniae*. In addition, HBMECs infected with *C. pneumoniae* increased amyloid production and processing as determined by Western immunoblotting for Aβ 1–42 (Fig. 2b) and Fluorescence Activated Cell Sorting (FACS). The processing of amyloid by these cell types, in addition to their activation to promote transmigration of peripheral monocytes through the endothelial barrier,[32] suggests

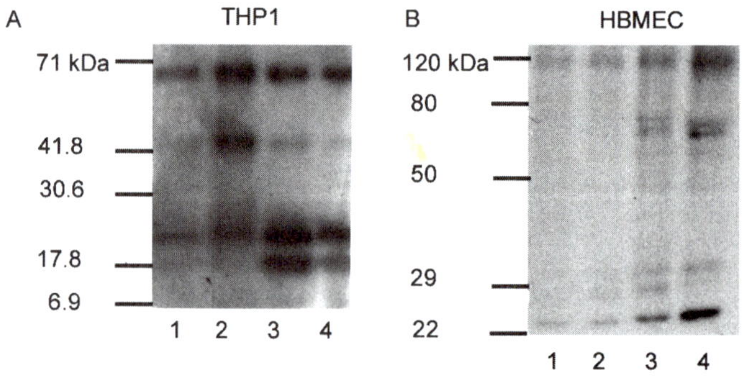

FIGURE 2. Processing of amyloid following acute *C. pneumoniae* infection *in vitro*. *C. pneumoniae* infection of (A) THP-1 monocytes and (B) human brain microvascular endothelial cells (HBMECs) resulted in an increase in Aβ immunoreactive peptides over the course of 72 h. Lane 1, uninfected; lane 2, 24 h postinfection; lane 3, 48 h postinfection; lane 4, 72 h postinfection. Note: Lower molecular mass bands appearing at 48 h (lane 3) and 72 h (lane 4) postinfection for both THP-1 (Aβ 1–40) and HBMECs (Aβ 1–42).

that angiopathy could arise following local amyloid production and processing in the brain. The influx of infected monocytes into the brain with ultimate production and processing of β-amyloid may provide a nidus of amyloid deposition that could promote further damage and/or "seeding" of further plaque development.

Entry of infected monocytes into the brain also may result in infection of resident microglia, astroglia, and neurons. Previous studies have shown that glial cells are infected in the AD brain,[20] and recent studies in our laboratory have shown that a human neuroblastoma cell line can be infected with *C. pneumoniae* (unpublished observation). The neuroblastoma cells appear to harbor the infection and develop a chronic/persistent infection with this organism. Intriguingly, the persistently infected nerve cells do not undergo staurosporine-stimulated apoptosis. Observations indicate that infection of neurons with *C. pneumoniae* result in inclusions containing organisms that localize close to the nucleus and Golgi apparatus and resemble those of granulovacuolar degeneration. Granulovacuolar degeneration occurring in AD has never been understood fully, but it occurs in pyramidal neurons typically in the hippocampus, and in neurons in the olfactory bulbs.[49] Further analysis *in vitro*, and of AD neurons *in situ*, is required to determine if *C. pneumoniae* infection of neurons results in the type of pathology found in AD.

8. ANIMAL MODEL FOR SPORADIC ALZHEIMER'S DISEASE

We have developed a mouse model to test whether *C. pneumoniae* can be a primary trigger for AD pathology following a noninvasive route of inoculation.[50] Since young, naïve, nontransgenic mice do not develop AD pathology, they are a suitable host for analyzing whether inoculation with this organism would lead to any pathological change in the brain. Young [3 months of age (mo)], female BALB/c mice were inoculated intranasally with an isolate of *C. pneumoniae* obtained from an AD brain. Uninfected control mice received vehicle only, which consisted of Hank's balanced salt solution. The intranasal route of inoculation was used for several reasons: the recognized olfactory deficits in AD,[36] the previous PCR findings of *C. pneumoniae* in human olfactory bulbs, and the direct access to the nasal olfactory neuroepithelium of this respiratory pathogen. Mice were sacrificed at 1, 2, and 3 months postinoculation by perfusion fixation followed by collection of the brain and olfactory bulbs. Representative areas of the brain and olfactory bulbs were analyzed for the presence of *C. pneumoniae*, inflammation, and pathology characteristic of AD, in particular amyloid plaques and neurofibrillary tangles.

Light microscopy and electron microscopy (Fig. 3) revealed the presence of *C. pneumoniae* in the olfactory bulbs of animals 1–3 months postinfection. Even more remarkable was the identification of Aβ 1–42 immunoreactive amyloid plaques in areas of the brain associated with AD pathology (e.g., hippocampus, entorhinal cortex) and in a region less affected in AD (e.g., cerebellum). Few plaques were first observed at 1 month postinfection in all regions. Increased numbers of plaques were observed at 3 months postinfection in areas typically

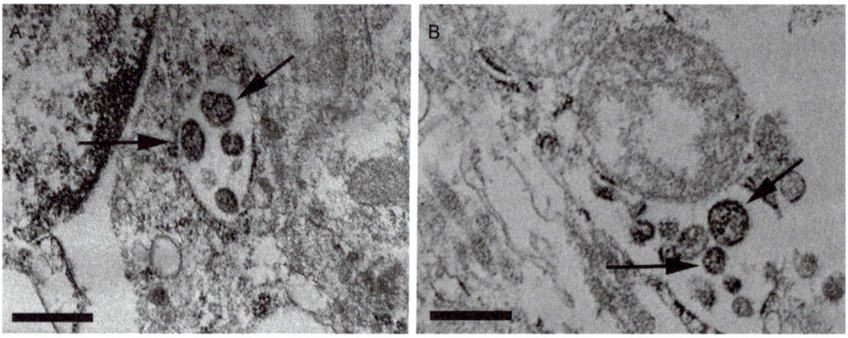

FIGURE 3. Olfactory bulbs from mice infected with *C. pneumoniae.* BALB/c mice were infected intranasally with *C. pneumoniae* and sacrificed at 3 months postinoculation. Profiles of *C. pneumoniae* (A, B, arrows) were observed in cells in the olfactory bulbs. (A) A typical inclusion containing chlamydial bodies. Bars = 0.4 μm.

affected in AD (Fig. 4). Evidence of inflammation was observed infrequently with reactive astrocytes located in some areas of amyloid plaque deposition as well as around some blood vessels. Analysis of the amyloid plaques following thioflavin S staining revealed fibrillogenesis in a few plaques (Fig. 5). Fibrillogenesis of amyloid peptides is a common finding in mature plaques in the AD brain. These data suggest that mature amyloid plaque formation in the brain can result from infection with *C. pneumoniae.* We observed in some areas of the brain intracellular immunolabeling for Aβ 1–42 in pyramidal neurons. However, whether inflammation is responsible for increased amyloid production/processing in the neurons is not known. Intracellular processing of the amyloid precursor protein into Aβ 1–42 recently has been recognized in the pyramidal neurons in the entorhinal cortex of AD brains, and is speculated to account for a portion of the dense-core amyloid plaques observed in AD cortices.[51] Our data suggest

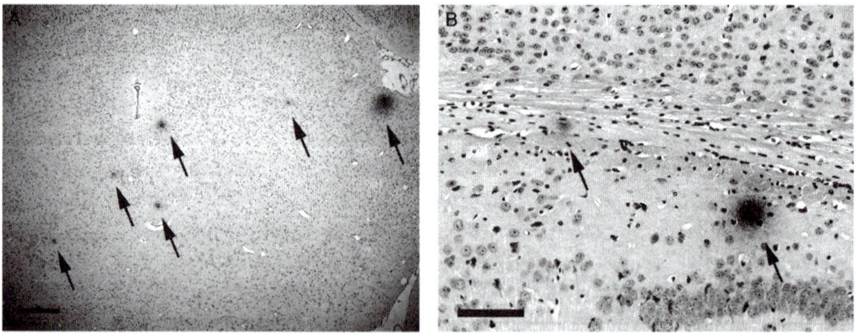

FIGURE 4. Aβ1–42 immunoreactive plaques in mouse brain. Mice were inoculated intranasally with *C. pneumoniae* and sacrificed at 3 months postinfection. Numerous Aβ 1–42 immunoreactive plaques were apparent within the brain (A, B, arrows). Note: Plaques were observed in regions typically affected by AD pathology. (B) Plaques in a section at the level of the hippocampal formation. Bars = 200 μm.

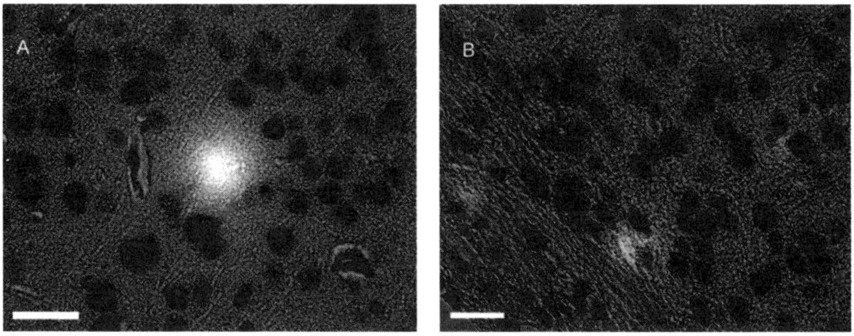

FIGURE 5. Amyloid plaques labeled with Thioflavin S. Amyloid plaques apparent at 3 months postinfection demonstrate fibrillogenic forms of amyloid. Bars = 50 μm.

that the initial intracellular accumulation of Aβ 1–42 may give rise to the extracellular accumulation of Aβ 1–42. Whether this intracellular accumulation could promote other neuropathology remains to be determined.

Silver staining (Gallyas) and immunolabeling for the tau protein failed to demonstrate neurofibrillary tangle pathology in the mouse brain at 1, 2, and 3 months postinfection. It is possible that the amount of time postinfection was not sufficient to initiate this type of pathology. Therefore, we have initiated experiments investigating whether long-term infection (e.g., >1 year) will result in tangle formation and increased plaque formation consistent with the progressive nature of AD. Alternatively, aged immunosenescent animals may be more susceptible to tangle formation following acute or chronic chlamydial infection. To address these questions, studies are now underway using this mouse model. This unique model should provide a foundation for testing how sporadic AD may be a consequence of infection and for testing intervention strategies designed to eradicate the infection, diminish inflammation, and improve/restore cognitive function.

9. SUMMARY

Alzheimer's disease is a complex disorder that invariably leads to extensive neurodegeneration and dementia. As hereditary forms of this disease are rare, there is much interest in determining risk and causative factors, resulting in the prominent sporadic form of this disease. Currently, there are accepted risk factors that include age, neurotrauma, ApoEε4 allelic genotype, and atherosclerosis. Although infection is accepted to be involved with nerve cell damage and symptoms of dementia as observed in HIV-1 associated dementia,[43] infection as a potential causative agent for AD is controversial. Intriguingly, *C. pneumoniae* infection has been associated with many of the accepted risk factors for AD, including age, atherosclerosis, and apoEε4. Furthermore, neurotrauma following head injury would facilitate blood-borne entry of *C. pneumoniae* into the CNS. Infection with *C. pneumoniae* stimulates a proinflammatory response as well as promotes the development of chronic inflammation. This may well account for neuronal damage and degeneration early in the disease process.

Our data demonstrate that the organism enters the brain following infection of the olfactory neuroepithelia and/or of the peripheral blood monocytes. Infection of both human monocytes and microvascular endothelial cells has been demonstrated to facilitate the entry of *C. pneumoniae* through the endothelial barrier as well as eliciting increased production and processing of the β-amyloid precursor protein into Aβ 1–40/42 forms. Implications of these interrelated events include infection in the CNS with a nidus for the production of amyloid peptides capable of "seeding" further amyloid plaque formation. Alternatively, infection of olfactory neuroepithelia and subsequently the olfactory bulbs, and deeper structures in the brain, may result in neuronal damage, leading to increased amyloid production and processing. As glial cells in AD brains have been shown to be infected with *C. pneumoniae*, glial responses leading to the production of inflammatory modulators such as cytokines and reactive oxygen species may promote neuroinflammation and further degeneration. Thus, multiple mechanisms and cascading events may be initiated following infection, thereby contributing to the neurodegeneration and symptomatology characteristically observed in AD.

Although the specific mechanisms of amyloid plaque deposition have yet to be elucidated, our research represents an initial "proof of principle" that infection with *C. pneumoniae* may be a "trigger" in the pathogenesis of sporadic AD. This proof of principle is based on the induction of plaque pathology, following infection, in the brains of normal, young, nontransgenic mice that have no predisposition toward the development of pathology consistent with AD. This model provides an opportunity to evaluate how infection may be influential in the development of sporadic AD as compared with other models using genetically modified animals that more closely resemble the development of familial AD.

10. FUTURE DIRECTIONS

The continued development of *in vitro* and *in vivo* models to determine how infection with *C. pneumoniae* leads to Alzheimer's pathology is one of the most promising directions to follow in the current research efforts. In the nontransgenic animal models, in particular, the observation of neurofibrillary tangle formation following infection will certainly strengthen the hypothesis that infection causes pathology similar to that observed in AD. Furthermore, increased amyloid plaque development as a function of multiple inoculations as well as increased infection time, and similar demonstrations in older animals, will certainly address the issues of progressive disease and associated pathology. In the event that older animals exhibit accelerated pathology with shorter duration of infection, issues of immunosenescence and the ability to resolve infections as a function of age will need to be addressed. In addition, behavioral and cognitive testing of animals following infection will help to correlate structural damage with functional change. Extensive understanding of the basic mechanistic features of *C. pneumoniae* infection in the brain will lead to final acceptance of the role that *C. pneumoniae* plays in sporadic AD.

The animal model will allow us to test different treatment modalities to determine the efficacy of bactericidal antibiotics to specifically eradicate *C. pneumoniae* infection in the brain. These studies will provide a basis for the eventual treatment of AD patients. Our immediate goals of this treatment would be to slow disease progression, and/or ameliorate cognitive function with a diminishment of symptoms. Great efforts will be required throughout the scientific community to increase our knowledge of how this infection interacts with host genetics and other environmental risk factors. Only with these efforts will we ultimately be able to diagnose, treat, eradicate, and prevent *C. pneumoniae*–induced sporadic AD.

ACKNOWLEDGMENTS. We thank the many families who donated tissues for our research efforts into Alzheimer's disease. We thank the following individuals for their contributions to the work presented herein: Drs. J. Todd Abrams, E. James Arking, Kerin Fresa-Dillon, Herve Gerard, Alan P. Hudson, Judy Whittum-Hudson, Tochi Okwuosa, Kevin Tomfohrde, Jennifer Ward, Ronnen Abramov, Jarrid Bernhardt, Gowri Harinarayanan, and Mr. Andrew Bowe, Ms. Anna Gibbard, Ms. Anita Ketty, Ms. Maria Roupas, Mr. David Steves, and Mr. Jonathan Zahler. The studies were supported by research grants AG10160, AI44055, and awards from the Foundation for Research Into Diseases of Aging (FRIDA), and the National Foundation for Infectious Diseases.

REFERENCES

1. Evans, D. A., Funkenstein, H. H., Albert, M. S., Scherr, P. A., Cook, N. R., Chown, M. J., Hebert, L. E., Hennekens, C. H., and Taylor, J. O., 1989, Prevalence of Alzheimer's disease in a community population of older persons. Higher than previously reported, *JAMA* **262:**2551–2556.
2. Brookmeyer, R., Gray, S., and Kawas, C., 1998, Projections of Alzheimer's disease in the United States and the public health impact of delaying disease onset, *Am. J. Public Health* **88:**1337–1342.
3. McKhann, G., Drachman, D., Folstein, M., Katzman, R., Price, D., and Stadlan, E. M., 1984, Clinical diagnosis of Alzheimer's disease: Report of the NINCDS-ADRDA Work Group under the auspices of Department of Health and Human Services Task Force on Alzheimer's Disease, *Neurology* **34:**939–944.
4. Wade, J. P., Mirsen, T. R., Hachinski, V. C., Fisman, M., Lau, C., and Merskey, H., 1987, The clinical diagnosis of Alzheimer's disease, *Arch. Neurol.* **44:**24–29.
5. Lee, M. K., Slunt, H. H., Martin, L. J., Thinakaran, G., Kim, G., Gandy, S. E., Seeger, M., Koo, E., Price, D. L., and Sisodia, S. S. (1996). Expression of presenilin 1 and 2 (PS1 and PS2) in human and murine tissues, *J. Neurosci.* **16:**7513–7525.
6. Goatc, A., Charticr-Harlin, M. C., Mullan, M., Brown, J., Crawford, F., Fidani, L., Giuffra, L., Haynes, A., Irving, N., James, L., *et al.*, 1991, Segregation of a missense mutation in the amyloid precursor protein gene with familial Alzheimer's disease, *Nature* **349:**704–706.
7. Levy-Lahad, E., Wijsman, E. M., Nemens, E., Anderson, L., Goddard, K. A., Weber, J. L., Bird, T. D., and Schellenberg, G. D., 1995, A familial Alzheimer's disease locus on chromosome 1, *Science* **269:**970–973.

8. Rogaev, E. I., Sherrington, R., Rogaeva, E. A., Levesque, G., Ikeda, M., Liang, Y., Chi, H., Lin, C., Holman, K., Tsuda, T., and *et al.*, 1995, Familial Alzheimer's disease in kindreds with missense mutations in a gene on chromosome 1 related to the Alzheimer's disease type 3 gene, *Nature* **376:**775–778.
9. Sherrington, R., Rogaev, E. I., Liang, Y., Rogaeva, E. A., Levesque, G., Ikeda, M., Chi, H., Lin, C., Li, G., Holman, K., *et al.*, 1995, Cloning of a gene bearing missense mutations in early-onset familial Alzheimer's disease, *Nature* **375:**754–760.
10. Selkoe, D. J., Podlisny, M. B., Joachim, C. L., Vickers, E. A., Lee, G., Fritz, L. C., and Oltersdorf, T., 1988, Beta-amyloid precursor protein of Alzheimer disease occurs as 110- to 135-kilodalton membrane-associated proteins in neural and nonneural tissues, *Proc. Natl. Acad. Sci. U.S.A.* **85:**7341–7345.
11. Schellenberg, G. D., 1995, Genetic dissection of Alzheimer disease, a heterogeneous disorder, *Proc. Natl. Acad . Sci. U.S.A.* **92:**8552–8559.
12. Corder, E. H., Saunders, A. M., Strittmatter, W. J., Schmechel, D. E., Gaskell, P. C., Small, G. W., Roses, A. D., Haines, J. L., and Pericak-Vance, M. A., 1993, Gene dose of apolipoprotein E type 4 allele and the risk of Alzheimer's disease in late onset families, *Science* **261:**921–923.
13. de Knijff, P., van den Maagdenberg, A. M., Frants, R. R., and Havekes, L. M., 1994, Genetic heterogeneity of apolipoprotein E and its influence on plasma lipid and lipoprotein levels, *Hum. Mutat.* **4:**178–194.
14. Roses, A. D., 1996, Apolipoprotein E alleles as risk factors in Alzheimer's disease, *Annu. Rev. Med.* **47:**387–400.
15. Bergem, A. L., 1994, Heredity in dementia of the Alzheimer type, *Clin. Genet.* **46**(1 Spec No)**:**144–1449.
16. Schaefer, E. J., and Ordovas, J. M., 1986, Metabolism of apolipoproteins A-I, A-II, and A-IV, *Methods Enzymol.* **129:**420–443.
17. Hofman, A., Ott, A., *et al.*, 1997, Atherosclerosis, apolipoprotein E, and prevalence of dementia and Alzheimer's disease in the Rotterdam Study, *Lancet* **349:**151–154.
18. Itzhaki, R. F., Lin, W. R., Shang, D., Wilcock, G. K., Faragher, B., and Jamieson, G. A., 1997, Herpes simplex virus type 1 in brain and risk of Alzheimer's disease, *Lancet* **349:**241–244. [see comments]
19. Miklossy, J., 1993, Alzheimer's disease—a spirochetosis? *Neuroreport* **4:**841–848.
20. Balin, B. J., Gerard, H. C., Arking, E. J., Appelt, D. M., Branigan, P. J., Abrams, J. T., Whittum-Hudson, J. A., and Hudson, A. P., 1998, Identification and localization of *Chlamydia pneumoniae* in the Alzheimer's brain, *Med. Microbiol. Immunol. (Berl)* **187:** 23–42.
21. Schumacher, H. R., Jr., Gerard, H. C., Arayssi, T. K., Pando, J. A., Branigan, P. J., Saaibi, D. L., and Hudson, A. P., 1999, Lower prevalence of *Chlamydia pneumoniae* DNA compared with *Chlamydia trachomatis* DNA in synovial tissue of arthritis patients, *Arthritis Rheum.* **42:**1889–1893.
22. Gerard, H. C., Wang, G. F., Balin, B. J., Schumacher, H. R., and Hudson, A. P., 1999, Frequency of apolipoprotein E (APOE) allele types in patients with Chlamydia-associated arthritis and other arthritides, *Microb. Pathog.* **26:**35–43.
23. Mahony, J. B., Woulfe, J., Munoz, D., Browning, D., Chong, S., Smieja, M., 2000a, *Chlamydia pneumoniae* in the Alzheimer's brain—Is DNA detection hampered by low copy number? Paper presented at *Fourth Meeting of the European Society for Chlamydia Research, Helsinki, Finland.*
24. Mahony, J. B., Woulfe, J., Munoz, D., Browning, D., Chong, S., Smieja, M., 2000b, Identification of *Chlamydiae pneumoniae* in the Alzheimer's brain. Paper presented at *World Alzheimer Congress 2000 Proceedings.*
25. Ossewaarde, J. M., Gielis-Proper, S. K., Meijer, A., Roholl, P. J. M., 2000, *Chlamydia pneumoniae* antigens are present in the brains of Alzheimer patients, but not in the brains of patients with other dementias, Paper presented at *Fourth Meeting of the European Society for Chlamydia Research, Helsinki, Finland.*

26. Nochlin, D., Shaw, C. M., Campbell, L. A., and Kuo, C. C., 1999, Failure to detect *Chlamydia pneumoniae* in brain tissues of Alzheimer's disease, *Neurology* **53:**1888.
27. Gieffers, J., Reusche, E., Solbach, W., and Maass, M., 2000, Failure to detect *Chlamydia pneumoniae* in brain sections of Alzheimer's disease patients, *J. Clin. Microbiol.* **38:**881–882.
28. Ring, R. H., and Lyons, J. M., 2000, Failure to detect *Chlamydia pneumoniae* in the late-onset Alzheimer's brain, *J. Clin. Microbiol.* **38:**2591–2594.
29. Taylor, G. S., Vipond, I. B., Paul, I. D., Matthews, S., Wilcock, G. K., and Caul, E. O., 2002, Failure to correlate *C. pneumoniae* with late onset Alzheimer's disease, *Neurology* **59:**142–143.
30. Grayston, J. T., Campbell, L. A., Kuo, C. C., Mordhorst, C. H., Saikku, P., Thom, D. H., and Wang, S. P., 1990, A new respiratory tract pathogen: *Chlamydia pneumoniae* strain TWAR, *J. Infect. Dis.* **161:**618–625.
31. Hudson, A. P., Gerard, H. C., Whittum-Hudson, J. A., Appelt, D. M., and Balin, B. J., 2000, *Chlamydia pneumoniae*, APOE genotype, and Alzheimer's disease, in: Chlamydia pneumoniae *and Chronic Disease* (J. L'age-Stehr, ed.), Springer-Verlag, New York, pp. 121–136.
32. MacIntyre, A., Abramov, R., Hammond, C. J., Hudson, A. P., Arking, E. J., Little, C. S., Appelt, D. M., Balin, B. J., 2003, *Chlamydia pneumoniae* infection of human brain endothelial cells and monocytes promotes the transmigration of monocytes through an *in vitro* blood brain barrier, *J. Neurosci. Res.*, **71:**740–750.
33. Airenne, S., Surcel, H. M., Alakarppa, H., Laitinen, K., Paavonen, J., Saikku, P., and Laurila, A., 1999, *Chlamydia pneumoniae* infection in human monocytes, *Infect. Immun.* **67:**1445–1449. [published erratum appears in *Infect. Immun.* (1999) **67:**6716]
34. Boman, J., Soderberg, S., Forsberg, J., Birgander, L. S., Allard, A., Persson, K., Jidell, E., Kumlin, U., Juto, P., Waldenstrom, A., and Wadell, G., 1998, High prevalence of *Chlamydia pneumoniae* DNA in peripheral blood mononuclear cells in patients with cardiovascular disease and in middle-aged blood donors, *J. Infect. Dis.* **178:**274–277.
35. MacIntyre, A., Hammond, C. J., Little, C. S., Appelt, D. M., and Balin, B. J., 2002, *Chlamydia pneumoniae* infection alters the junctional complex proteins of human brain microvascular endothelial cells, *FEMS Microbiol. Lett.* **217:**167–172.
36. Hyman, B. T., Van Hoesen, G. W., Kromer, L. J., and Damasio, A. R., 1986, Perforant pathway changes and the memory impairment of Alzheimer's disease, *Ann. Neurol.* **20:**472–481.
37. Rasmussen, S. J., Eckmann, L., Quayle, A. J., Shen, L., Zhang, Y. X., Anderson, D. J., Fierer, J., Stephens, R. S., and Kagnoff, M. F., 1997, Secretion of proinflammatory cytokines by epithelial cells in response to Chlamydia infection suggests a central role for epithelial cells in chlamydial pathogenesis, *J. Clin. Invest.* **99:**77–87.
38. Simon, A. K., Seipelt, E., Wu, P., Wenzel, B., Braun, J., and Sieper, J., 1993, Analysis of cytokine profiles in synovial T cell clones from chlamydial reactive arthritis patients: Predominance of the Th1 subset, *Clin. Exp. Immunol.* **94:**122–126.
39. Hauss-Wegrzyniak, B., Lukovic, L., Bigaud, M., and Stoeckel, M. E., 1998, Brain inflammatory response induced by intracerebroventricular infusion of lipopolysaccharide: An immunohistochemical study, *Brain Res.* **794:** 211–224.
40. Lue, L. F., Brachova, L., Civin, W. H., and Rogers, J., 1996, Inflammation, A beta deposition, and neurofibrillary tangle formation as correlates of Alzheimer's disease neurodegeneration, *J. Neuropathol. Exp. Neurol.* **55:**1083–1088.
41. Breitner, J. C., 1996, The role of anti-inflammatory drugs in the prevention and treatment of Alzheimer's disease, *Annu. Rev. Med.* **47:**401–411.
42. Wood, P. L., 1998, Role of CNS macrophages in neurodegeneration, in: *Neuroinflammation Mechanisms and Management* (P. L. Wood, ed.), Humana Press, Totowa, NJ, pp. 1–59.
43. Boven, L. A., Middel, J., Breij, E. C., Schotte, D., Verhoef, J., Soderland, C., and Nottet, H. S., 2000, Interactions between HIV-infected monocyte-derived macrophages and human brain microvascular endothelial cells result in increased expression of CC chemokines, *J. Neurovirol.* **6:**382–389.

44. Hu, J., and Van Eldik, L. J., 1999, Glial-derived proteins activate cultured astrocytes and enhance beta amyloid-induced glial activation, *Brain Res.* **842:**46–54.
45. Simpson, J. E., Newcombe, J., Cuzner, M. L., and Woodroofe, M. N., 1998, Expression of monocyte chemoattractant protein-1 and other beta-chemokines by resident glia and inflammatory cells in multiple sclerosis lesions, *J. Neuroimmunol.* **84:**238–249.
46. Weiss, J. M., Downie, S. A., Lyman, W. D., and Berman, J. W., 1998, Astrocyte-derived monocyte-chemoattractant protein-1 directs the transmigration of leukocytes across a model of the human blood–brain barrier, *J. Immunol.* **161:**6896–6903.
47. Persidsky, Y., 1999, Model systems for studies of leukocyte migration across the blood–brain barrier, *J. Neurovirol.* **5:**579–590.
48. Gieffers, J., Pohl, D., Treib, J., Dittmann, R., Stephan, C., Klotz, K., Hanefeld, F., Solbach, W., Haass, A., and Maass, M., 2001, Presence of Chlamydia pneumoniae DNA in the cerebral spinal fluid is a common phenomenon in a variety of neurological diseases and not restricted to multiple sclerosis, *Ann. Neurol.* **49:**585–589.
49. Esiri, M. M., and Wilcock, G. K., 1984, The olfactory bulbs in Alzheimer's disease, *J. Neurol. Neurosurg. Psychiatry* **47:**56–60.
50. Little, C. S., Hammond, C. J., MacIntyre, A., Balin, B. J., and Appelt, D. M., 2004, *Chlamydia pneumoniae* induces Alzheimer-like amyloid plaques in brains of BALB/c mice, *Neurobiol. Aging* **25:**419–429.
51. D'Andrea, M. R., Nagele, R. G., Wang, H. Y., and Lee, D. H., 2002, Consistent immunohistochemical detection of intracellular beta-amyloid42 in pyramidal neurons of Alzheimer's disease entorhinal cortex, *Neurosci. Lett.* **333:**163–166.

16

Chlamydia pneumoniae and Inflammatory Arthritis

JUDITH A. WHITTUM-HUDSON,
H. RALPH SCHUMACHER, and ALAN P. HUDSON

1. INTRODUCTION

According to the currently accepted diagnostic criteria of the American College of Rheumatology (ACR), reactive arthritis is a sterile immune-mediated pathogenesis process of the joint that follows bacterial infection of either the gastrointestinal or urogenital system.[1] The enteric organisms most closely associated with the disease include various species of the genera *Salmonella, Yersinia, Campylobacter, Shigella,* and others; the primary urogenital pathogen involved in the genesis of reactive arthritis is the obligate intracellular bacterium *Chlamydia trachomatis.*[2] Recent studies from many laboratories, however, have equivocated this official disease definition.[1] For example, while it does appear to be the case that synovial inflammation elicited by prior infection with *Salmonella, Yersinia,* and other enteric organisms involves primarily protein antigens in the joint rather than viable bacteria,[3–5] *C. trachomatis* has been demonstrated to be not only viable at that site, but also metabolically active.[6,7] Thus, as developed in detail below, our current understanding of the pathogenesis process, and host–pathogen interaction, in reactive arthritis clearly is incomplete.

Importantly, several groups have identified organisms in the inflamed joints of patients with reactive arthritis other than those canonical species specified by the official ACR definition. One approach employed by many investigators has been to use generalized polymerase chain reaction (PCR) primer systems targeting highly conserved DNA sequence regions separated by unique regions of the bacterial 16S rRNA genes. Results of the many studies structured in this fashion have been surprisingly congruent, demonstrating various species of

JUDITH A. WHITTUM-HUDSON and ALAN P. HUDSON • Wayne State University School of Medicine, Detroit, Michigan. RALPH SCHUMACHER • University of Pennsylvania School of Medicine and D.V.A. Medical Center, Philadelphia, Pennsylvania.

Pseudomonas, Mycoplasma, and other genera known to include both pathogenic and nonpathogenic organisms.[8] Interestingly, some of these studies also identified unusual bacterial species in the inflamed joint, including *Moraxella osloensis, Stenotrophomonas maltophilia,* and others, some of which are opportunistic pathogens of emerging interest.[9] However, it is not clear at this point what influence, if any, these organisms exert on the initiation or maintenance of synovial pathogenesis. A second approach to identification of noncanonical bacterial species in the joint has been to use PCR primer systems targeting a specific organism. One bacterium targeted in this manner in recent studies has been the respiratory pathogen *Chlamydia pneumoniae.* As developed in this chapter, *C. pneumoniae* has undergone relatively intense scrutiny in relation to its potential role in eliciting synovial inflammation and its resulting chronic reactive arthritis. The results of those studies indicate that this organism does indeed participate actively in the inflammatory process characteristic of this clinical entity, although many important details concerning how it does so remain to be elucidated.

2. REACTIVE ARTHRITIS

Reactive arthritis is an incompletely understood inflammatory process of the joint. Classic reactive arthritis produces an oligoarthritis that normally involves the lower extremities; the disease also usually produces enthesitis and may include sacroiliitis.[1,2] The HLA B27 allele is seen in patients with reactive arthritis somewhat more frequently than in the general population, but it is not invariably present.[2] The disease shows inflammatory changes such as increased leukocyte counts in joint effusions, as well as inflammation of the synovium. In many cases, episodes of active arthritis last weeks or longer, but finally result in remission. Approximately half of all individuals with acute reactive arthritis become chronic or remitting, either with or without evidence of reinfection by the associated or by other organisms.

As indicated above, the spectrum of organisms involved in reactive arthritis is probably larger than that suggested by the strict ACR definition. Indeed, more than 75% of patients seen in an early arthritis clinic showed similar unexplained oligoarthritis (occasionally polyarthritis) with no documented history of antecedent infection;[1,2,10] this issue is developed in more detail below. Patients with classic reactive arthritis frequently display a variety of extraarticular features, including iritis, conjunctivitis, urethritis, cervicitis, and several types of dermatitis.[1,2] However, most rheumatologists indicate that such additional features are not required for diagnosis.

Reiter's syndrome is a subset of reactive arthritis, also defined by ACR criteria. The synovitis in patients with early Reiter's syndrome is characterized by superficial inflammation with large numbers of polymorphonuclear cells and vascular congestion;[11] this is distinctly different from findings in the early rheumatoid arthritis synovium.[12,13] In chronic Reiter's syndrome, synovial changes tend to be similar to those seen in patients with rheumatoid arthritis, with higher levels of lymphocyte and plasma cell infiltration. Fibrin deposition may be seen in both acute and chronic disease. Electron microscopic studies show microvascular occlusion by platelet-fibrin thrombi, electron-dense protein-like deposits in vessel walls, and strands of infiltrating fibrin. While *C. trachomatis* has long been associated with development of Reiter's syndrome in some patients, no information is available at present concerning a possible role for *C. pneumoniae* in the genesis of this clinical entity. Thus the discussion following will focus primarily on classic reactive arthritis.

3. *Chlamydia trachomatis*, PERSISTENCE, AND REACTIVE ARTHRITIS

The role played by *C. pneumoniae* in the genesis of inflammatory arthritis is best understood in the context of, and in comparison to, that played by its more extensively studied sister-species *C. trachomatis*. As mentioned, this latter organism is one of the canonical bacterial species specified by ACR diagnostic criteria as etiologic for reactive arthritis. Early immunocytochemical and immunoelectron microscopic studies indicated that intact organisms and chlamydial antigens sometimes could be identified in the synovium of patients with reactive arthritis. For example, one electron microscopic study of patients having disease of less than 4 weeks' duration gave evidence for *C. trachomatis* within synovial mononuclear cells.[14] Keat *et al.* used monoclonal antibodies to demonstrate *C. trachomatis* elementary bodies in synovial tissue and/or synovial fluid of 5 of 8 patients with sexually acquired reactive arthritis; none of 8 controls showed the organism.[15,16] A study of 129 Reiter's syndrome patients showed elementary and reticulate bodies in 54% of urethral biopsies and synovial fluid samples. In early light and electron microscopic studies on synovial tissue of Reiter's syndrome patients, investigators found that 6 of 8 with early disease showed *Chlamydia*-like bodies at that site.[17] These were confirmed in two cases using anti–*C. trachomatis* antibodies.

A few studies have reported culture of *C. trachomatis* from joints,[18] but most groups have not been able to reproduce these results. Thus, although microscopy showed apparently intact *Chlamydia* and/or protein antigens from the organism in synovial materials from patients with inflammatory arthritis, virtually all attempts to demonstrate viability of those bacteria via culture have been unsuccessful. These negative culture results for *C. trachomatis* from relevant synovial materials engendered the current ACR definition of the disease, including *Chlamydia*-associated reactive arthritis, as a sterile inflammatory process of the synovium. We now know that synovial *C. trachomatis* cells are viable and metabolically active, and to a large extent we understand why synovial materials are culture-negative for the organism, as developed next.

The standard textbook description of the biphasic developmental cycle of *Chlamydiae* was derived primarily from the study of *in vitro* culture systems that used "permissive" host cells, *i.e.* eukaryotic cells that support passage of the organism through the entire cycle, such as the HeLa cell line. Over the last 25 years, however, it has become clear that under some circumstances the chlamydial developmental cycle can be arrested at a late point, obviating production and release of new elementary bodies. This state of arrested development under certain growth conditions and/or within certain host cell types is referred to as "persistence." The idea that *Chlamydiae* can engage in persistent infection is not new, although much of the evidence supporting the reality and clinical significance of persistent chlamydial infection in the joint and elsewhere has been generated during the last 15 years. A detailed review of the historical development of the idea of persistence in relation to chlamydial infections is beyond the scope of this chapter. However, much of the influential initial work in this area was done by the Byrne laboratory, based on studies of *C. trachomatis* infection of HeLa cells under treatment first with penicillin, later with low levels of IFNγ.[19] In those studies, it was shown that infected cultures so treated contain reticulate body-like forms displaying aberrant morphology within the inclusions, and that supernatants from treated cultures contained no, or extremely low levels of, new elementary bodies.[20] The Byrne group also showed that such aberrant chlamydial forms accumulate replicated and fully segregated copies of the bacterial chromosome in the absence of cell division.[21] Removing IFNγ from the medium releases *Chlamydia* from this block in completion of the developmental cycle, resulting in return to normal morphology and production of new elementary bodies.[21]

Detailed molecular genetic examination of persistently infecting *C. trachomatis*, both in an *in vitro* monocyte model of persistence,[22] and in synovial biopsy samples from patients with *C. trachomatis*–associated reactive arthritis, has revealed a number of unusual biochemical and physiologic characteristics which distinguish this state from that of actively growing organisms. Initial studies published some years ago demonstrated that synovial tissue, not synovial fluid, is the preferred site of residence for the organism in chronic reactive arthritis,[23] that synovial *C. trachomatis* cells are indeed viable and metabolically active,[24] and that those cells display a highly unusual profile of gene expression, which explains many of the puzzling aspects of chronic synovial infection by the organism. For example during persistent infection, *C. trachomatis* upregulates production of the Hsp60 protein, a powerful immunogen, while attenuating expression of *omp1*, the gene encoding the major outer-membrane protein;[24] this latter gene product is a critical component of the bacterial cell wall, and its loss during persistent infection largely explains the aberrant morphology of chronically infecting synovial *Chlamydia* observed in electron microscopy and other studies of persistent *C. trachomatis*.

Recent research has demonstrated that while ATP production *via* the chlamydial glycolytic and pentose phosphate pathways occurs at a high level in actively growing *C. trachomatis*, during persistent infection both *in vitro* and in relevant patient samples, expression of all genes encoding products required for function of those pathways is abrogated (Table I);[25] this engenders a significant decrease in the overall metabolic rate of persistent organisms. Importantly, further studies showed that, although genes such as *dnaA, polA, mutS*, and others required for chromosomal DNA replication and partition are expressed during persistence, genes whose products are required for the cytokinesis process, such as *ftsK* and *ftsW*, are not expressed;[26] this observation explains the highly attenuated production of new infectious elementary bodies by persistent *C. trachomatis*, in turn giving an explanation for the well-known culture-negativity

TABLE I
Expression of *C. trachomatis* Genes Encoding Proteins for the Glycolytic and Pentose Phosphate Pathways During Active (HEp-2) vs. Persistent (Monocyte) Infection[a]

	Hours Postinfection of HEp-2 Cells								
Gene	0	4	8	11	16	20	24	36	48
pyk[b]	–	–	–	+	+	+	+	+	+
gap[b]	–	–	–	+	+	+	+	+	+
tal[c]	–	–	–	+	+	+	+	+	+
gnd[c]	–	–	–	+	+	+	+	+	+
	Days Postinfection of Human Monocytes								
Gene	0	1	2	3	5	7			
pyk[b]	–	+	+	–	–	–			
gap[b]	–	+	+	–	–	–			
tal[c]	–	+	+	–	–	–			
gnd[c]	–	+	+	–	–	–			

[a]Determinations done by standard RT-PCR; see ref. 25 for details, original data, and patient analyses.
[b]Representative genes encoding products of the glycolytic pathway.
[c]Representative genes encoding products of the pentose phosphate pathway.

of synovial materials from patients with *Chlamydia*-associated arthritis. These and other data relevant to the biology and pathogenesis of persistent synovial *C. trachomatis* are reviewed in detail in refs. 5 and 7, but the important point is that chronic joint inflammation elicited by *C. trachomatis* is a function of persistent, not active, infection by the organism.

4. *Chlamydia pneumoniae* AND ARTHRITIS

Over the last 10 years several laboratories have reported the presence of DNA from *Chlamydia pneumoniae* in synovial materials of patients with an inflammatory joint disease resembling reactive arthritis, usually in the form of undifferentiated mono- or oligo-arthritis.[(27–30)] Although much evidence argues that it is extremely likely, and even though a tacit agreement appears to exist in the rheumatology community that it is so, unequivocal demonstration that the organism is etiologic for the observed joint disease in this situation remains to be obtained. Some systematic study of this question has been done, however. In one report from our group, *C. pneumoniae*–specific PCR screening assays were developed and used to survey a large number of synovial DNA samples; the samples chosen for analysis were known to be PCR-negative for *C. trachomatis.* Results of this survey indicated that 13% of the samples analyzed were PCR-positive for DNA from *C. pneumoniae,* a rate lower than that for *C. trachomatis* in our patient population but still significant.[(31)] Importantly, virtually all PCR-positive samples were from synovial tissue, suggesting that, as with *C. trachomatis*, the primary site of long-term residence for *C. pneumoniae* in the joint is the tissue, not synovial fluid. The same strategy used for the demonstration that synovial *C. trachomatis* cells are viable and metabolically active has been employed to assess those characteristics of *C. pneumoniae* in synovial tissue. This experimental approach involves using an RT-PCR assay targeting primary transcripts from the *C. pneumoniae* rRNA operons to assess those transcripts in RNA/cDNA from synovial tissues known to be PCR-positive for the organism. These studies confirmed that this organism, again like its sister-species, is viable and metabolically active during chronic infection of the joint.[32] Importantly, however, in none of the patients groups studied so far by any laboratory has it been possible to deduce a clear-cut set of extra-articular features that characterize inflammatory arthritis presumably induced by *C. pneumoniae*, in contrast to the situation for *C. trachomatis*–induced joint disease. This difference in disease phenotype almost certainly results from differences in the details of host–pathogen interaction for these two organisms, as developed below.

Culture of *Chlamydia pneumoniae* has not been attempted extensively from PCR-positive synovial tissue, and thus it is not yet clear that synovial *C. pneumoniae* reside in joint tissue in the persistent state. This seems likely to be the case, however, since the primary host for the organism, and its vehicle for dissemination from the respiratory system, is the monocyte/macrophage. Importantly, in collaboration with G.I. Byrne's laboratory, our group recently showed that the aberrant morphology, as well as all of the differential transcriptional attributes described above, which characterize persistent *C. trachomatis*, also obtain for persistent *C. pneumoniae*, when persistence is induced by IFNγ treatment of infected HEp-2 cells (Table II).[33] As yet unpublished observations also indicate that the transcriptional differences summarized for *C. trachomatis* in the human monocyte model of persistent infection also obtain for *C. pneumoniae* infection in the same model. Thus, as far as has been determined, both the molecular genetic and morphologic differences shown by chronically infecting synovial *C. trachomatis* also apply to synovial *C. pneumoniae*, suggesting that the latter, like the former, exist in that tissue in the persistent state.

However, not all aspects of long-term synovial infection by *C. pneumoniae* are identical to those shown by its sister-species. In relation to studies of *C. pneumoniae* in a different, nonarthritis, context,[34] it was demonstrated that individuals PCR-positive for this organism in synovial tissue have a prevalence of the ε4 allele type at the *APOE* locus on chromosome 19 which is more than 4-fold higher than that of the general population;[35] this was not the case for reactive arthritis patients PCR-positive for *C. trachomatis* DNA, or arthritis patients synovially infected with other organisms. This observation suggests that the gene product of the ε4 allele is somehow involved in pathogenesis by *C. pneumoniae*. Recent unpublished results from our group further indicate that host cells expressing this allele infect with standard strains of *Chlamydia pneumoniae* significantly more easily than do cells lacking it, although we do not yet understand by what means the ε4 product confers such increased infectivity on host cells.

TABLE II
Expression of *C. pneumoniae* Genes Encoding Proteins for DNA Replication and Cytokinesis During Active (HEp-2 − IFNγ) vs. Persistent (HEp-2 + IFNγ) Infection[a]

	Hours Postinfection of HEp-2 Cells							
Gene	0	12	24	36	48	60	72	96
dnaA[b]	−	+	+	+	+	+	+	+
polA[b]	−	+	+	+	+	+	+	+
ftsK[c]	−	+	+	+	+	+	+	+
ftsW[c]	−	+	+	+	+	+	+	+

	Expression at 48 h Postinfection		
Gene	no IFN-γ	0.15 ng/ml rIFN-γ	0.50 ng/ml rIFN-γ
dnaA[b]	+	+	+
polA[b]	+	+	+
ftsK[c]	+	+/−	−
ftsW[c]	+	+/−	−

[a]Determinations done by standard RT-PCR; see ref. 33 for details and all original data.
[b]Representative genes encoding products required for DNA replication.
[c]Representative genes encoding products required for cytokinesis.

5. *Chlamydia pneumoniae* AND CYTOKINES

The cytokine balance elicited during a host immune response to any infectious agent strongly influences whether that pathogen is eliminated with minimal damage to the host, or whether pathogen- or immune-mediated damage ensues.[36] The predominance of Th1- or Th2-associated cytokines, *i.e.*, IL-2, IFNγ, TNFα versus IL-4, IL-6, IL-10, produced by lymphocytes at sites of inflammation is often used as a general indicator of which CD4+ population is involved at that site, but macrophages/monocytes produce some T cell–associated cytokines, including IL-10, TNFα, IL-6. Th1 CD4+ cells drive delayed-type hypersensitivity responses, whereas Th2 CD4+ cells are associated with T-dependent antibody responses.[37] CD8+ T cells have other effector functions, including cytotoxicity for infected cells; they produce some of the same Th1- and Th2-like cytokines (*e.g.*, IFNγ, TNFα, IL-5). Studies have shown a shift over the course of chronic disease from a Th1- to a Th2-dominant cytokine pattern.[38] In addition to proinflammatory mediators such as IL-1 and TNFα, monocytes/macrophages produce a varied repertoire of cytokines, including IL-12 and TGFβ, some of which are upregulated in experimental or clinical *Chlamydia* infection.[39–40]

Many studies have indicated roles for cytokines in the growth, differentiation, and persistence of *Chlamydiae.*[19,36] As mentioned, if cultured in the presence of low levels of IFNγ, growth of both *C. trachomatis* and *C. pneumoniae* is inhibited *in vitro*, aberrant forms of intracellular organisms develop, and the profile of bacterial genes expressed becomes unusual.[19,33] The growth inhibition results from induction of indoleamine-2,3-dioxygenase, which deprives the pathogen of tryptophan by degrading it.[41] Moreover, IFNγ causes a selective transcriptional downregulation of several chlamydial genes, including *omp1*; LPS production is also attenuated, and other chlamydial genes, including *groEL*, are upregulated.[5,7] When IFNγ is removed from the growth medium, infectious organisms are again produced, LPS and major outer membrane protein levels return to normal, and the *groEL* gene product is decreased. TNFα, a proinflammatory cytokine produced by both T cells and macrophages, has similar, synergistic effects to those of IFNγ in this system.[41] Studies have shown that IFNα/β and NO also are induced during *in vitro* chlamydial infection of host cells. Interestingly, a recent report indicated that the amount of NO released by macrophages has important regulatory effects on chlamydial pathogenesis.[42] IFNα/β can modulate inflammation and immune responses and have been associated with some forms of arthritis. The presence of NO is consistent with recent experimental results showing that activated NO synthetase correlates with clearance of infection. *In vivo* sites of chlamydial infection show chronic inflammation and generally contain T cells, monocytes/macrophages, and at some sites B lymphocytes.[43]

It is not clear whether chlamydial infection elicits an inflammatory response because of upregulated cytokine production in infected or neighboring cells, or if the immune response to infected cells drives the inflammatory response via influx of cytokine-producing lymphocytes and macrophages. Nonetheless, synovial materials have been studied for the panel of cytokines present in *C. trachomatis*– and *C. pneumoniae*–induced inflammatory arthritis. After development of inflammation, Th1/Th2 CD4+ cells, as well as CD8+ cells and macrophages, have been detected in synovial fluid.[44] Many studies have indicated that, in patient materials from individuals with early disease, proinflammatory cytokines such as IL-12 and IFNγ are prominent.[39,44] However, in a large-scale study of the steady-state pattern of cytokine and chemokine mRNA production elicited by *C. trachomatis* during chronic synovial infection, a significant difference in the pattern of inflammatory mediators was identified compared with that of early disease. That is, in *C. trachomatis*–infected joint tissues from chronic reactive arthritis patients, IL-10 and IL-8 messengers strongly predominated, although low levels of mRNA encoding IFNγ, TNFα, and IL-15 were also identified; transcripts for the MCP-1, but not the RANTES, chemokine were present.[45] In the same study, it was shown that in synovial tissue samples from arthritis patients chronically infected at that site with *C. pneumoniae*, essentially only mRNA encoding IL-8, IL-1β, and RANTES were present (Fig. 1).

Thus, chronic synovial infection with *C. pneumoniae* shows important differences in host attributes and host response than does chronic infection at that site with *C. trachomatis*. Such differences almost certainly account for the observed differences in extra-articular features noted above for *C. trachomatis*– vs. *C. pneumoniae*–induced arthritis. Regardless, all available data indicate that joint infection with *C. pneumoniae* does engender a form of inflammatory, reactive arthritis.

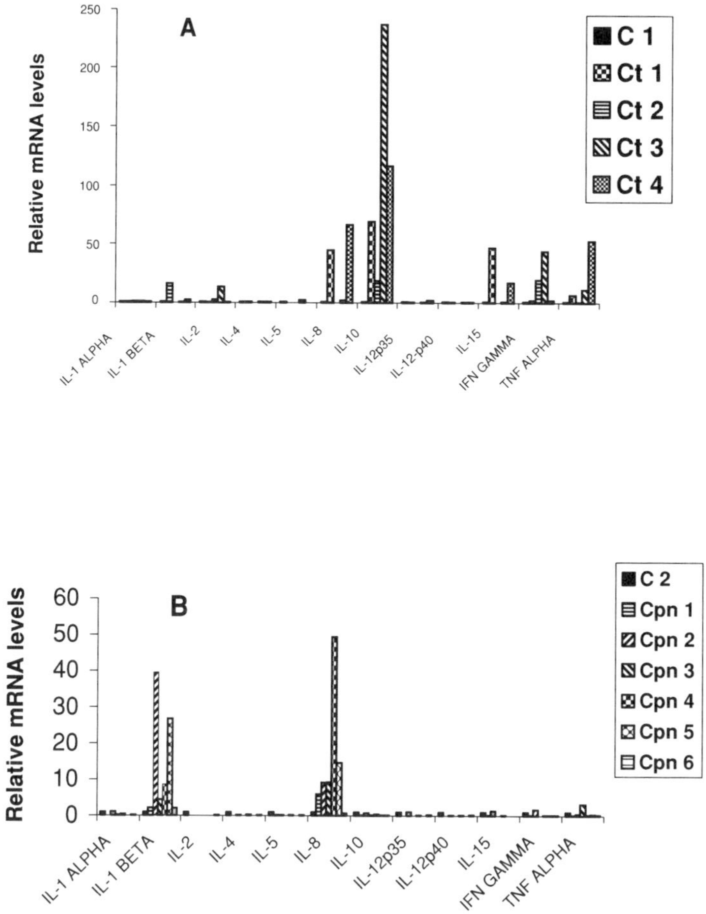

FIGURE 1. Relative levels of cytokine-encoding mRNA in synovial tissue chronically infected with *C. trachomatis* (panel A) or *C. pneumoniae* (panel B). Similar assays using RNA/cDNA from a patient with undifferentiated oligoarthritis but who was PCR-negative for all organisms examined showed no induction of cytokine transcripts (not shown). Assays were done by real time RT-PCR; see ref. 45 for all experimental details and complete data and discussion. C1, C2, control patients. Figure adapted with permission from ref. 45.

6. SUMMARY

Chlamydia pneumoniae is a pathogen of much current interest in relation to several chronic and acute human diseases. Associations of this organism with some clinical entities, such as multiple sclerosis and late-onset (sporadic) Alzheimer's disease, remain highly controversial, but a role for *C. pneumoniae* in the genesis of inflammatory arthritis has gained much credence over the last several years. While the published data briefly summarized in this chapter clearly support a role for *C. pneumoniae* in this disease, a good deal remains to be learned concerning the mechanisms by which disease generation is initiated and/or maintained by the bacterium. For example, studies of *C. trachomatis* and its pathogenic role in reactive arthritis convincingly show that this bacterium resides long-term in synovial tissue in the persistent, rather than the actively growing, state, and that joint pathogenesis by this organism must be understood in terms of host–pathogen interactions specific to that biologic state. Currently available data concerning the biology of *C. pneumoniae* in the joint, while strongly suggestive of persistent infection, do not unequivocally demonstrate that this organism, like its sister-species, undergoes persistence during chronic infection of the synovium. Further, some data suggest that molecular genetic and other details of persistent infection and host–pathogen interaction differ somewhat between *C. trachomatis* and *C. pneumoniae*. Those potential differences must be defined carefully, since the lack of clear extraarticular features for the inflammatory arthritis apparently induced by the latter organism probably relate to those differences. Moreover, the role of the *APOE* ε4 gene product must be elucidated, since some observations indicate that it may be important, specifically in the *C. pneumoniae*–associated pathogenesis process. Thus, even though it is likely that this unusual organism does indeed elicit synovial pathogenesis, much work remains to be done before its role in joint disease is fully understood.

ACKNOWLEDGMENTS. The writing of this chapter was supported by NIH grants AR-42541 and AI-44055 to A.P.H. and AI-44493 and AR-48331 to J.A.W.-H.

REFERENCES

1. Klippel, J. H., Weyand, C. M., and Wortmann, R. L., 1997, *Primer on the Rheumatic Diseases*, 11th ed., Arthritis Foundation Press, Atlanta.
2. Schumacher, H. R., 2000, *Chlamydia*-associated arthritis, *Isr. Med. Assoc. J.* **2:**532–535.
3. Granfors, K., Jalkanen, S., von Essen, R., Lahesmaa-Rantala, R., Isomaki, O., Pekola-Heino, K., Merilahti-Palo, R., Saario, R., Isomaki, H., and Toivanen, A., 1989, *Yersinia* antigen in synovial-fluid cells from patients with reactive arthritis, *N. Engl. J. Med.* **320:** 216–221.
4. Köhler, L., Zeidler, H., and Hudson, A. P., 1998, Aetiologic agents in reactive arthritis: Their molecular biology and phagocyte–host interactions, *Baillière's Clin. Rheumatol.* **12:**589–609.
5. Inman, R. D., Whittum-Hudson, J. A., Schumacher, H. R., and Hudson, A. P., 2000, *Chlamydia*-associated arthritis, *Curr. Opin. Rheumatol.* **12:**254–262.
6. Gérard, H. C., Branigan, P. J., Schumacher, H. R., and Hudson, A. P., 1998, Synovial *Chlamydia trachomatis* in patients with reactive arthritis/Reiter's syndrome are viable but show aberrant gene expression, *J. Rheumatol.* **25:**734–742.
7. Villareal, C., Whittum-Hudson, J. A., and Hudson, A. P., 2002, Persistent *Chlamydiae* and chronic arthritis, *Arthritis Res.* **4:**5–9.
8. Gray, J., Marsh, P. J., and Walker, D. J., 1994, A search for bacterial DNA in RA synovial fluids using polymerase chain reaction, *Br. J. Rheumatol.* **33:**997–999; See also Wilbrink *et al.*, 1998, *Arthritis Rheum.* **41:**535–543; Li *et al.*, 1996, *Arthritis Rheum.* **39:**950–958.
9. Gérard, H. C., Wang, Z., Wang, G. F., El-Gabalawy, H., Goldbach-Mansky, R., Li, Y., Majeed, W., Zhang, H., Ngai, N., Hudson, A. P., and Schumacher, H. R., 2001, Chromosomal DNA from a variety of bacterial species is present in synovial tissue in patients with various forms of arthritis, *Arthritis Rheum.* **44:**1689–1697.

10. Zeidler, H., Werdier, D., Klauder, A., Brinkmann, S., Viswat, M., Mones, M. L., Hulsemann, J. L., and Keck, E., 1987, Undifferentiated arthritis and spondyloarthropathy as a challenge for prospective follow-up, *Clin. Rheumatol.* **6**(Suppl. 2)**:**112–120.
11. Weinberger, H., Ropes, M., Kulka, J. P., and Bauer, W., 1962, Reiter's syndrome: Clinical and pathological observations, *Medicine* **41:**35–91.
12. Schumacher, H. R., 1975, Synovial membrane and fluid morphological alterations in early rheumatoid arthritis: Microvascular injury and virus-like particles, *Ann. N. Y. Acad. Sci.* **256:**39–64.
13. Schumacher, H. R., Bartista, B. B., Krauser, R. E., Mathur, A. K., and Gall, E. P., 1994, Histological appearance of the synovium in early rheumatoid arthritis, *Sem. Arthritis Rheum.* **23:**3–10.
14. Norton, W. L., Lewis, D., and Ziff, M., 1966, Light and electron microscopic observation on the synovitis of Reiter's disease, *Arthritis Rheum.* **9:**747–757.
15. Keat, A. C., 1986, *Chlamydia trachomatis* infection in human arthritis, in: *Chlamydial Infections* (D. Oriel Ridgway, G., Schachter, J., Taylor-Robinson, D., and Ward, M., eds.), Cambridge University Press, New York, pp. 269–279.
16. Keat, A., Thomas, B., Dixey, J., Osborn, M., Sonnex, C., and Taylor-Robinson, D., 1987, *Chlamydia trachomatis* and reactive arthritis: The missing link, *Lancet* **I:**72–74.
17. Schumacher, H. R., Magge, S., Cherian, P. V., Sleckman, J., Rothfuss, S., Clayburne, G., and Sieck, M., 1988, Light and electron microscopic studies on the synovial membrane in Reiter's syndrome. Immunocytochemical identification of chlamydial antigen in patients with early disease, *Arthritis Rheum.* **31:**937–946.
18. Schachter, J., Marshall, J. B., Jones, J. P., Engleman, E. P., and Meyer, K. F., 1966, Isolation of *Bedsoniae* from the joint of patients with Reiter's syndrome, *Proc. Soc. Exp. Biol. Med.* **122:**283–285.
19. Beatty, W. L., Morrison, R. P., and Byrne, G. I., 1995, Persistent *Chlamydiae*, from cell culture to a paradigm for chlamydial pathogenesis, *Microbiol. Rev.* **58:**685–699.
20. Beatty, W. L., Byrne, G. I., and Morrison, R. P., 1993, Morphologic and antigenic characterization of interferon gamma mediated persistent *Chlamydia trachomatis* infection, *Proc. Natl. Acad. Sci. U.S.A.* **90:**3998–4002.
21. Beatty, W. L., Morrison, R. P., and Byrne, G. I., 1995, Reactivation of persistent *Chlamydia trachomatis* infection in cell culture, *Infect. Immun.* **63:**199–205.
22. Gérard, H. C., Köhler, L., Branigan, P. J., Zeidler, H., Schumacher, H. R., and Hudson, A.P., 1998, Viability and gene expression in *Chlamydia trachomatis* during persistent infection of cultured human monocytes, *Med. Microbiol. Immunol.* **187:**115–120.
23. Branigan, P. J., Gérard, H. C., Schumacher, H. R., and Hudson, A. P., 1996, Comparison of synovial tissue and fluid as sources for nucleic acids for detection of *C. trachomatis* by polymerase chain reaction, *Arthritis Rheum.* **39:**1740–1746.
24. Gérard, H. C., Branigan, P. J., Schumacher, H. R., and Hudson, A. P., 1998, Synovial *Chlamydia trachomatis* in patients with reactive arthritis/Reiter's syndrome are viable but show aberrant gene expression, *J. Rheumatol.* **25:**734–742.
25. Gérard, H. C., Freise, J., Rudy, D., Wang, Z., Roberts, G., Krauss-Opatz, B., Köhler, L., Zeidler, H., Schumacher, H. R., Whittum-Hudson, J. A., and Hudson, A. P., 2002, *Chlamydia trachomatis* genes whose products are related to energy metabolism are expressed differentially in active vs. persistent infection, *Microb. Infect.* **4:**13–22.
26. Gérard, H. C., Krauße-Opatz, B., Rudy, D., Wang, Z., Rao, J. P., Zeidler, H., Schumacher, H. R., Whittum-Hudson, J. A., Köhler, L., and Hudson, A. P., 2001, Expression of *Chlamydia trachomatis* genes required for DNA synthesis and cell division in active vs. persistent infection, *Mol. Microbiol.* **41:**731–741.
27. Beaudreuil, J., Hayem, G., Meyer, O., and Khan, M. F., 1995, Reactive arthritis ascribed to *Chlamydia pneumoniae*, *Rev. Rheum.* **3:**222–224.
28. Braun, J., Laitko, S., Treharne, J., Eggens, U., Wu, P., Distler, A., and Sieper, J., 1994, *Chlamydia pneumoniae*—a new causative agent of reactive arthritis and undifferentiated oligoarthritis, *Ann. Rheum. Dis.* **53:**100–105.

29. Cascina, A., Marone Bianco, A., Mangiarotti, P., Montecucco, C. M., and Meloni, F., 2002, Cutaneous vasculitis and reactive arthritis following respiratory infection due to *Chlamydia pneumoniae*: Report of a case, *Clin. Exp. Rheumatol.* **20:**845–847.
30. Wilkinson, N. Z., Kingsley, G. H., Sieper, Braun, J., and Ward, M. E., 1996, Detection of *Chlamydia pneumoniae* and *Chlamydia trachomatis* in the synovium of patients with a range of rheumatic diseases, in *: Proceedings of the European Society for Chlamydia Research* (A. Starry, ed.), Societa Editrice Esculapio, Bologna, p. 189.
31. Schumacher, H. R., Gérard, H. C., Arayssi, T. K., Pando, J. A., Branigan, P. J., Saaibi, D. L., and Hudson, A. P., 1999, *Chlamydia pneumoniae* is present in synovial tissue of arthritis patients with lower prevalence than that of *C. trachomatis*, *Arthritis Rheum.* **42:**1889–1893.
32. Gérard, H. C., Schumacher, H. R., El-Gabalawy, H., Goldback-Mansky, R., and Hudson, A. P., 2000, *Chlamydia pneumoniae* infecting the human synovium are viable and metabolically active, *Microb. Pathog.* **29:**17–24.
33. Byrne, G. I., Ouellette, S. P., Wang, Z., Rao, J. P., Lu, L., Beatty, W. L., and Hudson, A. P., 2001, *Chlamydia pneumoniae* expresses genes required for DNA replication but not cytokinesis during persistent infection of HEp-2 cells, *Infect. Immun.* **69:**5423–5429.
34. Balin, B. J., Gérard, H. C., Arking, E. J., Appelt, D. M., Branigan, P. J., Abrams, J. T., Whittum-Hudson, J. A., and Hudson, A. P., 1998, Identification and localization of *Chlamydia pneumoniae* in the Alzheimer's brain, *Med. Microbiol. Immunol.* **187:**23–42.
35. Gérard, H. C., Wang, G. F., Balin, B. J., Schumacher, H. R., and Hudson, A. P., 1999, Frequency of apolipoprotein E (*APOE*) allele types in patients with *Chlamydia*-associated arthritis and other arthritides, *Microb. Pathog.* **26:**35–43.
36. Ward, M. E., 1999, Mechanisms of *Chlamydia*-induced disease, in: *Chlamydia—Intracellular Biology, Pathogenesis, and Immunity* (R. S. Stephens, ed.), ASM Press, Washington, D.C., pp. 171–210.
37. Mosmann, T. R., and Coffman, R. L., 1989, TH1 and TH2 cells: Different patterns of lymphokine secretion lead to different functional properties, *Annu. Rev. Immunol.* **7:**145–173.
38. Carter, L. L., and Dutton, R. W., 1996, Type 1 and Type 2: A fundamental dichotomy for all T-cell subsets, *Curr. Opin. Immunol.* **8:**336–342.
39. Kotake, S., Schumacher, H. R., Arayssi, T. K., Gérard, H. C., Branigan, P. J., Hudson, A. P., Yarboro, C. H., Klippel, J. H., and Wilder, R. L., 1999, IFNγ, IL-10, and IL-12 p40 gene expression in synovial tissues from patients with recent-onset *Chlamydia*-associated arthritis, *Infect. Immun.* **67:**2682–2686.
40. Van Voorhis, W. C., Barrett, L. K., Cosgrove Sweeney, Y. T., Kuo, C. C., and Patton, D. L., 1996, Analysis of lymphocyte phenotype and cytokine activity in the inflammatory infiltrates of the upper genital tract of female macaques infected with *Chlamydia trachomatis*, *J. Infect. Dis.* **174:**647–650.
41. Malinverni, R., 1996, The role of cytokines in chlamydial infections, *Curr. Opin. Infect. Dis.* **9:**150–156.
42. Huang, J., DeGraves, F. J., Lenz, S. D., Gao, D., Feng, P., Schlapp, T., and Kaltenbach, B., 2002, The quantity of nitric oxide released by macrophages regulates *Chlamydia*-induced disease, *Proc. Natl. Acad. Sci. U.S.A.* **99:**3914–3919.
43. Whittum-Hudson, J. A., Taylor, H. R., Farazdaghi, M., and Prendergast, R. A., 1986, Immunohistochemical study of the local inflammatory response to chlamydial ocular infection, *Invest. Ophthalmol. Vis. Sci.* **27:**64–69.
44. Simon, A. K., Seipelt, E., Wu, P., Wenzel, B., Braun, J., and Sieper, J., 1993, Analysis of cytokine profiles in synovial T cell clones from chlamydial reactive arthritis patients: Predominance of the Th1 subset, *Clin. Exp. Immunol.* **94:**122–126.
45. Gérard, H. C., Wang, Z., Whittum-Hudson, J. A., El-Gabalawy, H., Goldbach-Mansky, R., Bardin, T., Schumacher, H. R., and Hudson, A. P., 2002, Cytokine and chemokine mRNA produced in synovial tissue chronically infected with *Chlamydia trachomatis* and *Chlamydia pneumoniae*, *J. Rheumatol.* **29:**1827–1835.

17

Role of *Chlamydia pneumoniae* as an Inducer of Asthma

DAVID L. HAHN

1. INTRODUCTION

1.1. Definition of Induction

Chlamydia pneumoniae (Cpn) could have three distinguishable causal effects as an asthma inducer. First, an acute infection (or reactivation of a latent infection) could cause an acute worsening of preexisting asthma. Indeed, it is generally accepted that acute Cpn infection can cause some acute asthma *exacerbations*[1,2] but most evidence suggests that other organisms [mainly respiratory viruses, and, to a lesser extent, *Mycoplasma pneumoniae* (Mpn[3])] are associated with the majority of asthma exacerbations in both children[4] and adults.[5] Second, because of its propensity to produce persistent infection, Cpn chronic lung infection could cause worsening of established asthma over time. Again, Cpn has been associated with asthma severity[6] and a possible *promoting* role for Cpn in asthma is a focus of active investigation.[7] Third, an acute primary or secondary Cpn infection, in a previously asymptomatic nonasthmatic individual, could cause acute wheezing that develops into chronic asthma. That Cpn can *initiate* asthma is a radical idea for which there is some evidence[8] but how important a role Cpn plays as an asthma initiator remains to be determined. It is conceivable that acute Cpn infection, as an asthma initiator, could contribute to a substantial number of cases. This chapter focuses on the substantial body of evidence favoring a promoting role for Cpn, and also reviews the existing evidence for Cpn as an asthma initiator.

DAVID L. HAHN • East Clinic Division of Dean Medical Center, 1821 S. Stoughton Road, Madison, Wisconsin 53716

1.2. Definition of Asthma

Asthma can be defined either by diagnostic or functional criteria. This distinction is important. In the United States, clinicians try to make clear-cut diagnostic distinctions between asthma, chronic bronchitis, and emphysema whereas in European countries, particularly the Netherlands, the tendency is to view these diseases as a continuum. This debate between "splitters"[9] and "lumpers"[10] has not been resolved to everyone's satisfaction. American asthma specialists have tended to view asthma as a noninfectious allergic (atopic) condition beginning in childhood and also have tended to diagnose older patients with asthma symptoms as chronic obstructive pulmonary disease (COPD), particularly if a history of smoking was present.[11] It is now apparent from population-based epidemiological studies that (1) neither atopy[12] nor allergen exposure[13] are primary causes for childhood asthma, (2) half of asthma may be related to nonatopic (neutrophilic) rather than to atopic (eosinophilic) inflammation,[14,15] and (3) adult-onset asthma (AOA) is as frequent as childhood-onset asthma.[16] Of further interest are data showing that smoking is associated with Cpn infection as well as COPD.[17,18]

The natural history of obstructive airways syndromes may evolve over a lifetime,[19,20] airway-disease-related variables (age, sex, smoking, pulmonary function, and markers for atopy) have unimodal and continuous distributions across clinical diagnoses,[21] and a significant minority of patients with asthma-like symptoms cannot be placed in the classical diagnostic categories.[10,21] To avoid diagnostic bias, it is probably more scientifically accurate to view atopy and smoking, along with other factors associated with asthma (e.g. bronchial hyper responsiveness), as covariates rather than as diagnostic attributes. For purposes of population-based, patient-outcome-oriented primary care clinical research, I favor the use of a functional definition for asthma based solely on cardinal symptoms (cough, wheeze, chest tightness, shortness of breath) and objective evidence for reversible airways obstruction determined by lung function testing.

Persistent symptomatic reversible airway obstruction is a common entity encountered in primary care settings.[16,21] The majority of cases are diagnosed as asthma, but the "overlap" syndromes [chronic asthmatic bronchitis and asthma with chronic airways obstruction (AS-CAO[22])] are often difficult to distinguish from asthma in clinical practice, respond to similar treatments, are linked to asthma in epidemiological studies and may represent later stages in the natural history of reversible airway obstruction.[10,19,22] The concept that asthma and COPD might have a common underlying etiopathology characterized by a unique host response to exogenous stimuli was proposed by Orie[23] and has become known as the "Dutch Hypothesis." The Dutch refer to obstructive airways syndromes (asthma and COPD) as *chronic nonspecific lung disease* (CNSLD),[10] and chronic Cpn infection has been proposed as an etiologic factor in CNSLD.[20] As well as reviewing the evidence for a role in the initiation and promotion of asthma, this chapter also reviews the evidence suggesting a role for Cpn in the development of lung remodeling, a cardinal feature of COPD.

1.3. Importance of Asthma

Asthma is an important cause of respiratory morbidity (symptoms, impaired quality of life, medication side effects), health care utilization (clinic and emergency room visits, hospitalizations, medication costs), and mortality. National costs for asthma in the USA were over $6 billion in 1990[24] and are steadily increasing since asthma prevalence leaped almost 20%, from 10.4 million to 12.4 million, between 1990 and 1992[25,26] and continues to rise. The history of an "infectious initiation"[27] and markers for chlamydial infection[28] may be significant risk factors for the development of COPD from nonatopic AOA. If COPD sequelae are included in the equation, costs of asthma are much greater than currently reported, since more than 16 million adults have COPD, and COPD accounts for approximately 110,000 deaths per year, $18 billion in direct health care costs and almost $10 billion in indirect costs.[29,30]

The incidence of asthma is greatest in childhood but the prevalence of active asthma is equally distributed between children and adults because of the longer period of time available for accrual of new cases of adult asthma[31] and the greater likelihood for asthma remission to occur in childhood-onset compared to AOA.[32] These facts account for the data that, nationally, patients 18 years or older account for 72% of direct costs, 61% of indirect costs, and 66% of overall costs of asthma care, excluding medications.[24] In one clinical setting, adults 19 years or older accounted for 51% of the costs of asthma treatment, including medications.[33] Death from asthma is a rare event in children, but asthma mortality increases steadily with age such that the mortality rate in the elderly population (age 70 and older) can be more than 40 times the death rate in children aged 14 or less.[34] It has been hypothesized that this age-related asthma mortality is due to the premature development of fixed obstruction known to accompany long-standing asthma.[35] These utilization data correlate with the clinical picture of asthma derived from numerous observations: compared with childhood-onset asthma, adult-onset asthma is associated with fewer markers of atopy,[36] more likely to affect women,[37] more clinically severe,[38,39] less likely to remit,[40,41] and associated with more fixed obstruction.[22,42,43]

1.4. Current Asthma Treatments Are Palliative, Not Curative

Current asthma treatment is based on a paradigm of asthma as a noninfectious atopic condition whose "root cause" is inflammation.[44,45] It is now well established that chronically administered antiinflammatory medications, primarily inhaled corticosteroids (ICS), ameliorate asthma symptoms and improve prebronchodilator FEV_1 (forced expiratory volume in 1 s).[46–49] However, the therapeutic effects of ICS treatment are not maintained upon discontinuation,[47] implying that ICS treatment is for the most part suppressive (palliative), not curative. The hoped-for additional effect that ICS treatment prevents the accelerated development of fixed obstruction in childhood asthma is not supported by evidence.[50,51] A randomized controlled trial found that ICS administration failed to slow the decline of postbronchodilator FEV_1 in adult asthma,[49,52] and an uncontrolled, before–after observational trial in adults found that postbronchodilator FEV_1 actually declined more rapidly after the addition of ICS.[35] Any new asthma therapy that is superior to antiinflammatory drugs could have significant impacts on asthma morbidity, mortality, and costs of care.

2. *Chlamydia pneumoniae* AS A PROMOTER OF ASTHMA SEVERITY

Cpn-specific total Ig and the isotypes IgG and IgA antibodies have been associated with asthma in many studies of adults,[53] but serologic studies in childhood asthma have been mixed.[54,55] Conversely, several studies using culture or polymerase chain reaction (PCR) testing have shown high prevalence of Cpn in the upper airways of children with asthma,[54,56,57] but Cpn is rarely detected in upper-airway samples of adult asthmatics, although Cpn has been detected in adult asthmatic lower airways.[58,59] Furthermore, many studies have found that Cpn antibodies were associated with adult asthma severity.[6,7,54,57,60–64] It is generally recognized that children do not develop antibodies against Cpn as readily as adults and therefore serology is a less reliable marker of possible chronic infection in children as compared with adults. Conversely, positive serology in adults could indicate previous exposure rather than persistent infection so that serologic associations in the absence of organism detection in the lung are not conclusive evidence for infection. Thus, currently available serologic evidence upon which most of the Cpn–asthma associations are based serves as a basis for more detailed investigation but is not in itself proof of causation. This further evidence must include results of antichlamydial antibiotic treatment studies in asthma that are confounded by potential antiinflammatory effects of the agents employed and by uncertainty about whether persistent Cpn infection can be eradicated by currently available antibiotic treatments. Thus, it will be challenging to produce conclusive evidence that chronic Cpn infection causes or contributes to persistent asthma.

2.1. *Chlamydia pneumoniae*–Asthma Serologic Associations

A 1999 review of 18 controlled epidemiological studies containing over 4000 cases and controls reported significant associations between Cpn and asthma in 15 of 18 studies.[53] An additional 12 positive case–control studies were presented at the 2000 Meeting of the European Society for Chlamydia Research and at the 2000 European Respiratory Society Meeting.[63,65–75] Some of these studies found significant associations with PCR positivity in airway secretions as well as with Cpn-specific antibodies. Overall, the 27 positive studies included both children and adults with asthma; some reported stronger associations with nonatopic asthma than with atopic asthma, but differentiation on the basis of atopy was not universal. Some studies that measured only IgG antibodies failed to detect significant differences between cases and controls,[76–78] whereas studies that included IgA antibodies were more often positive.[7,53,79–82] A number of recently published seroepidemiologic studies add further evidence for a significant association between asthma and Cpn-specific IgA[7,79–81] and also with chlamydial heat shock protein–60 (Hsp60) antibodies.[63,66,83] The latter observation, of an association between asthma and chlamydial Hsp60 antibodies, is intriguing because of the established link between chlamydial Hsp60 seroreactivity and the known chlamydia-caused diseases—trachoma, pelvic inflammatory disease, and tubal infertility.[84–86]

Several studies have found evidence that acute Cpn infection, when present in acute exacerbations of asthma[82,87] or of chronic bronchitis[88] is associated with increased sputum neutrophil and eosinophil cationic protein levels,[82] asthma exacerbation severity,[87] and chronic bronchitis exacerbation frequency.[88] A larger number of studies have found significant associations between markers of possible chronic infection and markers of asthma severity (Table I).

Cpn-specific antibodies, in the absence of acute infection, have been associated with asthma severity as measured by asthma symptom frequency,[61] number of exacerbations,[56] quantity of inhaled corticosteroid medication use,[62] and symptom severity classification.[7] Especially Cpn-specific IgA antibodies, both in serum[6,7,62,63] and in sputum[63] or nasal washings,[56] have been associated with asthma severity, suggesting that this short-half-life antibody may be a useful marker for chronic infection.[79] Two preliminary treatment studies, one in children[54] and one in adults,[61] also suggest that macrolide treatment for Cpn-associated asthma is more successful in less severe disease and before the occurrence of lung remodeling.

2.2. Results of Antibiotic Treatment Directed against Chlamydial Infection in Asthma

2.2.1. *Macrolide Antibiotics Have Been Used in Asthma, but Indications and Mechanisms Are Unclear*

A role for triacetyloleandomycin (TAO), a macrolide antibiotic, in the treatment of asthma was first suggested in 1959.[90] It remains unclear to what extent antiasthma effects of TAO and other macrolides may be related to "steroid sparing" activity,[91–93] inhibition of theophylline clearance,[94] direct antiinflammatory effects,[95–97] or to other mechanisms such as direct antiviral activity.[98] Early[99] and more recent[100] clinical observations indicated that TAO and erythromycin might improve symptoms and bronchial hyperresponsiveness, respectively, in patients who were not receiving corticosteroids or theophylline. These observations suggest that effects on steroid and theophylline metabolism alone cannot account for the antiasthma properties of macrolides. Of additional interest is a preliminary report of prolonged remissions in severe asthma after withdrawal of long-term TAO therapy.[101] Because remission of severe asthma is unusual, this observation begs an explanation, as antiinflammatory or antiviral effects of macrolides are expected to wane soon after treatment discontinuation. An antibiotic effect was first suggested by clinical observations documenting

TABLE I
Studies Finding Associations Between *Chlamydia pneunoniae* Infection and Asthma Severity

Reference	Asthma population/ Study design	Results
Emre *et al.*[54]	12 Cpn culture-positive children/Case series	Macrolide treatment was less effective in more severe asthma.
Hahn *et al.*[60]	Grp1: 12 adults with persistent asthma; Grp2: 30 adults with intermittent asthma; Grp3: 89 with nonwheezing respiratory illnesses/Case–control	Cpn seropositivity (total Cpn-specific Ig titers > 1:16) was strongly associated with asthma severity (100% of Grp1, 80% of Grp2, 53% of Grp3, $P < 0.001$). Antibody geometric mean titer (GMT) was also associated with severity (Grp1 = 76, Grp2 = 29, Grp3 = 19, $P = 0.0001$).
Hahn[61]	46 Cpn seropositive adults (some culture positive)/ Before–after trial	Macrolide treatment was less effective in asthma patients with a coexisting component of fixed obstruction.
Cook *et al.*[6]	1518 nonasthmatic hospital controls, 123 acute asthma patients with exacerbations admitted to hospital (chronic bronchitis and COPD excluded), 46 severe chronic asthma outpatients/Case–control	Chronic/recent infection (IgG64-256 or IgA $\geq$ 8) present in 12.7% of controls, 14.6% of exacerbations, and 34.8% of severe asthma (Odds ratio 3.99, 95% confidence interval 3.6–9.9, for severe asthma versus controls).
Cunningham *et al.*[56]	108 children with asthma, aged 9–11, enrolled in a 13-month longitudinal community-based study/ Prospective cohort	Cumulative frequency of Cpn PCR+ in nasal washings was 45% (independent of symptom status). Children with multiple exacerbations were more likely to remain PCR+ ($P < 0.02$). Amount of Cpn-specific secretory IgA was more than 7 times higher in children with four or more exacerbations versus 1 exacerbation ($P < 0.02$).
Black *et al.*[62]	619 adult asthma patients/Nested case–control	Use of high-dose inhaled corticosteroids (ICS) was associated with increased Cpn-specific IgG ($P = 0.04$), and IgA ($P = 0.0001$) seropositivity compared to use of low-dose ICS. In patients with IgG $\geq$ 1:64 and/or IgA $\geq$ 1:16, there was an inverse association between IgG and FEV_1 ($P = 0.04$) and IgA antibodies were associated with a higher daytime asthma symptom score ($P = 0.04$).

TABLE I
(*Cont.*)

Reference	Asthma population/ Study design	Results
Huittinen *et al.*[63]	105 mild–moderate asthmatics and 33 healthy controls/ Case–control	Serum IgA heat shock protein–60 (Hsp60) antibodies (40% cases, 22% controls, $P < 0.05$), sputum IgA (51% cases vs. 25% controls, $P < 0.01$) and sputum IgA Hsp60 (41% vs. 22%, $P < 0.05$) correlated with asthma severity.
Schmidt *et al.*[89]	106 children (66 male, 40 female, ages 1 month to 17 years) undergoing bronchoscopy for therapy-resistant obstructive symptoms and/or recurrent or chronic bronchitis/ pneumonia, without identification of any other cause/Nested case–control	52% were PCR+ on BAL (half were strongly positive and half were weakly positive; the investigators suggest that weak positives might indicate chronic infection). Compared to PCR– children, PCR+ children had less eosinophilia of the nasal mucosa, less total serum IgE antibodies, and worse pulmonary function. However, weak positive PCR patients had the highest rates of allergic sensitization, reduction in lung volume, and the most obstruction.
Von Hertzen *et al.*[7]	116 adults with asthma from a chest clinic (31 men, 85 women; 13 severe, 54 moderate, 49 mild asthma) and 50 blood donor controls (31 men, 19 women)/ Case–control	Severe and moderate asthma were significantly associated with elevated IgA antibody titers to Cpn: ORs were 5.6 (95% CI 1.3–24) for severe asthma and 5.7 (2–16) for moderate asthma. cHsp60 antibodies were more frequent and of higher titer among asthmatics than controls but the differences were not significant.

prolonged asthma improvement and even remission after microbiologic eradication of Cpn in infected asthma patients.[54,58,61] The first randomized, controlled trial of a macrolide to treat asthma[102] documented a significant treatment effect on morning peak expiratory flow rate (PEFR) immediately after treatment but this result could have been due either to an antiinflammatory or to an antibiotic effect.[103] It is now recognized that macrolides possess significant antiinflammatory effects that may be useful to treat a variety of chronic inflammatory lung diseases.[104] The dramatic beneficial effect of low-dose, long-term macrolide treatment in diffuse panbronchiolitis (DPB), a chronic inflammatory lung disease prevalent in Japan,[105] is an example. Cpn had been continuously isolated from a patient with DBP,[106] but chlamydial infection does not appear to contribute significantly to the etiology,[107] establishing macrolides as having a profound nonantibiotic effect in DPB that requires continuous therapy to maintain benefit. Another example is cystic fibrosis, in which significant macrolide treatment benefit has been demonstrated and interpreted as due to an antiinflammatory effect;[108] however, some cases of CF have been associated with Cpn infection.[109] Current controversy surrounds the issue whether beneficial effects of macrolide treatment for asthma and related conditions are due to antiinflammatory[110] or to antibiotic[111] mechanisms, but all parties agree that future randomized, controlled trials are warranted to resolve this issue.[103,111] Distinguishing an antiinflammatory mechanism from an antibiotic one is important, among other reasons, because of implications for length of treatment.

2.2.2. *Antichlamydial Antibiotic Treatment for Asthma*

Table II summarizes results of preliminary treatment studies that support the need for further randomized trials. Table III illustrates that the positive treatment effect of antibiotics may be equivalent to or greater than that of inhaled corticosteroids (ICS).

2.2.2a. Randomized Controlled Trials

Evidence for a promoting role for chronic Cpn infection in asthma requires randomized treatment trials for confirmation.[115] No randomized, controlled trials (RCTs) in children and two RCTs of antichlamydial treatment for adult asthma have been published[102,116] (Table I). Kraft *et al.*[114] reported that 31 of 55 (56%) adults with asthma were PCR positive for *M. pneumoniae* (Mpn), Cpn, or both in bronchoalveolar lavage (BAL) fluid or bronchial biopsy and that only PCR-positive subjects randomized to 6 weeks of clarithromycin treatment had improved pulmonary function and reduced expression of IL-5, suggesting an antibiotic rather than an antiinflammatory effect. Black *et al.*[102] randomized 232 adults with asthma and serologic evidence of previous Cpn exposure to 6 weeks of roxithromycin therapy and reported that this macrolide treatment was associated with a significant improvement in morning peak expiratory flow rate (PEFR) at the end of treatment. Although the level of improvement in morning PEFR in treated subjects was maintained, the significant difference from the control group was no longer present at 3 and 6 months because of steady gradual improvement in the control group. Therefore, the results of this study leave unanswered the question whether Cpn is important in asthma.[103] The Black *et al.*[102] trial contained methodological deficiencies, including lack of (1) monitoring for changes in asthma controller medications (to control for "medication confounding": the potential for changing doses of controller medication in the usual care setting), (2) performance of direct organism detection (indicating whether subjects were actually infected at baseline, whether treatment eradicated infection, and whether cessation of treatment was followed by reestablishment of infection), and (3) inclusion of a control group of asthmatic patients with asthma of similar severity but without evidence of infection (to determine whether observed effects were a result of antiinflammatory activity).[103] My group recently completed a multisite randomized, placebo-controlled pilot feasibility trial employing 6 weekly doses of azithromycin in 45 adult primary care outpatients with persistent asthma and found consistent 3-month posttreatment trends favoring azithromycin for all measures of asthma-specific symptoms, rescue medication use, and quality of life

TABLE II
Treatment Results Supporting a Causal Role for Cpn Infection in Asthma

Reference	Asthma population/ Study design	Results
Emre *et al.*[54]	12 Cpn culture-positive children/Case series	Lasting improvement in symptoms and pulmonary function associated with microbiologic eradication of Cpn.
Hahn[61]	46 Cpn seropositive adults (some culture positive)/ Before–after trial	25 had lasting improvement or complete remission in symptoms and improved pulmonary function.
Hahn *et al.*[112]	1 adolescent & 2 adults with severe, steroid-dependent asthma and serologic evidence suggesting Cpn infection/Case series	Significant improvement after prolonged antibiotics; all three able to discontinue oral steroids.
Hahn *et al.*[58]	Case of nonatopic AOA/Case report	Improvement in asthma symptoms, FEV_1 and quality-of-life documented after microbial eradication of Cpn from BAL[a] fluid.
Espositio *et al.*[113]	71 wheezing children of whom 24 had Mpn or Cpn infection/ Nonrandomized treatment	Macrolide treatment significantly decreased relapse rate in Mpn/Cpn infected children (0 vs. 69% relapsed within 3 months) although 2/3 Cpn PCR+ cases remained PCR+ after treatment.
Black *et al.*[102]	232 Cpn seropositive adults/RCT[b]	Roxithromycin group (6-week treatment) had persisting increase in PEF[c] that was statistically significant only at 6 weeks.
Kraft *et al.*[114]	55 adult asthmatics of whom 31 were PCR positive for Mpn or Cpn/RCT[b]	Only PCR+ asthmatics receiving clarithromycin (6-week treatment) had significantly improved pulmonary function and decreased IL-5 cytokine expression in BAL.
Hahn *et al.* (unpublished)	45 adult asthmatics/pilot RCT[b]	Consistent trends in favor of azithromycin (6-week treatment) for all measures of asthma symptoms, rescue medication use and asthma-specific quality of life 3 months posttreatment.

[a]Bronchoalveolar lavage.
[b]Randomized, controlled trial.
[c]Peak expiratory flow.

(Hahn, D. L., Plane, M. B., personal communication). Positive effects 3 months after treatment cessation are consistent with an antibiotic mechanism. Our preliminary results need to be confirmed by larger trials before firm conclusions can be drawn, however.

In summary, elevated antibody titers to Cpn, but not to other organisms, are associated with asthma.[64] Many studies find that Cpn titers are associated with markers for asthma severity and that the association is stronger for long-standing

TABLE III
Comparison of Open-Label Results of Azithromycin vs. Blinded Result of Inhaled Corticosteroids on FEV_1 in Asthma

		Control Groups			Treated Groups		
	Study duration, months.	No. of subjects	Baseline $FEV_1{}^a$	% change	No. of subjects	Baseline $FEV_1{}^a$	% change
RCTs: inhaled corticosteroids							
Haahtela *et al.*[48]	24	53	87%	0	50	87%	+4
Juniper *et al.*[46]	12	16	90%	0	16	92%	0
Kerstjens *et al.*[49]	30	183	63%	−1	91	65%	+10†
Vathenen *et al.*[47]	1 1/2	16	98%	−6	18	96%	+8
Azithromycin							
Hahn[61]	6	–	–	–	46	68%	+12

[a]Percent predicted †maximum response at 3 months posttreatment, then FEV_1 declined 0.033L/year (same as in control group).

than for recent-onset asthma.[117] These data have been interpreted to indicate that Cpn is primarily a promoter rather than an initiator of asthma.[64] It has also been acknowledged, however, that the data do not exclude a role for Cpn in asthma initiation.[7]

3. *Chlamydia pneumoniae* AS AN ASTHMA INITIATOR

3.1. Asthma Is Often Associated with Preceding Respiratory Illnesses

Patients developing AOA often recall that their asthma symptoms started after an acute respiratory illness such as acute bronchitis, pneumonia, or an influenza-like illness.[27] Clinical observations[118,119] and prospective epidemiological studies[120–122] also support an association between bronchitis/pneumonia and subsequent AOA. While these observations have often been interpreted to suggest that the preceding illnesses were actually misdiagnosed asthma symptoms or merely viral exacerbations of previously unrecognized asthma, a third possibility is that acute infectious illnesses might actually play a role in asthma initiation. Epidemiological associations of respiratory illness and subsequent asthma also pertain to children[123] and adolescents.[124] The risk of developing asthma has been associated with close household contact with nonrelatives (friends and spouses).[125] This observation could be explained by an infectious agent as a cause for asthma.[126]

3.2. New-Onset Asthma after Acute Cpn Infection

Soon after the discovery of the TWAR organism (now called *C. pneumoniae*) several case reports of asthma initiation after acute Cpn infection were published.[127–129] Subsequently, I and my colleague, Roberta McDonald, performed a prospective study that included 10 adult outpatients presenting in primary care clinical practice with a first-ever reported wheezing attack, representing all such cases encountered over a 10-year time period.[8] All 10 patients met serologic criteria for an acute primary ($n = 8$) or an acute secondary ($n = 2$) Cpn infection using criteria that included a fourfold or greater titer rise and/or presence of Cpn-specific IgM antibody. Six of the 10 had persistent wheezing symptoms that eventually met American Thoracic Society criteria for chronic asthma ($n = 5$) or chronic bronchitis ($n = 1$). The latter patient was Cpn-culture-positive during the chronic bronchitic phase of his illness.[8] After 6 months to 2 years of persisting symptoms, all subjects were treated with prolonged courses of antibiotics with antichlamydial activity, and asthma symptoms resolved completely in every case.[61] Further details on these patients are presented in Table IV. These data linking serologic evidence of acute Cpn infection with new-onset asthma that appeared to respond to antimicrobial treatment strongly suggest the possibility that Cpn can initiate asthma in previously asymptomatic individuals. Attempts to replicate these findings are of high priority to determine the quantitative contribution of Cpn infection to new-onset asthma. In my opinion, this research must include prospective studies in primary care settings where most acute lower respiratory illnesses and new-onset asthma present.

3.3. Infection Should Be Added to the List of Possible Causes for Worldwide Increases in Asthma

A consensus exists that asthma prevalence and mortality have been increasing worldwide in recent decades, but the reasons for these changes in frequency and severity of disease are unknown.[130] None of the proposed risk factors (e.g., changes in atopy, smoking, air pollution, poor housing, better-insulated housing, etc.) appear to explain these temporal trends.[131] Strachan[131] has pointed out that quite large changes in the level of relatively powerful causal agents are required to explain documented increases in asthma prevalence. An infectious disease pandemic represents one candidate as a powerful causal agent. A role for Cpn infection as a cause for worldwide increases in asthma is plausible and has been suggested.[132] Ecologic data show that Cpn infection[133] and asthma[134] in all age groups and in both sexes have increased simultaneously in Finland over recent decades. Considering the possibility that worldwide increases in asthma are due to an infectious disease pandemic, further research into the Cpn–asthma association assumes great importance.

4. SUMMARY: *Chlamydia pneumoniae* AS AN ASTHMA INDUCER

A growing body of evidence (based on culture isolation, polymerase chain reaction (PCR) detection, serologic studies, and treatment results) links Cpn infection with asthma primarily in adults,[53,59,135] and in children as well.[54,56] Nonatopic asthma has been associated with an infectious initiation in general[27] and with Cpn infection in particular.[8,117,128] Cpn respiratory tract infections can initiate asthma[8] and are present in up to one-half of adults with asthma.[116] It is even possible that the contribution of Cpn infection to asthma is so great

TABLE IV
Clinical Data[a] in 10 Patients with *Chlamydia pneumoniae* Infection and *de novo* Wheezing[b]

Age, Sex	Date	Total Ig	IgM	IgG	IgA	Clinical description
1. 37, M	10/13/88 (4 days)[c]	128	16			Bronchitis with wheezing
	11/14/88	128	0			
	9/13/89	128	0			No asthma
2. 39, M	11/21/88 (5 days)	0	16			Laryngitis, bronchitis with wheezing
	12/23/88	256	128			
	4/17/89	64	16			
	6/29/89	32	0			No asthma
3. 59, F	3/9/89 (7 days)	16	32			Pneumonia with wheezing
	4/3/89	256	512			
	7/28/89	64	64			
	11/27/89	32	0			No asthma
4. 35, F	8/22/89 (24 days)	512	0			Bronchitis with wheezing[d]
	10/16/89	4096	0			
	9/18/91	256	0			No asthma
5. 55, M	1/26/89 (4 days)	64	16			COPD, pneumonia with wheezing
	1/10/92	512	0	256	256	Asthma diagnosed
	2/6/92	256	0	256	256	
6. 47, F	3/21/89 (48 days)	256	16			Bronchitis with wheezing
	4/21/89	256	0			Persistent wheezing
	9/16/89	256	0			Asthma diagnosed
	2/3/92	256	0	256	16	
	3/5/92	512	0	256	16	
	2/18/93	256	0	512	16	
	5/26/93	256	0	256	16	
	1/17/96	256	0		32	
7. 51, M	4/24/90 (35 days)	256	64			COPD, bronchitis with wheezing
	5/21/90	128	128			Persistent wheezing
	11/11/91	256	0	128	16	Asthma diagnosed
	12/11/91	256	0	128	16	
	3/6/92	256	0	128	16	
	2/19/93	256	0	128	16	
8. 39, F	10/29/93 (3 days)	32	0		<8	Bronchitis with wheezing
	3/24/94	128	0		16	Asthma diagnosed
9. 56, M	4/11/94 (70 days)	512	256		≥ 64	Community-acquired pneumonia with wheezing
	6/17/94	1024	32		≥ 64	Chronic bronchitis diagnosed[d]
	3/31/95	512	8		64	
10. 35, F	7/7/94 (66 days)	1024	128		≥ 64	Bronchitis with wheezing
	8/5/94	1024	64		≥ 64	Persistent wheezing
	10/17/94	1024	16		≥ 64	
	12/19/94	1024	8		≥ 64	Asthma diagnosed

[a]Reprinted with permission of *Ann. Allergy Asthma Immunol.* 1998;339–344. Copyright 1998.
[b]Defined as the first-ever wheezing episode experienced by the patient. Missing IgG and IgA results were due to unavailability of sera.
[c]Days post illness onset.
[d]Culture positive.

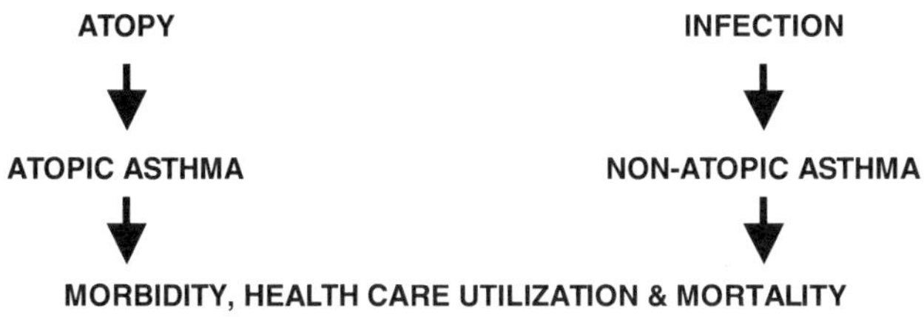

FIGURE 1. Schematic illustration of current model of dual etiologies—atopy and infection—producing atopic and nonatopic asthma syndromes, respectively.

that infection has played a role in causing worldwide increases in reactive airway diseases over the past several decades. The emerging concept of infection as a potential contributing cause for asthma is illustrated in Fig. 1.

5. *Chlamydia pneumoniae* AND LUNG REMODELING

5.1. Infection May Also Be Related to Airway Remodeling in Asthma and COPD

The combination of adult-onset nonatopic asthma and Cpn seroreactivity has been associated specifically with an accelerated decline in lung function, suggesting that Cpn infection could contribute to the development of lung remodeling and COPD (Fig. 2). Ten Brinke *et al.* studied 101 adults with severe asthma recruited from outpatient pulmonary clinics at 10 Dutch hospitals.[(28)] Fifty-one percent (52 subjects) had early-onset asthma (before age 18) and 49% had AOA, with 24 subjects (24%) having adult-onset nonatopic asthma associated with IgG antibody positivity against Cpn. This latter group exhibited a decline in lung function that was 4 times greater than the other three groups combined (2.3% per year for Cpn-positive nonatopic AOA versus 0.5% per year for other subjects). Early-onset and adult-onset study groups were well matched in age, sex, asthma severity, pulmonary function, serologic parameters, and smoking status. The rate of decline of lung function in this cross-sectional, retrospective study was comparable to that reported in prospective studies, and recall bias could not explain the differences predicted by serological status.[(28)] The authors suggest that "The striking association between seropositivity to *C. pneumoniae* and increased decline in lung function in AOA is compatible with the hypothesis that this respiratory pathogen might be involved in airway remodeling."[(28)]

Additional studies[(53)] have described associations between Cpn serology and COPD, and it has been hypothesized that Cpn infection augments smoking-associated inflammation in COPD.[(136)] Cpn has been documented by PCR, immunohistochemistry (IHC), and/or electron microscopy (EM) in COPD,[(137,138)] and also in emphysema,[(139)] lung tissue. These observations are compatible with the idea that chronic chlamydial lung infection may relate to the pathogenesis of established chlamydial diseases such as trachoma, pelvic inflammatory disease and tubal infertility that involve inflammation, and fibrosis and scarring

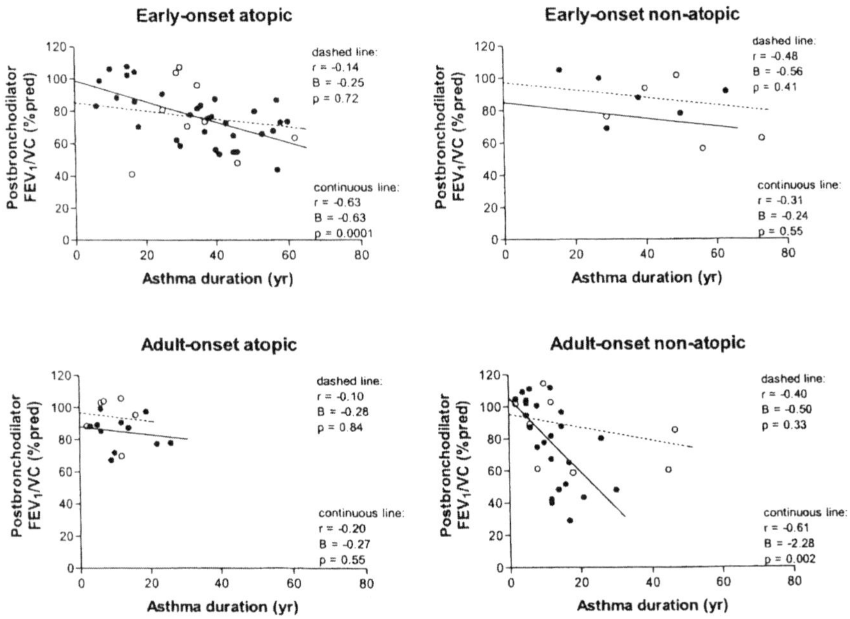

FIGURE 2. Relationship between asthma duration and postbronchodilator FEV_1/VC (percent predicted [% pred]) in different subgroups of patients with severe asthma according to age of onset of asthma, atopic status, and *C pneumoniae* IgG seropositivity (filled circles and continuous lines) versus seronegativity (open circles and dashed lines). Only in the subgroup of patients with adult-onset nonatopic asthma was a significant difference ($P = 0.03$) in the slopes of the regression lines observed. B, Slope of the regression line, with corresponding P value. Reprinted from *J. All. Clin. Immunol.* **107**, ten Brinke *et al.*, "Persistent airflow limitation in adult-onset nonatopic asthma is associated with serologic evidence of *Chlamydia pneumoniae* infection," pp. 449–54, Copyright (2001), with permission from Elsevier Science.

resulting from an immunopathologic host response to chronic infection. A recent randomized trial of long-term, low-dose erythromycin in COPD reported profound beneficial effects of the macrolide on incidence of mild and severe COPD exacerbations but provided no information about the underlying mechanism of action.[140] These epidemiological and histologic studies are complemented by a variety of *in vitro* and *in vivo* pathogenesis studies that are reviewed below.

5.2. Cpn Produces Cytokines Linked to Asthma and Lung Remodeling

In vitro studies have demonstrated that Cpn infection of relevant lung cells can produce cytokines associated with asthma inflammation.[141] Potential pathogenic factors include release of reactive oxygen species, TNF-alpha, IL-1beta, and IL-8 from *in vivo* and *ex vivo* Cpn infected alveolar macrophages[142] and IL-8, prostaglandin-E2, ICAM-1, cyclooxygenase-2 and NF-kB upregulation in human airway epithelial cells, in association with transepithelial migration

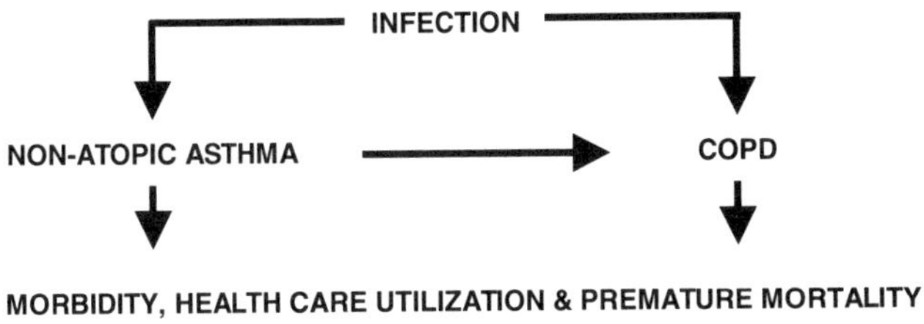

FIGURE 3. An expanded model illustrating the proposed pathways whereby infection influences nonatopic asthma and COPD to produce morbidity, health care utilization, and mortality.

of PMNs.[143] Relevance of these results to clinical asthma is highlighted by additional findings that IL-8 and neutrophil inflammation have been associated specifically with non atopic asthma, whereas eosinophil and mast cell inflammation are present in both atopic and non atopic asthma.[14] It is intriguing to note that a bronchoscopic study of 55 adults with asthma found that mast cells were significantly more prevalent in the 31 subjects who were PCR positive for Mpn and/or Cpn, compared to PCR negative subjects.[116] *In vitro* experiments also demonstrate that Cpn can induce factors relevant to the process of lung remodeling: human bronchial smooth muscle cells infected by Cpn produce IL-6, basic fibroblast growth factor and interferon-beta[144,145] and Cpn-infected human macrophages express matrix metalloproteinase[146] and 92-kD gelatinase.[147] Collectively, these factors may contribute to the development of airway smooth muscle hypertrophy, myofibroblast proliferation and excess extracellular matrix deposition and/or degradation that characterize stages of asthma lung remodeling[148] and airway damage in COPD.

5.3. Summary: *Chlamydia pneumoniae* and Lung Remodeling

Taken together, the studies reviewed in this section lay the groundwork for the hypothesis that persistent Cpn infection in lung tissue can accelerate the process of lung remodeling, the hallmark of COPD. Based on this evidence, an expanded model for the role of infection in the development of obstructive airways disease is presented in Figure 3. Since nonatopic asthma (i.e., the asthma syndrome most likely to be caused by infection) may account for up to 50% of cases, Cpn infection could potentially have a major public health impact on asthma, and ultimately perhaps also on the treatment and/or prevention of other obstructive airways diseases such as chronic bronchitis and emphysema (COPD).

6. CONCLUSION: *Chlamydia pneumoniae*, CHRONIC NONSPECIFIC LUNG DISEASE (CNSLD) AND THE "DUTCH HYPOTHESIS"

According to Vermeire *et al.*,[9] the CIBA Symposium proposed chronic nonspecific lung disease (CNSLD) in 1959 as an umbrella term for chronic bronchitis, asthma, emphysema and irreversible or persistent obstructive lung disease. In 1961 Orie[23] proposed the "Dutch Hypothesis" which stated that CNSLD represented different expressions of a single disease entity characterized by an hereditary predisposition to develop allergy and bronchial hyper reactivity in response to environmental factors. Prior to the discovery of the Cpn–asthma association, the pros and cons of the "Dutch Hypothesis" were fully debated without the hypothesis being proven or disproven.[9,10] It should be obvious to the reader of this review that the discovery of Cpn as a potential factor in asthma and COPD casts new light on the importance of examining the concept of CNSLD as a pathophysiologic entity.

It is now well established that acute Cpn infection can cause acute bronchitis and pneumonia[149,150] and additional evidence presented herein suggests that lower respiratory tract illnesses caused by acute Cpn infection can develop into asthma and chronic bronchitis.[8,151] Chronic Cpn infection has also been associated with a wide variety of chronic upper-airway illnesses[152,153] as well as with the spectrum of acute and chronic lower-airway conditions including acute bronchitis,[154] asthma and COPD.[81] Taken together, these data suggest a role for Cpn in the entire spectrum of respiratory illnesses embracing the natural history of CNSLD.[19,20,155] Just as early identification and treatment of genital chlamydial infection of women is required to prevent the occurrence of pelvic inflammatory disease and tubal infertility, and timely treatment of eye infection in children is required to prevent blinding trachoma, early identification and treatment of Cpn infection in chronic airways disease will be important to prevent the development of chronic sequelae, if Cpn is confirmed as a treatable cause for even a subset of CNSLD.

ACKNOWLEDGMENTS. I would like to thank Mary Beth Plane for a critical review of a previous draft of this chapter.

REFERENCES

1. Allegra, L., Blasi, F., Centanni, S., Cosentini, R., Denti, F., Raccanelli, R., Tarsia, P., and Valenti, V., 1994, Acute exacerbations of asthma in adults: Role of *Chlamydia pneumoniae* infection, *Eur. Respir. J.* **7:**2165–2168.
2. Clementsen, P., Permin, H., and Norn, S., 2002, *Chlamydia pneumoniae* infection and its role in asthma and chronic obstructive pulmonary disease, *J. Invest. Allergol. Clin. Immunol.* **12:**73–79.
3. Lieberman, D., Lieberman, D., Printz, S., Ben-Yakov, M., Lazarovich, Z., Ohana, B., Friedman, M. C., Dvoskin, B., Leinonen, M., and Boldur, I., 2003, Atypical pathogen infection in adults with acute exacerbation of bronchial asthma, *Am. J. Respir. Crit. Care Med.* **167:**406–410.
4. Johnston, S. L., Pattemore, P. K., Sanderson, G., Smith, S., Lampe, F., Josephs, L., Symington, P., O'Toole, S., Myint, S. H., Tyrrell, D. A. J., and Holgate, S. T., 1995, Community study of role of viral infections in exacerbations of asthma in 9–11 year old children, *Br. Med. J.* **310:**1225–1228.
5. Nicholson, K. G., Kent, J., and Ireland, D. C., 1993, Respiratory viruses and exacerbations of asthma in adults, *Br. Med. J.* **307:**982–986.

6. Cook, P. J., Davies, P., Tunnicliffe, W., Ayres, J. G., Honeybourne, D., and Wise, R., 1998, *Chlamydia pneumoniae* and asthma, *Thorax* **53:**254–259.
7. von Hertzen, L., Vasankari, T., Liippo, K., Wahlström, E., and Puolakkainen, M., 2002, *Chlamydia pneumoniae* and severity of asthma, *Scand. J. Infect. Dis.* **34:**22–27.
8. Hahn, D. L., and McDonald, R., 1998, Can acute *Chlamydia pneumoniae* infection initiate chronic asthma? *Ann. Allergy Asthma Immunol.* **81:**339–344.
9. Vermeire, P. A., and Pride, N. B., 1991, A "splitting" look at chronic nonspecific lung disease (CNSLD): Common features but diverse pathogenesis, *Eur. Respir. J.* **4:**490–496.
10. Sluiter, H. J., Koëter, G. H., de Monchy, J. G. R., Postma, D. S., de Vries, K., and Orie, N. G. M., 1991, The Dutch hypothesis (chronic non-specific lung disease) revisited, *Eur. Respir. J.* **4:**479–489.
11. Yunginger, J. W., Reed, C. E., O'Connell, E. J., Melton, L. J., O'Fallon, W. M., and Silverstein, M. D., 1992, A community-based study of the epidemiology of asthma: Incidence rates, 1964–1983, *Am. Rev. Respir. Dis.* **146:**888–894.
12. Pearce, N., Pekkanen, J., and Beasley, R., 1999, How much asthma is really attributable to atopy? *Thorax* **54:**268–272.
13. Pearce, N., Douwes, J., and Beasley, R., 2000, Is allergen exposure the major primary cause of asthma? *Thorax* **55:**424–431.
14. Amin, K., Lúdvíksdóttir, D., Janson, C., Nettelbladt, O., Björnsson, E., Roomans, G. M., Boman, G., Sevéus, L., and Venge, P., 2000, Inflammation and structural changes in the airways of patients with atopic and nonatopic asthma, *Am. J. Respir. Crit. Care Med.* **162:**2295–2301.
15. Douwes, J., Gibson, P., Pekkanen, J., and Pearce, N., 2002, Non-eosinophilic asthma: Importance and possible mechanisms, *Thorax* **57:**643–648.
16. Hahn, D. L., and Beasley, J. W., 1994, Diagnosed and possible undiagnosed asthma: A Wisconsin Research Network (WReN) study, *J. Fam. Pract.* **38:**373–379.
17. Hahn, D. L., and Golubjatnikov, R., 1992, Smoking is a potential confounder of the *Chlamydia pneumoniae*–coronary artery disease association, *Arterioscler. Thromb.* **12:**945–947.
18. Karvonen, M., Tuomilehto, J., Pitkäniemi, J., Naukkarinen, A., and Saikku, P., 1994, Importance of smoking for *Chlamydia pneumoniae* seropositivity, *Int. J. Epidemiol.* **23:**1315–1321.
19. Hahn, D. L., 2002, Evaluation and management of acute bronchitis, in: *20 Common Problems in Respiratory Disorders* (W. J. Hueston, eds.), New York, McGraw-Hill, pp. 141–153.
20. Hahn, D. L., 2002, *Chlamydia pneumoniae* and the "Dutch Hypothesis," *Chest* **122:**1510–1512.
21. Brand, P. L., Kerstjens, H. A., Postma, D. S., Sterk, P. J., Quanjer, P. H., Sluiter, H. J., Dijkman, G. H., van Herwaarden, C. L., Hilvering, C., and Jansen, H. M., 1992, Long-term multicentre trial in chronic nonspecific lung disease: Methodology and baseline assessment in adult patients. Dutch CNSLD Study Group, *Eur. Respir. J.* **5:**21–31.
22. Burrows, B., 1991, Epidemiologic evidence for different types of chronic airflow obstruction, *Am. J. Respir. Dis.* **143:**1452–1455.
23. Orie, N. G. M., Sluiter, H. J., Vries, K. D., Tammeking, G. J., and Witkop, J., 1961, The host factor in bronchitis, in: *Bronchitis: An International Symposium, 27–29 April 1960, University of Gronigen, the Netherlands* (N. G. M. Orie and H. J. Sluiter, eds.), Assen, Netherlands; Springfield, Illinois, Royal Vangorcum; Charles C. Thomas, pp. 44–59.
24. Weiss, K. B., Gergen, P. J., and Hodgson, T., 1992, An economic evaluation of asthma in the United States, *New Engl. J. Med.* **326:**862–866.
25. Adams, P. F., and Benson, V., 1991, Current estimates from the National Health Interview Survey, National Center for Health Statistics, *Vital Health Statistics* **10**.
26. Centers for Disease Control and Prevention. Vital and Health Statistics, Current Estimates from the National Health Interview Survey, 1992 (U.S. Department of Health and

Human Services, Public Health Service, National Center for Health Statistics), 1994, DHHS Publication No. PHS 94–1517.

27. Hahn, D. L., 1995, Infectious asthma: A reemerging clinical entity? *J. Fam. Pract.* **41**:153–157.
28. ten Brinke, A., van Dissel, J. T., Sterk, P. J., Zwinderman, A. H., Rabe, K. F., and Bel, E. H., 2001, Persistent airflow limitation in adult-onset nonatopic asthma is associated with serologic evidence of *Chlamydia pneumoniae* infection, *J. Allergy Clin. Immunol.* **107**:449–454.
29. Bach, P. B., Brown, C., Gelfand, S. E., and McCrory, D. C., 2001, Management of acute exacerbations of Chronic Obstructive Pulmonary Disease: A summary and appraisal of the published evidence, *Ann. Intern. Med.* **134**:600–620.
30. Sin, D. D., Stafinski, T., Ng, Y. C., Bell, N. R., and Jacobs, P., 2002, The impact of chronic obstructive pulmonary disease on work loss in the United States, *Am. J. Respir. Crit. Care Med.* **165**:704–707.
31. Cookson, J. B., 1987, Prevalence rates of asthma in developing countries and their comparison with those of Europe and North America, *Chest* **91**:97S–103S.
32. Burrows, B., 1987, The natural history of asthma, *J. Allergy Clin. Immunol.* **80**:375S–377S.
33. Stempel, D. A., Hedblom, E. C., Durcanin-Robbins, J. F., and Sturm, L. L., 1996, Use of a pharmacy and medical claims database to document cost centers for 1993 annual asthma expenditures, *Arch. Fam. Med.* **5**:36–40.
34. Juel, K., and Pederson, P. A., 1992, Increasing asthma mortality in Denmark 1969–88 not a result of a changed coding practice, *Ann. Allergy* **68**:180–182.
35. Dompeling, E., Van Schayck, C. P., Van Grunsven, P. M., Van Herwaarden, C. L. A., Akkermans, R., Molema, J., Folgering, H., and Van Weel, C., 1993, Slowing the deterioration of asthma and chronic obstructive pulmonary disease observed during bronchodilator therapy by adding inhaled corticosteroids, *Ann. Intern. Med.* **118**:770–778.
36. Burrows, B., Martinez, F., Halonen, M., Barbee, R. A., and Cline, M. G., 1989, Association of asthma with serum IgE levels and skin-test reactivity to allergens, *N. Engl. J. Med.* **320**:271–277.
37. de Marco, R., Locatelli, F., Sunyer, J., and Burney, P., 2000, Differences in incidence of reported asthma related to age in men and women. A retrospective analysis of the data of the European Respiratory Health Survey, *Am. J. Respir. Crit. Care Med.* **162**:68–74.
38. Toogood, J. H., Jennings, B., Baskerville, J., and Lefcoe, N. M., 1984, Personal observations on the use of inhaled corticosteroid drugs for chronic asthma, *Eur. J. Respir. Dis.* **65**:321–338.
39. Cline, M. G., Lebowitz, M. D., and Burrows, B., 1993, Determinants of the percent predicted FEV_1 in asthma, *Am. J. Respir. Dis.* **147**(part 2 of 2 parts):A380.
40. Bronnimann, S., and Burrows, B., 1986, A prospective study of the natural history of asthma. Remission and relapse rates, *Chest* **90**:480–484.
41. Rönmark, E., Jönsson, E., and Lunbbäck, B., 1999, Remission of asthma in the middle aged and elderly: Report from the Obstructive Lung Disease in Northern Sweden study, *Thorax* **54**:611–613.
42. Rijcken, B., Schouten, J. P., Rosner, B., and Weiss, S. T., 1991, Is it useful to distinguish between asthma and chronic obstructive pulmonary disease in respiratory epidemiology? *Am. J. Respir. Dis.* **143**:1456–1457.
43. Frazier, E. A., Vollmer, W. M., Wilson, S. R., Hayward, A. D., and Buist, A. S., 1997, Characteristics of older asthmatics with moderate–severe disease, *Am. J. Respir. Crit. Care Med.* **155**(part 2 of 2 parts):A286.
44. Expert Panel Report. National Asthma Education Program: Guidelines for the diagnosis and management of asthma. Office of Prevention, Education, and Control. National Heart. Lung, and Blood Institute. National Institutes of Health. Bethesda, Maryland 20892. Publication No. 91–3042.
45. Expert Panel Report II, 1997, Guidelines for the diagnosis and management of asthma. US Department of Health and Human Services. Public Health Service. National Institutes of Health, National Heart, Lung, and Blood Institute, p. 136.

46. Juniper, E. F., Kline, P. F., Vanzieleghem, M. A., Ramsdale, E. H., O'Byrne, P., and Hargreave, F. E., 1990, Effect of long-term treatment with an inhaled corticosteroid (budesonide) on airway hyperresponsiveness and clinical asthma in nonsteroid-dependent asthmatics, *Am. Rev. Respir. Dis.* **142:**832–836.
47. Vathenen, A. S., Knox, A. J., Wisniewski, A., and Tattersfield, A. E., 1991, Time course of change in bronchial reactivity with an inhaled corticosteroid in asthma, *Am. Rev. Respir. Dis.* **143:**1317–1321.
48. Haahtela, T., Järvinen, M., Kava, T., Kiviranta, K., Koskinen, S., Lehtonen, K., Nikander, K., Persson, T., Reinikainen, K., Selroos, O., Sovijärvi, A., Stenius-Aarniala, B., Svahn, T., Tammivaara, R., and Laitinen, L. A., 1991, Comparison of a β_2-agonist, terbutaline, with an inhaled corticosteroid, budesonide, in newly detected asthma, *N. Engl. J. Med.* **325:**388–392.
49. Kerstjens, H. A. M., Brand, P. L. P., Hughes, M. D., Robinson, N. J., Postma, D. S., Sluiter, H. L., Bleecker, E. R., Dekhuijzen, P. N. R., DeJong, P. M., Mengelers, H. J. J., Overbeek, S. E., and Schoonbrood, D. F. M. E., 1992, A comparison of bronchodilator therapy with or without inhaled corticosteroid therapy for obstructive airways disease, *N. Engl. J. Med.* **327:**1413–1419.
50. Hahn, D. L., van Schayck, C. P., Dompeling, E., and Folgering, H., 1993, Effect of inhaled steroids on the course of asthma, *Ann. Intern. Med.* **119:**1051–1052.
51. Aronson, N., Lefevre, F., and Piper, M., September 2001, Management of Chronic Asthma. Evidence Report/Technology Assessment Number 44. (Prepared by Blue Cross and Blue Shield Association Technology Evaluation Center under Contract No. 290–97-0015.). Rockville, MD, Agency for Healthcare Research and Quality, AHRQ Publication No. 01-E044.
52. Hahn, D. L., Kerstjens, H. A. M., Brand, P. L. P., and Postma, D. S., 1993, Bronchodilator therapy with or without inhaled corticosteroid therapy for obstructive airways disease, *N. Engl. J. Med.* **328:**1044–1045.
53. Hahn, D. L., 1999, *Chlamydia pneumoniae*, asthma and COPD: What is the evidence? *Ann. Allergy Asthma Immunol.* **83:**271–292.
54. Emre, U., Roblin, P. M., Gelling, M., Dumornay, W., Rao, M., Hammerschlag, M. R., and Schachter, J., 1994, The association of *Chlamydia pneumoniae* infection and reactive airway disease in children, *Arch. Pediatr. Adolesc. Med.* **148:**727–732.
55. Mills, G. D., Lindeman, J. A., Fawcett, J. P., Herbison, G. P., and Sears, M., 2000, *Chlamydia pneumoniae* serological status is not associated with asthma in children or young adults, *Int. J. Epidemiol.* **29:**280–284.
56. Cunningham, A. F., Johnston, S. L., Julious, S. A., Lampe, F. C., and Ward, M. E., 1998, Chronic *Chlamydia pneumoniae* infection and asthma exacerbations in children, *Eur. Respir. J.* **11:**345–349.
57. Johnston, S. L., 1997, Influence of viral and bacterial respiratory infections on exacerbations and symptom severity in childhood asthma, *Pediatr. Pulmonol.* **16:**88–89.
58. Hahn, D. L., Middleton, K. M., Campbell, L. A., and Wang, S.-P., 1999, Eradication of *Chlamydia pneumoniae* from bronchoalveolar lavage (BAL) fluid associated with asthma improvement: Case report, *Ann. Allergy Asthma Immunol.* **84:**115.
59. Martin, R. J., Kraft, M., Chu, H. W., Berns, E. A., and Cassell, G. H., 2001, A link between chronic asthma and chronic infection, *J. Allergy Clin. Immunol.* **107:**595–601.
60. Hahn, D. L., and Golubjatnikov, R., 1994, Asthma and chlamydial infection: A case series, *J. Fam. Pract.* **38:**589–595.
61. Hahn, D. L., 1995, Treatment of *Chlamydia pneumoniae* infection in adult asthma: A before–after trial, *J. Fam. Pract.* **41:**345–351.
62. Black, P. N., Scicchitano, R., Jenkins, C. R., Blasi, F., Allegra, L., Wlodarczyk, J., and Cooper, B. C., 2000, Serological evidence of infection with *Chlamydia pneumoniae* is related to the severity of asthma, *Eur. Respir. J.* **15:**254–259.
63. Huittinen, T., Harju, T., Paldanius, M., Wahlström, E., Rytilä, P., Kinnula, V., Saikku, P., and Leinonen, M., 2000, *Chlamydia pneumoniae* HSP60 antibodies in adults with stable

asthma, in: *Proceedings: Fourth Meeting of the European Society for Chlamydia Research*, Helsinki, Finland, Esculapio, Bologna, Italy, p. 185.

64. von Hertzen, L. C., 2002, Role of persistent infection in the control and severity of asthma: Focus on *Chlamydia pneumoniae*, *Eur. Respir. J.* **19:**546–556.
65. Ramos, M., Arrieta, L., Samaniego, J., Garcia, A. R., Quitano, J. A., and Castañeda, A., 2000, *Chlamydia pneumoniae* infection in asthma and coronary heart patients: Serology study, in: *Proceedings: Fourth Meeting of the European Society for Chlamydia Research*, Helsinki, Finland, Esculapio, Bologna, Italy, p. 166.
66. Roblin, P. M., Witkin, S. S., Weiss, S. M., Gelling, M., and Hammerschlag, M. R., 2000, Immune response to *Chlamydia pneumoniae* in patients with asthma: Role of heat shock proteins (HSPs), in: *Proceedings: Fourth Meeting of the European Society for Chlamydia Research*, Helsinki, Finland, Esculapio, Bologna, Italy, p. 209.
67. Anwar, M., and Badawi, H., 2000, Role of *Chlamydia pneumoniae* in the pathogenesis of childhood respiratory tract infection (RTI) and asthma, in: *Proceedings: Fourth Meeting of the European Society for Chlamydia Research*, Helsinki, Finland, Esculapio, Bologna, Italy, p. 242.
68. Gomes, J. P., Rocha, M. G., Barona, T., Carvalhas, M. E., Borrego, M. J., and Catry, M. A., 2000, *Chlamydia pneumoniae* infection and acute exacerbations of asthma in children, in: *Proceedings: Fourth Meeting of the European Society for Chlamydia Research*, Helsinki, Finland, Esculapio, Bologna, Italy, p. 262.
69. Esposito, S., Blasi, F., Arioso, C., Morelli, N., Forloni, M., Droghetti, R., and Allegra, L., 2000, Recurrent wheezing in children: Role of *Chlamydia pneumoniae*, in: *Proceedings: Fourth Meeting of the European Society for Chlamydia Research*, Helsinki, Finland, Esculapio, Bologna, Italy, p. 281.
70. Müller, C. E., Schmidt, S. M., Bruns, R., and Wierbitzky, S. K. W., 2000, *Chlamydia pneumoniae* and asthma? No eosinophil inflammation or atopy but lower vital capacity in chronic bronchitis and respiratory *Chlamydia pneumoniae* infection, in: *Proceedings: Fourth Meeting of the European Society for Chlamydia Research*, Helsinki, Finland, Esculapio, Bologna, Italy, p. 282.
71. Petitjean, J., Vincent, F., Le Moël, G., Fradin, S., Vabret, A., Brun, J., and Freymuth, F., 2000, *Chlamydia pneumoniae* and acute exacerbation of chronic obstructive pulmonary disease or asthma in adults, in: *Proceedings: Fourth Meeting of the European Society for Chlamydia Research*, Helsinki, Finland, Esculapio, Bologna, Italy, p. 285.
72. Sirmatel, F., Dikensoy, O., and Sirmatel, O., 2000, The association of *Chlamydia pneumoniae* with late onset asthma, in: *Proceedings: Fourth Meeting of the European Society for Chlamydia Research*, Helsinki, Finland, Esculapio, Bologna, Italy, p. 292.
73. Atis, S., Öztürk, C., and Çalikoglu, M., 2000, Serology of *Chlamydia pneumoniae* in relation to asthma and atopy, *Eur. Respir. J.* **16**(Suppl. 31):20S (Abstract P310).
74. Shi, Y., Zheng, W., and Xia, X., 2000, Clinical study of *Chlamydia pneumoniae* infection in asthma patients, *Eur. Respir. J.* **16**(Suppl. 31):20S (Abstract P311).
75. Kocabas, A., Avsar, M., and Koksal, F., 2000, *Chlamydia pneumoniae* infection in adults with asthma, *Eur. Respir. J.* **16**(Suppl. 31):20S (Abstract P313).
76. Weiss, S., Quist, J., Roblin, P., Sokolovskaya, N., Hammerschlag, M., and Schachter, J., 1995, The relationship between *Chlamydia pneumoniae* and bronchospasm in adults (Abstract K39), in: *Abstracts of the 35th Interscience Conference on Antimicrobial Agents and Chemotherapy (ICAAC)*, San Francisco, California, American Society for Microbiology, p. 294.
77. Larsen, F. O., Norn, S., Mordhorst, C. H., Skov, P. S., Milman, N., and Clementsen, P., 1998, *Chlamydia pneumoniae* and possible relationship to asthma. Serum immunoglobulins and histamine release in patients and controls, *APMIS* **106:**928–934.
78. Routes, J. M., Nelson, H. S., Noda, J. A., and Simon, F. T., 2000, Lack of correlation between *Chlamydia pneumoniae* antibody titers and adult-onset asthma, *J. Allergy Clin. Immunol.* **105:**391–392.

79. Hahn, D. L., Peeling, R. W., Dillon, E., McDonald, R., and Saikku, P., 2000, Serologic markers for *Chlamydia pneumoniae* in asthma, *Ann. Allergy Asthma Immunol.* **84:**227–233.
80. Gencay, M., Rüdiger, J. J., Tamm, M., Solér, M., Perruchoud, A. P., and Roth, M., 2001, Increased frequency of *Chlamydia pneumoniae* antibodies in patients with asthma, *Am. J. Respir. Crit. Care Med.* **163:**1097–1100.
81. Falck, G., Gnarpe, J., Hansson, L.-O., Svärdsudd, K., and Gnarpe, H., 2002, Comparison of individuals with and without specific IgA antibodies to *Chlamydia pneumoniae.* Respiratory morbidity and the metabolic syndrome, *Chest* **122:**1587–1593.
82. Wark, P. A. B., Johnston, S. L., Simpson, J. L., Hensley, M. J., and Gibson, P. G., 2002, *Chlamydia pneumoniae* immunoglobulin A reactivation and airway inflammation in acute asthma, *Eur. Respir. J.* **20:**834–840.
83. Huittinen, T., Hahn, D., Wahlstrom, E., Saikku, P., and Leinonen, M., 2001, Host immune response to *Chlamydia pneumoniae* heat shock protein 60 is associated with asthma, *Eur. Respir. J.* **17:**1078–1082.
84. Brunham, R. C., Maclean, I. W., Binns, B., and Peeling, R. W., 1985, *Chlamydia trachomatis*: Its role in tubal infertility, *J. Infect. Dis.* **152:**1275–1282.
85. Peeling, R. W., Kimani, J., Plummer, F., Maclean, I., Cheang, M., Bwayo, J., and Brunham, R. C., 1997, Antibody to chlamydial Hsp60 predicts an increased risk for chlamydial pelvic inflammatory disease, *J. Infect. Dis.* **175:**1153–1158.
86. Peeling, R. W., Bailey, R. L., Conway, D. J., Holland, M. J., Campbell, A. E., Jallow, O., Whittle, H. C., and Mabey, D. C. W., 1998, Antibody response to the 60-kDa chlamydial heat-shock protein is associated with scarring trachoma, *J. Infect. Dis.* **177:**256–259.
87. Cosentini, R., Blasi, F., Tarsia, P., Capone, P., Papetti, M. C., Canetta, C., Graziadei, G., and Allegra, L., September 14–18, 2002, *Chlamydia pneumoniae* and severe asthma exacerbations, in: *12th European Respiratory Society Annual Congress*, Stokholm, Sweden, Abstract 1454.
88. Blasi, F., Damato, S., Cosentini, R., Tarsia, P., Raccanelli, R., Centanni, S., and Allegra, L., 2002, *Chlamydia pneumoniae* and chronic bronchitis: Association with severity and bacterial clearance following treatment, *Thorax* **57:**672–676.
89. Schmidt, S. M., Müller, C. E., Bruns, R., and Wiersbitzky, S. K. W., 2001, Bronchial *Chlamydia pneumoniae* infection, markers of allergic inflammation and lung function in children, *Pediatr. Allergy Immunol.* **12:**257–265.
90. Kaplan, M. A., and Goldin, M., 1959, The use of triacetyloleandomycin in chronic infectious asthma, in: *Antibiiotic Annual 1958–1959* (H. Welch and F. Marti-Ibauez, eds.), New York, Interscience Publishers, pp. 273–276.
91. Selenke, W., Longo, G., Glode, J., and Townley, R., 1969, Glucocorticoid sparing effects of certain macrolide antibiotics, *J. Allergy* **43:**156–157.
92. Ong, K. S., Grieco, M. H., and Rosner, W., 1978, Enhancement by oleandomycin of the inhibitory effect of methylprednisolone on phytohemagglutinin-stimulated lymphocytes, *J. Allergy Clin. Immunol.* **62:**115–118.
93. Szefler, S. J., Rose, J. Q., Ellis, E. F., Spector, S. L., Green, A. W., and Jusko, W. J., 1980, Effect of troleandomycin on methylprednisolone disposition, *J. Allergy Clin. Immun.* **65:** 181.
94. Weinberger, M., Hudgel, D., Spector, S., and Chidsey, C., 1977, Inhibition of theophylline clearance by troleandomycin, *J. Allergy Clin. Immunol.* **59:**228–231.
95. Miyachi, Y., Yoshioka, A., Imamura, S., and Niwa, Y., 1986, Effect of antibiotics on the generation of reactive oxygen species, *J. Invest. Dermatol.* **86:**449–453.
96. Næss, A. and Solberg, C. O., 1988, Effects of two macrolide antibiotics on human leukocyte membrane receptors and functions, *Acta Pathol. Microbiol. Immunol. Scand.* **96:**503–508.
97. Anon, 1991, Antibiotics as biological response modifiers, *Lancet* **337:**400–402.

98. Suzuki, T., Yamaya, M., Sekizawa, K., Hosoda, M., Yamada, N., Ishizuka, S., Yoshino, A., Yasuda, H., Takahashi, H., Nishimura, H., and Sasaki, H., 2002, Erythromycin inhibits rhinovirus infection in cultured human tracheal epithelial cells, *Am. J. Respir. Crit. Care Med.* **165:**1113–1118.

99. Fox, J. L., 1961, Infectious asthma treated with triacetyloleandomycin, *Penn. Med. J.* **64:**634–635.

100. Miyatake, H., Taki, F., Taniguchi, H., Suzuki, R., Takagi, K., and Satake, T., 1991, Erythromycin reduces the severity of bronchial hyperresponsiveness in asthma, *Chest* **99:** 670–673.

101. German, D., Serwonska, M., and Strub, M., 1992, Response of asthmatics to withdrawal from macrolide antibiotic (Ma)–Medrol (Me) therapy, *J. Allergy Clin. Immunol.* **89:**Abstracts 341.

102. Black, P. N., Blasi, F., Jenkins, C. R., Scicchitano, R., Mills, G. D., Rubinfeld, A. R., Ruffin, R. E., Mullins, P. R., Dangain, J., Cooper, B. C., Bem David, D., and Allegra, L., 2001, Trial of roxithromycin in subjects with asthma and serological evidence of infection with *Chlamydia pneumoniae*, *Am. J. Respir. Crit. Care Med.* **164:**536–541.

103. Johnston, S. L., 2001, Is *Chlamydia pneumoniae* important in asthma? The first controlled trial of therapy leaves the question unanswered, *Am. J. Respir. Crit. Care Med.* **164:**513–514.

104. Jaffe, A., and Bush, A., 2001, Anti-inflammatory effects of macrolides in lung disease, *Pediatr. Pulmonol.* **31:**464–473.

105. Yanagihara, K., Kadoto, J., and Kohno, S., 2001, Diffuse panbronchiolitis—pathophysiology and treatment mechanisms, *Int. J. Antimicrob. Agents* **18**(Suppl. 1):83–87.

106. Miyashita, N., Matsumoto, A., Kubota, Y., Nakajima, M., Niki, Y., and Matsushima, T., 1996, Continuous isolation and characterization of *Chlamydia pneumoniae* from a patient with diffuse panbronchiolitis, *Microbiol. Immunol.* **40:**547–552.

107. Miyashita, N., Niki, Y., Nakajima, M., Kawane, H., and Matsushima, T., 1998, *Chlamydia pneumoniae* infection in patients with diffuse panbronchiolitis and COPD, *Chest* **114:**969–971.

108. Wolter, J., Seeney, S., Bell, S., Bowler, S., Masel, P., and McCormack, J., 2002, Effect of long term treatment with azithromycin on disease parameters in cystic fibrosis: A randomized trial, *Thorax* **57:**212–216.

109. Emre, U., Bernius, M., Roblin, P., Gaerlan, P. F., Summersgill, J. T., Steiner, P., Schacter, J., and Hammerschlag, M., 1996, *Chlamydia pneumoniae* infection in patients with cystic fibrosis, *Clin. Infect. Dis.* **22:**819–823.

110. Garey, K. W., Rubinstein, I., Gotfried, M. H., Khan, I. J., Varma, S., and Danziger, L. H., 2000, Long-term clarithromycin decreases prednisone requirements in elderly patients with prednisone-dependent asthma, *Chest* **118:**1826–1827.

111. Kroegel, C., Rödel, J., Mock, B., Garey, K. J., and Rubinstein, I., 2001, *Chlamydia pneumoniae*, clarithromycin, and severe asthma, *Chest* **120:**1035–1036.

112. Hahn, D., Bukstein, D., Luskin, A., and Zeitz, H., 1998, Evidence for *Chlamydia pneumoniae* infection in steroid-dependent asthma, *Ann. Allergy Asthma Immunol.* **80:**45–49.

113. Esposito, S., Blasi, F., Arioso, C., Fioravanti, L., Fagetti, L., Droghetti, R., Tarsia, P., Allegra, L., and Principi, N., 2000, Importance of acute *Mycoplasma pneumoniae* and *Chlamydia pneumoniae* infections in children with wheezing, *Eur. Respir. J.* **16:**1142–1146.

114. Kraft, M., Cassell, G. H., Pak, J., and Martin, R. J., 2002, *Mycoplasma pneumoniae* and *Chlamydia pneumoniae* in asthma, *Chest* **121:**1782–1788.

115. Hahn, D. L., 1996, Intracellular pathogens and their role in asthma: *Chlamydia pneumoniae* in adult patients, *Eur. Respir. Rev.* **6:**224–230.

116. Kraft, M., Hamid, Q., Cassell, G. H., Gaydos, C. A., Duffy, L. B., Rex, M. D., Pak, J., and Martin, R. J., 2001, Mycoplasma and chlamydia cause increased airway inflammation that is responsive to clarithomycin, *Am. J. Respir. Crit. Care Med.* **163**(part 2 of 2 parts): A551.

117. von Hertzen, L., Töyrylä, M., Gimishanov, A., Bloigu, A., Leinonen, M., Saikku, P., and Haahtela, T., 1999, Asthma, atopy and *Chlamydia pneumoniae* antibodies in adults, *Clin. Exp. Allergy* **29:**522–528.
118. Williamson, H. A., and Schultz, P., 1987, An association between acute bronchitis and asthma, *J. Fam. Pract.* **24:**35–38.
119. Williamson, H. A., 1987, Pulmonary function tests in acute bronchitis: Evidence for reversible airway obstruction, *J. Fam. Pract.* **25:**251–256.
120. Burrows, B., Knudson, R. J., and Leibowitz, M., 1977, The relationship of childhood respiratory illness to adult obstructive airway disease, *Am. Rev. Respir. Dis.* **115:**751–760.
121. Sherman, C. B., Tosteson, T. D., Tager, I. B., Speizer, F. E., and Weiss, S. T., 1990, Early childhood predictors of asthma, *Am. J. Epidemiol.* **132:**83–95.
122. Jónsson, J. S., Gíslason, T., Gíslason, D., and Sigurdsson, J. A., 1998, Acute bronchitis and clinical outcome three years later: Prospective cohort study, *Br. Med. J.* **317:**1433.
123. Infante-Rivard, C., 1993, Childhood asthma and indoor environmental factors, *Am. J. Epidemiol.* **137:**834–844.
124. Dodge, R. R., Burrows, B., Lebowitz, M. D., and Cline, M. G., 1993, Antecedent features of children in whom asthma develops during the second decade of life, *J. Allergy Clin. Immunol.* **92:**744–749.
125. Smith, J. M., and Knowler, L. A., 1965, Epidemiology of asthma and allergic rhinitis. I. In a rural area. II. In a university-centered community, *Am. Rev. Respir. Dis.* **92:** 16–38.
126. Smith, J. M., 1994, Asthma and atopy as diseases of unknown cause. A viral hypothesis possibly explaining the epidemiologic association of the atopic diseases and various forms of asthma, *Ann. Allergy* **72:**156–162.
127. Frydén, A., Kihlström, E., Maller, R., Persson, K., Romanus, V., and Anséhn, S., 1989, A clinical and epidemiological study of "ornithosis" caused by *Chlamydia psittaci* and *Chlamydia pneumoniae* (strain TWAR), *Scand. J. Infect. Dis.* **21:**681–691.
128. Hahn, D. L., Dodge, R., and Golubjatnikov, R., 1991, Association of *Chlamydia pneumoniae* (strain TWAR) infection with wheezing, asthmatic bronchitis and adult-onset asthma, *JAMA* **266:**225–230.
129. Thom, D. H., Grayston, J. T., Campbell, L. A., Kuo, C.-C., Diwan, V. K., and Wang, S.-P., 1994, Respiratory infection with *Chlamydia pneumoniae* in middle-aged and older adult outpatients, *Eur. J. Clin. Microbiol. Infect. Dis.* **13:**785–792.
130. Lewis, S., 1998, ISAAC-a hypothesis generator for asthma? *Lancet* **351:**1220–1221.
131. Strachan, D. P., 1995, Time trends in asthma and allergy: Ten questions, fewer answers, *Clin. Exp. Allergy* **25:**791–794.
132. Bone, R. C., 1991, Chlamydial pneumonia and asthma: A potentially important relationship, *JAMA* **266:**265.
133. Puolakkainen, M., Ukkonen, P., and Saikku, P., 1989, The seroepidemiology of Chlamydiae in Finland over the period 1971 to 1987, *Epidemiol. Infect.* **102:**287–295.
134. Klaukka, T., Peura, S., and Martikainen, J., 1991, Why has the utilization of antiasthmatics increased in Finland? *J. Clin. Epidemiol.* **44:**859–863.
135. Kraft, M., Cassell, G. H., Henson, J. E., Watson, H., Williamson, J., Marmion, B. P., Gaydos, C. A., and Martin, R. J., 1998, Detection of *Mycoplasma pneumoniae* in the airways of adults with chronic asthma, *Am. J. Respir. Crit. Care Med.* **158:**998–1001.
136. von Hertzen, L., 1998, *Chlamydia pneumoniae* and its role in chronic obstructive pulmonary disease, *Ann. Med.* **30:**27–37.
137. Hahn, D., Campbell, L. A., and Kuo, C.-C., 2000, Failure of four and six weeks of treatment to eradicate evidence of *Chlamydia pneumoniae* from human lung and vascular tissue: Pathology case reports, in: *Proceedings: Fourth Meeting of the European Society for Chlamydia Research*, Helsinki, Finland, Esculapio, Bologna, Italy, p. 395.
138. Wu, L., Skinner, S. J. M., Lambie, N., Vuletic, J. C., Blasi, F., and Black, P. N., 2000, Immunohistochemical staining for *Chlamydia pneumoniae* is increased in lung tissue

from subjects with chronic obstructive pulmonary disease, *Am. J. Respir. Crit. Care Med.* **162:**1148–1151.

139. Theegarten, D., Mogilevski, G., Anhenn, O., Stamatis, G., Jaeschock, R., and Morgenroth, K., 2000, The role of chlamydia in the pathogenesis of pulmonary emphysema. Electron microscopy and immunofluorescence reveal corresponding findings as in atherosclerosis, *Virchows Arch.* **437:**190–193.
140. Suzuki, T., Yanai, M., Yamaya, M., Satoh-Nakagawa, T., Sekizawa, K., Ishida, S., and Sasaki, H., 2001, Erythromycin and common cold in COPD, *Chest* **120:**730–733.
141. Leinonen, M., 1993, Pathogenetic mechanisms and epidemiology of *Chlamydia pneumoniae*, *Eur. Heart J.* **14**(Suppl. K):56–71.
142. Redecke, V., Dalhoff, K., Bohnet, S., Braun, J., and Maass, M., 1998, Interaction of *Chlamydia pneumoniae* and human alveolar macrophages: Infection and inflammatory response, *Am. J. Respir. Cell. Mol. Biol.* **19:**721–727.
143. Jahn, H.-U., Krüll, M., Wuppermann, F. N., Klucken, A. C., Rosseau, S., Seybold, J., Hegemann, J. H., Jantos, C. A., and Suttorp, N., 2000, Infection and activation of airway epithelial cells by *Chlamydia pneumoniae*, *J. Infect. Dis.* **182:**1678–1687.
144. Rödel, J., Woytas, M., Groh, A., Schmidt, K.-H., Hartmann, M., Lehmann, M., and Straube, E., 2000, Production of basic fibroblast growth factor and interleukin 6 by human smooth muscle cells following infection with *Chlamydia pneumoniae*, *Infect. Immun.* **68:**3635–3641.
145. Rödel, J., Assefa, S., Prochnau, D., Woytas, M., Hartmann, M., Groh, A., and Straube, E., 2001, Interferon-γ production by *Chlamydia pneumoniae* in human smooth muscle cells, *FEMS Immunol. Med. Microbiol.* **32:**9–15.
146. Kol, A., Sukhova, G. K., Lichtman, A. H., and Libby, P., 1998, Chlamydial heat shock protein 60 localizes in human atheroma and regulates macrophage tumor necrosis factor-alpha and matrix metalloproteinase expression, *Circulation* **98:**300–307.
147. Vehmaan-Kreula, P., Puolakkainen, M., Sarvas, M., Welgus, H. G., and Kovanan, P. T., 2001, *Chlamydia pneumoniae* proteins induce secretion of the 92-kDa gelatinase by human monocyte-derived macrophages, *Arterioscler. Thromb. Vasc. Biol.* **21:**e1–e8.
148. Redington, A. E., Roche, W. R., Madden, J., Frew, A. J., Djukanovic, R., Holgate, S. T., and Howarth, P. H., 2001, Basic fibroblast growth factor in asthma: Measurement in bronchoalveolar lavage fluid basally and following allergen challenge, *J. Allergy Clin. Immunol.* **107:**384–387.
149. Grayston, J. T., 1992, Infections caused by *Chlamydia pneumoniae* strain TWAR, *Clin. Infect. Dis.* **15:**757–763.
150. Grayston, J. T., Aldous, M., Easton, A., Wang, S.-P., Kuo, C.-C., Campbell, L. A., Altman, J., 1993, Evidence that *Chlamydia pneumoniae* causes pneumonia and bronchitis, *J. Infect. Dis.* **168:**1231–1235.
151. Hahn, D. L., 1994, Acute asthmatic bronchitis: A new twist to an old problem, *J. Fam. Pract.* **39:**431–435.
152. Falck, G., Heyman, L., Gnarpe, J., and Gnarpe, H., 1995, *Chlamydia pneumoniae* and chronic pharyngitis, *Scand. J. Infect. Dis.* **27:**179–182.
153. Falck, G., Engstrand, I., Gad, A., Gnarpe, J., Gnarpe, H., and Laurila, A., 1997, Demonstration of *Chlamydia pneumoniae* in patients with chronic pharyngitis, *Scand. J. Infect. Dis.* **29:**585–589.
154. Falck, G., Heyman, L., Gnarpe, J., and Gnarpe, H., 1994, *Chlamydia pneumoniae* (TWAR): A common agent in acute bronchitis, *Scand. J. Infect. Dis.* **26:**179–187.
155. Hahn, D. L., Azenabor, A. A., Beatty, W. L., and Byrne, G. I., 2002, *Chlamydia pneumoniae* as a respiratory pathogen, *Front. Biosci.* **7:**E66–E76.

18

Respiratory Tract Infections Caused by *C. pneumoniae* in Pediatric Patients

KAZUNOBU OUCHI

1. INTRODUCTION

The growing data have been indicating that *Chlamydia pneumoniae* is a common and important respiratory pathogen in children as well as in adults all over the world.[1] This organism causes both upper and lower respiratory tract infections, often mild and self-limiting. *C. pneumoniae*, like *Mycoplasma pneumoniae*, has been recognized as a main cause of "atypical" pneumonia in children.[1] *C. pneumoniae* causes not only an acute infection but also a chronic infection in children and may trigger exacerbations in their reactive airway disease.[2] However, there are a number of unsolved issues in the diagnosis and treatment for *C. pneumoniae* infections because of its persistence in nature.

In this chapter, I review on the role of *C. pneumoniae* in pediatric respiratory infections, including carrier status, upper respiratory tract infection, lower respiratory tract infection, diagnosis, and treatment.

2. RESPIRATORY DISEASE ASSOCIATED WITH *Chlamydia pneumoniae*

2.1. Carrier Status

Healthy carrier status or asymptomatic carriage of *C. pneumoniae* has been well documented in adults and children. Healthy carrier rate of *C. pneumoniae*

KAZUNOBU OUCHI • Department of Pediatrics, Kawasaki Medical School, Kurashiki, Okayama, Japan.

by isolation and/or PCR in the nasopharynx are 1 to 6% in children as reported in several studies.[3–5] The natural course of *C. pneumoniae* infection or long-term influence of the persistence of *C. pneumoniae* is not really understood. So far there is no useful typing method among strains of *C. pneumoniae*, so we cannot differentiate recurrence from reinfection of *C. pneumoniae*. Further study is really needed to elucidate the natural course of *C. pneumoniae* infection and the role of its persistence by using the sensitive and reliable typing method.

2.2. Upper Respiratory Tract Infection

Infection with *C. pneumoniae* has been implicated in upper respiratory tract infection. Sore throat and hoarseness, respectively, were found in 80 and 30% of the 20 University of Washington students shown to have *C. pneumoniae* infection by serology, so upper respiratory symptoms such as pharyngitis are common clinical findings of acute *C. pneumoniae* respiratory tract infection in adult.[6] *C. pneumoniae* infection was found in only 2% of adults with common cold by serology and is thought to be a less frequent cause in the etiology of the common cold than are viral diseases, main causes of the common cold.[7] However, upper respiratory tract infections with *C. pneumoniae* in children are quite less clear and only a few data have been reported. Recently, an Italian group reported that they found 21.3% incidence for *C. pneumoniae* infection in children with symptomatic pharyngitis.[8] A Swedish group also found *C. pneumoniae* in 21.0% of children with nasopharingitis.[9] The same Swedish group found a 15% positive rate for *C. pneumoniae* in the adenoid tissue from children.[9] These data seemingly suggest the emerging pathological role of *C. pneumoniae* in children with upper respiratory tract infections. However there are complicated and unsolved problems in this field. Which is better to diagnose *C. pneumoniae* infection, detection of *C. pneumoniae* or serology for diagnosis? In case of detection, is it always associated with the illness? Healthy carrier rates of *C. pneumoniae* by isolation and/or PCR in the nasopharynx are 1 to 6% in children as reported in several studies.[3–5] The lack of correlation between positive culture results and serology has been often reported.[3,10] This problem of *C. pneumoniae* is similar to that of *Streptococcus pyrogens*, of which healthy carriers in the nasopharynx are common. Presence of *C. pneumoniae* or *S. pyrogens* does not necessarily mean the cause of clinical findings. Furthermore it is very difficult to distinguish the symptomatic illness of *C. pneumoniae* infection from its healthy carrier status because of the poor specific antibody response in young infants and children. However these rates of healthy carrier status for *C. pneumoniae*, which are quite lower when compared with those obtained in children with upper respiratory tract infections, [3,8,9] implicate that *C. pneumoniae*, like *S. pyrogens*, has a pathogenic role in children.

The data on the association of *C. pneumoniae* infection and otitis media are contradictory. Detection rates of *C. pneumoniae* from middle-ear effusion of patients with acute otitis media are 0 to 67%.[4,11,12] However, to date little is known about the pathologic role of *C. pneumoniae* in not only acute purulent and secretory otitis media but also acute and chronic sinusitis.

2.3. Lower Respiratory Tract Infection

2.3.1. Bronchitis

Acute bronchitis is the most common clinical finding among lower respiratory tract infections due to *C. pneumoniae*. The rate of *C. pneumoniae* infection in patients with acute bronchitis has been reported to be varied and it ranges from 2 to 41%,[8,13–16] depending on the diagnostic criteria of acute bronchitis, the diagnostic methods used, and the patient populations selected. More than 3 times as many patients with acute bronchitis were seen as patients with community-acquired pneumonia among children. Typical symptoms of acute bronchitis due to *C. pneumoniae* are unproductive cough or the production of slightly yellowish mucoid sputum, sore throat, pain on swallowing, and hoarseness. These symptoms progress gradually in nature. Such patients sometimes present a pertussis-like illness. In a Swedish study, *C. pneumoniae* was found to be as much a main cause of persistent cough as were *Bordetella pertussis* and *M. pneumoniae*.[16] If appropriate treatment is given, the symptoms disappear very quickly even if they have persisted for several months.

2.3.2. Respiratory Tract Infection with Wheezing

It has been well known that infectious agents, such as many viruses and atypical pathogens, trigger asthmatic exacerbations.[2,17–22] Respiratory viruses such as the rhinovirus are thought to be a major contributor. However, these respiratory viruses cause only acute infections. On the contrary, *C. pneumoniae* cause not only acute infections but also chronic infections, often persisting for years.[3–5] *C. pneumoniae* may cause the persistent inflammation, which is the basic nature of bronchial asthma, through chronic respiratory tract infection. Several researchers have found specific anti–*C. pneumoniae* IgE in the sera of children with *C. pneumoniae* infection and wheezing.[17] These data suggested that a type 1 allergy contributes to the pathogenesis of asthma associated with *C. pneumoniae* infection, similar to RS virus infection.[14] According to age, *C. pneumoniae* infection was found more frequently in exacerbations of childhood asthma than in adult asthma (Table I).[2,17–22] Several studies found that macrolides antibiotic treatment of children with *C. pneumoniae* infection and wheezing showed dramatic clinical improvement in their reactive airway disease, especially among those with eradication of *C. pneumoniae*.[13,17,18] In these studies, clinical improvement was also better in childhood asthma than in adult asthma. A role of *C. pneumoniae* in exacerbations of asthma may be more significant in early childhood than in adulthood.

Recently, inhaled steroid drugs have been recommended to treat moderate to severe asthma patients. Many *in vitro* and animal data have indicated that steroid drugs exacerbate *C. pneumoniae* infection.[23,24] Our data also showed that inhaled steroid drugs enhance *C. pneumoniae* infection *in vitro* (Fig. 1). Further studies are needed to elucidate the role of oral and inhaled steroid drugs in patients with asthma and *C. pneumoniae* infection in the future. Furthermore, the optimal length of antibiotic therapy and the mechanism of favorable effect of antibiotic treatment in asthmatic children with *C. pneumoniae* infection should

TABLE I
Detection of *C. pneumoniae* Infection in Exacerbations of Asthma in Children and Adults: Results of Representative Studies

Author (ref.)	Location	Subjects (n)	Age (year)	Methods	Prevalence of *C. pneumoniae* infection (%)
Cunningham *et al.*[2]	Southhampton, UK	96	9–11	PCR[a]	45
Emre *et al.*[17]	New York, USA	118	5–16	Culture	11
Esposite *et al.*[18]	Milan, Italy	71	2–14	PCR/Serology	16
Kamesaki *et al.*[19]	Osaka, Japan	33	0.4–16	PCR/Culture/ Serology	42
Allegra *et al.*[20]	Milan, Italy	74	17–54 (Adult)	Serology	9
Cook *et al.*[21]	Birmingham, UK	123	16–82 (Adult)	Serology	6
Miyashita *et al.*[22]	Kurashiki, Japan	168	16–80 (Adult)	PCR/Culture/ Serology	9

[a] PCR: Polymerase chain reaction.

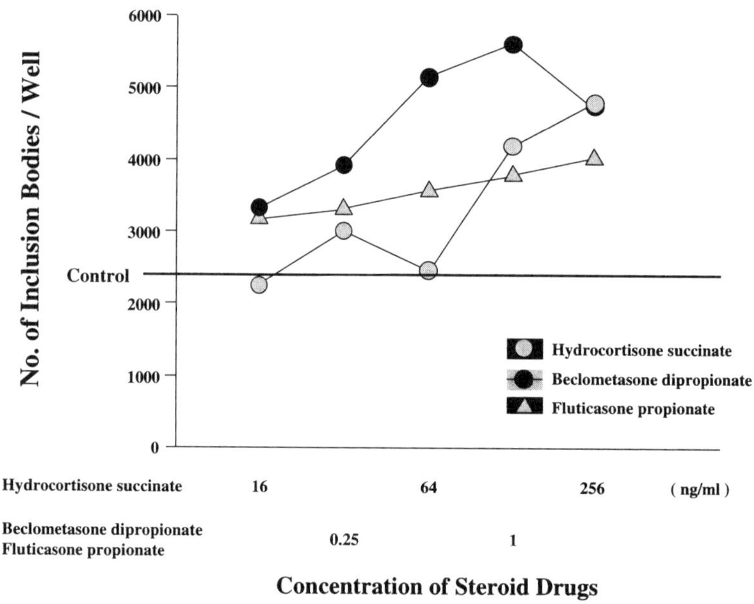

FIGURE 1. Increase of inclusion bodies of *C. pneumoniae* in the presence of glucocorticoids. *C. pneumoniae* inoculated HEp-2 cells incubated with fluticasone propionate, beclometasone dipropionate, or hydrocortisone succinate, compared with control cultures in which no drug was added throughout the 72-h culture period. Inoculated doses of *C. pneumoniae* incubated with glucocorticoids were the same as the control.

be established, because it is well known that macrolides have antiinflammatory activities in addition to their antibacterial activities.

2.3.3. Pneumonia

The role of *C. pneumoniae* in children with community-acquired pneumonia has been getting as well known as that in adults, although published data has been quite less for children than adults. Its proportion ranged between 2 and 28%,[13,25–34] depending on the diagnostic methods used and the patient populations, especially geographical location and the age group selected.[25–34] Most of these studies were based on serology alone,[25–27,32] except some which were based on both serology and detection of *C. pneumoniae* by isolation and/or polymerase chain reaction (PCR) using nasopharyngeal specimens.[13,28,30–34] Only one study was done using transthoracic needle aspiration.[29] Although pneumonia due to *C. pneumonae* was formerly thought to be less frequent in children aged less than 5 years, recent studies based on both serology and detection of *C. pneumoniae* have indicated that it has been frequent in children aged less than 5 years.[13,30–34] Clinical findings of pneumonia due to *C. pneumonaie* and *M. pneumoniae* are quite similar and usually regarded as classical atypical pneumonia, often mild and self-limiting. Thus, *C. pneumonaie* is a main cause of mild to moderate community-acquired pneumonia and ranged usually more than 10% in outpatient settings. On the contrary, *C. pneumoniae* causes severe community-acquired pneumonia less frequently and ranged usually less than 5% in inpatient settings (Table II).

However, there are several differences between the clinical findings for children with *C. pneumoniae* infection and those with *M. pneumoniae* infection. In our experience, *M. pneumoniae* tends to cause pneumonia rather than bronchitis, while *C. pneumoniae* tends to cause bronchitis rather than pneumonia.[34] Children with *M. pneumoniae* infection develop high fever and are accompanied by high cold agglutination titer more frequently than those with *C. pneumoniae* infection. Children with *C. pneumoniae* infection develop wheezing more frequently than those with *M. pneumoniae* infection.

Typical pattern of the chest X-ray of pneumonia with *C. pneumoniae* infection is bronchopneumonia with multiple infiltrates. The infiltrates are often segmental with small nodular infiltrates. Segments of various different lobes can be involved and there is a general preference for both lower lobes. Pleural effusion is rare and mostly not extensive. The chest X-ray of pneumonia patients with *M. pneumoniae* infection shows rather dense shadows, with more interstitial infiltrate and reactive pleurisy than of those with *C. pneumoniae*. However, we cannot predict causative agents on the basis of presenting manifestations, radiological features, and laboratory data except specific diagnostic tests.

3. DIAGNOSIS

Despite the substantial evidence of *C. pneumoniae* as a respiratory pathogen, progress has been hampered by the lack of standardized diagnostic

TABLE II

Role of *C. pneumoniae* infections in community-acquired pneumonia in children. Results of representative studies.

Author (ref.)	Location	Subjects (n)	Age (year)	Setting	Methods	Prevalence of *C. pneumoniae* infection (%)
Nohynek *et al.*[25]	Helsinki, Finland	135	0.2–15	I[a]	Serology	1
Korppi *et al.*[26]	Multicenter, Finland	195	117(<2 years of age) 78(<2 years of age)	I	Serology	1
Juven *et al.*[27]	Turku, Finland	254	0.1–16	I	Serology	3
Esposito *et al.*[28]	Multicenter, Italy	203	2–14	I	PCR/Serology	9
Principi *et al.*[13]	Multicenter, Italy	418	2–14	I	PCR/Serology	11
Vuori-Holopaninen *et al.*[29]	Oulu, Finland	34	0.8–14	I	Lung Tap/ PCR	3
Block *et al.*[30]	Multicenter, USA	260	3–12	O[b]	Culture/ Serology	28
Harris *et al.*[31]	Multicenter, USA	456	0.5–16	O	Culture/ Serology	15
Heiskanen-Kosma *et al.*[32]	Multicenter, Finland	201	0.3–14	O	Serology	10
Wubblel *et al.*[33]	Dallas, USA	174	0.5–16	O	PCR/Culture/ Serology	6
Ouchi *et al.*[34]	Shimonoseki, Japan	305	0.1–14	O>I	Serology	15

[a]I: Inpatient. [b]O: Outpatient.

methods, including serological testing, culture, DNA amplification, and tissue diagnostics[35]. The lack of standardized methods has resulted in a broader range of the reported role of *C. pneumoniae* infection and made it difficult for us to compare those results. Recent recommendations state as follows:[35] With regard to serological testing, only the use of microimmunofluorescence (MIF) test is recommended. Standardized definition for acute infection is based on a serum IgM titer ≧1:16 or a 4-fold increase in serum IgG titer, and that for possible acute infection is based on a serum IgG titer ≧1:512. Presumed past infection is on a serum IgG titer ≧1:16. The use of single IgG titers for determining acute infection and IgA for determining chronic infection are discouraged. Confirmation of a positive culture result requires propagation of the isolate and/or confirmation by use of PCR. There have been many reports using different PCR methods; however, only four PCR assays have been recommended in published reports.

There are several other problems in diagnosing *C. pneumoniae* infection with standardized diagnostic methods. It is very difficult to take good sputum specimens from children. So most studies were based on serology alone or on both serology and the detection of *C. pneumoniae* by isolation or PCR using nasopharyngeal specimens. Only one study was done by use of transthoracic needle aspiration.[29] It is very important to detect *C. pneumoniae* from patients with pneumonia for diagnosis, however only limited data are available for the correlation between nasopharyngeal specimens and sputum specimens in children. *C. pneumoniae* is well known to infect persistently in the nasopharynx either with some symptoms or without.[3–5] Therefore, it is not easy to differentiate symptomatic illness due to *C. pneumoniae* infection from symptomatic illness due to some other causative organism with asymptomatic *C. pneumoniae* persistent infection by only detection of *C. pneumoniae* from the nasopharynx. Cultures are important for examining the viability of *C. pneumoniae*, and providing isolates for biological characteristics and antimicrobial susceptibility. However, PCR holds more promise as a rapid diagnostic test and a sensitive test than does culture.

To date, only the MIF test has been recommended for serology.[35] The kinetics of the antibody response is relevant in *C. pneumoniae* infection.[8,14] On primary infection, the IgM response should appear as early as 3 weeks after onset of illness, and the IgG response at 6–8 weeks. On reinfection, the IgM response may be absent and the IgG response appears rapidly, within 1–2 weeks. Because of the absence of antibodies several weeks after the onset of infection, the antibody response may be missed if convalescent sera are obtained soon. Another difficulty of serological MIF testing is the absence of MIF antibodies in people with culture-confirmed infection. This is rather rare in adults, but has been often noted in younger children.[8] When sera from culture-positive but MIF-negative children with respiratory infection were examined by immunoblotting, over 89% were found to have antibodies to several *C. pneumoniae* proteins but only 24% reacted with the major outer membrane protein (MOMP).[14] The MOMP does not appear to be immunodominant in the immune response to *C. pneumoniae* infection, although it has been shown to be immunodominant in *C. trachomatis* infection. Even though we use the immunoblots of acute-phase and convalescent paired sera for serology, there may be more or less poor correlation between serology, culture, and PCR because of the poor serological response in younger children and the asymptomatic persistent infection.

There is also a problem of some interlaboratory variation in the performance of the MIF test.[35] Therefore, we need good enzyme immunoassays for detection of *C. pneumoniae* antibodies for their technical accessibility, objective evaluation, and easy standardization, and further careful comparative tests will need to be done for recommendation.

4. TREATMENT

There have been limited published data describing the clinical and bacteriological response to antibiotic therapy of respiratory infection due to *C. pneumoniae*. *C. pneumoniae* is susceptible *in vitro* to macrolides, tetracyclines, and fluoroquinolones. However, only macrolides can be safely used regardless of age because of the side effects of tetracyclines and fluoroquinolones in children.[8,14] Optimal dose and duration of therapy are uncertain, although prolonged treatment (more than 2 weeks') has been recommended in spite of a lack of well-controlled studies.[1] In several open, uncontrolled studies investigating patients with *C. pneumoniae* infection, clinical response did not differ between those patients who had treatment regimen active against *C. pneumoniae* and those who were treated with antibiotics that have no or poor activity against *Chlamydia in vitro*. Lack of standardization in diagnosis or of well-controlled studies may be the main reason for this basic confusion. Overall, the use of macrolide antibiotics led to 80–100% clinical and around 80% microbiological eradication from the nasopharynx in several recent open, uncontrolled studies.[14,30,31] Treatment with erythromycin for 14 days, clarithromycin for 10 days, or azithromycin for 5 days showed similar clinical and bacteriological responses in these studies. Persistent infection of *C. pneumoniae* after completion of macrolide treatment was not due to development of resistance to the antibiotics because the minimal inhibitory concentrations and the minimal bactericidal concentrations of isolates of *C. pneumoniae* from patients before and after treatment remained within the range susceptible to the antibiotics.[14] The reason why persistence of *C. pneumoniae* after antibiotic treatment active against *C. pneumoniae* remains unanswered. However, this persistent infection of *C. pneumoniae* may be due to the persistent form of *Chlamydia* which is induced *in vitro* with gamma-interferon, tryptophan depletion and so on, and becomes less susceptible to antibiotics.[36] In human natural infections, the active infection form and the persistent form might exist together and convert to either way in the local circumstance *in vivo*; however, to date no one has proved the persistent form of *C. pneumoniae* in natural infection.

We really need large-scale, carefully randomized, controlled trials that examine the effectiveness of antibiotic treatment for *C. pneumoniae* infections, including microbiological efficacy.

5. CONCLUSION

The growing and substantial data have indicated that *C. pneumoniae* has a more significant role than previously thought as causes of respiratory tract infections in children, especially younger children. *C. pneumoniae* frequently causes not only upper and lower respiratory tract infections but also triggers wheezing in children with reactive airway disease. *C. pneumoniae* causes persistent infections in nature with or without symptoms, and this evidence makes it very difficult to interpret healthy carrier status and diagnose acute infections. Furthermore, another difficulty in diagnosis is the absence of antibodies in people with culture-confirmed infection, which is rather rare in adults, but has been often noted in younger children.

The lack of standardized methods is another problem and has resulted in a broader range of the reported role of *C. pneumoniae* infection and makes it difficult for us to compare those results. We must follow recent recommendation statements for future trials. Furthermore, rapid and more effective diagnostic methods are necessary to diagnose *C. pneumoniae* infection in clinical settings.

There have been limited published data describing the clinical and bacteriological response of respiratory infection due to *C. pneumoniae* to antibiotic therapy. Optimal dose and duration of therapy are uncertain. We really need large-scale, carefully randomized, controlled trials that examine the effectiveness of antibiotic treatment for *C. pneumoniae* infections, including microbiological efficacy.

REFERENCES

1. Grayston, J. T., Campbell, L. A., Kuo, C. C., Mordhorst, C. H., Saikku, P., Thom, D. H., and Wang, S. P., 1990, A new respiratory tract pathogen: *Chlamydia pneumoniae* Strain TWAR, *J. Infect. Dis.* **161**:618–625.
2. Cunningham, A. F., Johnston, S. L., Julious, S. A., Lampe, F. C., and Ward, M. E., 1998, Chronic *Chlamydia pneumoniae* infection and asthma exacerbations in children, *Eur. Respir. J.* **11**:345–349.
3. Hyman, C. L., Roblin, P. M., Gaydos, C. A., Quinn, T. C., Schachter, J., and Hammerschlag, M. R., 1995, Prevalence of asymptomatic nasopharyngeal carriage of *Chlamydia pneumoniae* in subjectively healthy adults: Assessment by polymerase chain reaction–enzyme immunoassay and culture, *Clin. Infect. Dis.* **20**:1174–1178.
4. Block, S. L., Hammerschlag, M. R., Hedrick, J., Tyler, R., Smith, A., Roblin, P., Gaydos, C., Quinn, T. C., Palmer, R., and McCarty, J., 1997, *Chlamydia pneumoniae* in acute otitis media, *Pediatr. Infect. Dis. J.* **16**:858–862.
5. Miyashita, N., Niki, Y., Nakajima, M., Fukano, H., and Matsushima, T., 2001, Prevalence of asymptomatic infection with *Chlamydia pneumoniae* in subjectively healthy adults, *Chest* **119**:1416–1419.
6. Grayston, J. T., Kuo, C. C., Wang, S. P., and Altman, J., 1986, A new *Chlamydia psittaci* strain, TWAR, isolated in acute respiratory tract infections, *N. Engl. J. Med.* **315**:161–168.
7. Makela, M. J., Puhakka, T., Ruuskanen, O., Leinonen, M., Saikku, P., Kimpimaki, M., Blomqvist, S., Hyypia, T., and Aristila, P., 1998, Viruses and bacteria in the etiology of the common cold, *J. Clin. Microbiol.* **36**:539–542.
8. Principi, N., and Esposito, S., 2001, Emerging role *of Mycoplsma pneumoniae* and *Chlamydia pneumoniae* in pediatric respiratory tract infections, *Lancet Infect. Dis.* **1**:334–344.
9. Normann, E., Gnarpe, J., Gnarpe, H., and Wettergren, B., 1998, *Chlamydia pneumoniae* in children with acute respiratory tract infections, *Acta Pediatr.* **87**:23–27.
10. Gaydos, C. A., Roblin, P. M., Hammerschlag, M. R., Hyman, C. L., Eiden, J. J., Schachter, J., and Quinn, T. C., 1994, Diagnostic utility of PCR-enzyme immunoassay, culture, and serology for detection of *Chlamydia pneumoniae* in symptomatic and asymptomatic patients, *J. Clin. Microbiol.* **32**:903–905.
11. Goo, Y. A., Hori, M. K., Voorhies, J. H., Kuo, C. C., Wang, S. P., and Campbell, L. A., 1995, Failure to detect *Chlamydia pneumoniae* in ear fluids from children with otitis media with effusion, *Pediatr. Infect. Dis. J.* **14**:1000–1001.
12. Falck, G., Engstrand, I., Gnarpe, J., and Gnarpe, H., 1998, Association of *Chlamydia pneumoniae* with otitis media in children, *Scand. J. Infect. Dis.* **30**:377–380.
13. Principi, N., Esposito, S., Blasi, F., Allegra, L., and the Mowgli Study Group, 2001, Role of *Mycoplasma pneumoniae* and *Chlamydia pneumoniae* in children with community-acquired lower respiratory tract infections, *Clin. Infect. Dis.* **32**:1281–1289.
14. Hammerschlag, M. R., 2000, *Chlamydia pneumoniae* and the lung, *Eur. Respir. J.* **16**:1001–1007.
15. Bent, S., Saint, S., Vittinghoff, E., and Grady, D., 1999, Antibiotics in acute bronchitis: A meta-analysis, *Am. J. Med.* **107**:62–67.
16. Hallander, H. O., Gnarpe, J., Gnarpe, H., and Olln, P., 1999, *Bordetella pertussis, Bordetella parapertussis, Mycoplasma pneumoniae, Chlamydia pneumoniae* and persistent cough in children, *Scand. J. Infect. Dis.* **31**:281–286.

17. Emre, U., Sokolovskaya, N., Roblin, P. M., Schachter, J., and Hammerschlag, M. R., 1995, Detection of anti-*Chlamydia pneumoniae* IgE in children with reactive airway disease, *J. Infect. Dis.* **172:**265–267.
18. Esposito, S., Blasi, F., Arosio, C., Fioravanti, L., Fagetti, L., Droghetti, R., Tarsia, P., Allegra, L., and Principi, N., 2000, Importance of acute *Mycoplasma pneumoniae* and *Chlamydia pneumoniae* infections in children with wheezing, *Eur. Respir. J.* **16:**1142–1146.
19. Kamesaki, S., Suehiro, Y., Shinomiya, K., Matsushima, H., and Ouchi, K., 1998, *Chlamydia pneumoniae* infection in children with asthma exacerbation, *Allergy* **47:**667–673. (in Japanese)
20. Allegra, L., Blasi, F., Centanni, S., Cosentini, R., Denti, F., Raccanelli, R., Tarsia, P., and Valenti, V., 1994, Acute exacerbations of asthma in adults: Role of *Chlamydia pneumoniae* infection, *Eur. Respir. J.* **7:**2165–2168.
21. Cook, P. J., Davies, P., Tunnicliffe, W., Ayres, J. G., Honeybourne, D., and Wise, R., 1998, *Chlamydia pneumoniae* and asthma, *Thorax* **53:**254–259.
22. Miyashita, N., Kubota, Y., Nakajima, M., Niki, Y., Kawane, H., and Matsushima, T., 1998, *Chlamydia pneumoniae* and exacerbations of asthma in adults, *Ann. Allergy Asthma Immunol.* **80:**405–409.
23. Malinverni, R., Kuo, C. C., Campbell, L. A., and Grayston, J. T., 1995, Reactivation of *Chlamydia pneumoniae* lung infection in mice by cortisone, *J. Infect. Dis.* **172:**593–594.
24. Tsumura, N., Emre, U., Roblin, P. M., and Hammerschlag, M. R., 1996, Effect of hydrocortisone succinate on growth of *Chlamydia pneumoniae in vitro*, *J. Clin. Microbiol.* **34:**2379–2381.
25. Nohynek, H., Eskola, J., Laine, E., Halonen, P., Ruutu, P., Saikku, P., Kleemola, M., and Leinonen, M., 1991, The causes of hospital-treated acute lower respiratory tract infection in children, *Am. J. Dis. Child.* **145:**618–622.
26. Korppi, M., Heiskanen-Kosma, T., Jalonen, E., Saikku, P., Leinonen, M., Halonen, P., and Makela, P. H., 1993, Aetiology of community-acquired pneumonia in children treated in hospital, *Eur. J. Pediatr.* **152:**24–30.
27. Juven, T., Mertsola, J., Waris, M., Leinonen, M., Meurman, O., Roivainen, M., Escola, J., Saikku, P., and Ruuskanen, O., 2000, Etiology of community-acquired pneumonia in 254 hospitalized children, *Pediatr. Infect. Dis. J.* **19:**293–298.
28. Esposito, S., Blasi, F., Bellini, F., Allegra, L., and Principi, N., and the Mowgli Study Group, 2001, *Mycoplasma pneumoniae* and *Chlamydia pneumoniae* infections in children with pneumonia, *Eur. Respir. J.* **17:**241–245.
29. Vuori-Holopaninen, E., Salo, E., Saxen, H., Hedman, K., Hyypia, T., Lahdenpera, R., Leinonen, M., Tarkka, E., Vaara, M., and Peltola, H., 2002, Etiological diagnosis of childhood pneumonia by use of transthoracic needle aspiration and modern microbiological methods, *Clin. Infect. Dis.* **34:**583–590.
30. Block, S. L., Hedrick, J., Hammerschlag, M. R., Cassell, G. H., and Craft, J. C., 1995, *Mycoplasma pneumoniae* and *Chlamydia pneumoniae* in pediatric community-acquired pneumonia: Comparative efficacy and safety of clarithromycin vs. erythromycin ethylsuccinate, *Pediatr. Infect. Dis. J.* **14:**471–477.
31. Harris, J. S., Kolokathis, A., Campbell, M., Cassell, G. H., and Hammerschlag, M. R., 1998, Safety and efficacy of azithromycin in the treatment of community-acquired pneumonia in children, *Pediatr. Infect. Dis. J.* **17:**865–871.
32. Heiskanen-Kosma, T., Jokinen, C., Kurki, S., Heiskanen, L., Juvonen, H., Kallinen, S., Sten, M., Tarkiainen, A., Ronnberg, P. R., Kleemola, M., Makela, P. H., and Leinonen, M., 1998, Etiology of childhood pneumonia: Serologic results of a prospective, population-based study, *Pediatr. Infect. Dis. J.* **17:**986–991.
33. Wubbel, L., Muniz, L., Ahmed, A., Trujillo, M., Carubelli, C., McCoig, C., Abramo, T., Leinonen, M., and McCracken, G. H., 1999, Etiology and treatment of community-acquired pneumonia in ambulatory children, *Pediatr. Infect. Dis. J.* **18:**98–104.

34. Ouchi, K., Komura, H., Fujii, M., Matsushima, H., Maki, T., Hasegawa, K., and Nonaka, Y., 1999, *Chlamydia pneumoniae* infection and *Mycoplasma pneumoniae* infection in pediatric patients, *Kansenshogaku Zasshi* **73:**1177–1182.
35. Dowell, S. F., Peeling, R. W., Boman, J., Carlone, G. M., Fields, B. S., Guarner, J., Hammerschlag, M. R., Jackson, L. A., Kuo, C. C., Maass, M., Messmer, T. O., Talkington, D. F., Tondella, M. L., and Zaki, S. R., 2001, The *C. pneumoniae* workshop participants, standardizing *Chlamydia pneumoniae* assays: Recommendations from the Centers for Disease Control and Prevention (USA) and the Laboratory Center for Disease Control (Canada), *Clin. Infect. Dis.* **33:**492–503.
36. Beatty, W. L., Byrne, G. I., and Morrison, R. P., 1994, Repeated and persistent infection with *Chlamydia* and the development of chronic inflammation and disease, *Trends Microbiol.* **2:**94–98.

Subject Index

Abdominal aortic aneurysm (AAA), 113, 116, 123, 128, 177–178
Aberrant body (AB), 23–24
ABI-1131, 47
ABI-1657, 47
ACADEMIC study, 126
ACES. *See* Azithromycin and Coronary Events Study
Acquired immunity, 83, 84–88
Activator Protein-1, 123
Acute bronchitis, 57, 188, 248, 265
Adenosine triphosphate (ATP), 14, 33, 57, 230
Adenosine triphosphate/adenosine diphosphate (ATP/ADP) translocase, 14, 105
Adhesion molecules, 121, 139, 154
Adults
  asthma in, 239, 240, 241, 242, 246, 248, 249, 251
  CAP in, 58, 60–62
Alcmaeon of Croton, 153
Alzheimer's disease, 6, 211–223, 235
  animal models of, 219–221
  C. pneumoniae and inflammation, 216–218
  C. pneumoniae association with, 213–214
  C. pneumoniae nervous system entry, 214–216
  risk factors for, 212–213
Amoxicillin, 45–46, 127, 181
Amoxicillin-clavulanate, 50, 59
Ampicillin, 24
β-Amyloid, 211, 212, 217, 219–220, 222
β-Amyloid precursor protein (βAPP), 212, 218, 220, 222
Angina. *See* Unstable angina
Animal models. *See also* specific animals
  of Alzheimer's disease, 219–221
  antibiotics and, 163–166, 167, 174–175
  of atherosclerosis, 119–121, 153–168, 174–175
  of cardiovascular disease, 135–136
  of cell-mediated immunity, 84–86
  of lung infection, 156–157
  of myocarditis, 192–193
ANTIBIO study, 127
Antibiotics, 4, 24, 45–52. *See also* specific types
  in animal models, 163–166, 167, 174–175
  asthma and, 45, 49, 50–51, 243–248, 249, 265–267
  atherosclerosis and, 45, 52, 123–130, 131, 163–166, 167, 174–175, 176, 177
  CAP and, 50, 59
  in cardiac clinical intervention studies, 51–52
  cardiovascular disease and, 5–6
  coronary artery disease and, 51–52, 173–185
  peripheral vascular disease and, 176–178
  resistance to, 47–48, 52, 184
  respiratory infections and, 49–51, 269–271
  susceptibility to, 45–48
Antibodies, 3, 88–90
Antigen-presenting cells (APC), 83, 84, 86, 106, 108

Antigens
  myelin, 200
  neural, 200
  vaccines and, 101, 104–106
Antigen-specific immune response, 81, 82, 83
Antiinflammatory cytokines, 83, 84
AP-1, 140
APOE gene product, 232, 236
Apolipoprotein E (apoE) ϵ4 allele, 212, 213–214
Apolipoprotein E knockout (aopE-KO) mice, 155, 157–158, 163, 168, 175
Apoptosis, 103, 123, 193
Arthritis, 45, 228. *See also* Inflammatory arthritis; Reactive arthritis
Asthma, 1, 4, 6, 7, 23, 57, 239–254
  antibiotics and, 45, 49, 50–51, 243–248, 249, 265–267
  atopic, 240, 253
  C. pneumoniae as an initiator of, 248–249
  C. pneumoniae as a promoter of, 242–248
  definition of, 240
  definition of induction, 239
  importance of, 241
  lung remodeling and, 251–253
  nonatopic, 4, 240, 241, 251, 253
  treatments for, 241–242
  vaccines and, 100
Asthma with chronic airway obstruction (AS-CAO), 240
Atherosclerosis, 5–6, 7, 23, 30, 113–132, 139, 141, 142, 143, 212, 214. *See also* Cardiovascular disease; Coronary artery disease; Coronary heart disease
  animal models of, 119–121, 153–168, 174–175
  antibiotics and, 45, 52, 123–130, 131, 163–166, 167, 174–175, 176, 177
  association between infection and, 113–115
  blood *C. pneumoniae* and, 65, 66, 75–76, 77
  C. pneumoniae risk factors, 130
  evidence linking *C. pneumoniae* to, 116–130
  immunological mechanisms in, 121–123
  microorgranisms implicated in, 114t
  molecular mechanisms in, 121–123
  pathological specimen findings, 117–118, 119t
Atorvastatin, 184
ATP. *See* Adenosine triphosphate
Atypical pneumonia, 4
Avians, 154, 156, 167
AZACS. *See* Azithromycin in Acute Coronary Syndromes
Azalides, 46
Azithromycin
  asthma and, 246–247, 248t
  atherosclerosis and, 120, 123, 126, 127, 129, 163, 166, 174, 175
  CAP and, 59
  cardiovascular disease and, 6, 140
  coronary artery disease and, 52, 178–179, 180, 181, 183–184
  peripheral vascular disease and, 178
  resistance to, 47, 48
  respiratory infections and, 50, 270
  susceptibility to, 46
Azithromycin and Coronary Events Study (ACES), 128, 183–184
Azithromycin in Acute Coronary Syndromes (AZACS) study, 127, 183
Azithromycin in Coronary Artery Disease Elimination of Myocardial Infection with Chlamydia (ACADEMIC) study, 126

Basic fibroblast growth factor (bFGF), 121, 123, 253
B cells, 3, 7, 86, 103, 107–108, 193
Beta-lactam antibiotics, 45, 48, 49, 59, 63
Binary fission, 29
Biology of *Chlamydia*, 2–4
Blood, 70–77
  C. pneumoniae location in, 75
  immunostaining of, 66, 75, 76–77
  PCR of, 65–66, 70–76, 77
B lymphocytes, 234
Bordetella pertussis, 265
Bronchitis, 1
  acute, 57, 188, 248, 265
  antibiotics and, 49
  chronic, 4, 57, 100, 240, 243, 249, 253–254
  in pediatric patients, 265, 267

Cadherins, 216
Campylobacter, 227
CAP. *See* Community-acquired pneumonia
Cap-1, 105
Cardiac clinical interventions studies, 51–52
Cardiovascular disease, 135–149, 214. *See also* Atherosclerosis; Coronary artery disease; Coronary heart disease
  blood *C. pneumoniae* and, 75–76
  host cell signaling and, 139–142

Cardiovascular disease (*cont.*)
insights into, 5–6
strain heterogeneity in, 144–148
Caspase-3, 123
Catenin, 216
Cats, 154, 156
CD3 cells, 75
CD4 cells, 3, 83, 84–86, 87
Alzheimer's disease and, 217
inflammatory arthritis and, 233, 234
multiple sclerosis and, 200, 201
vaccines and, 101, 102, 103, 106
CD8 cells, 3, 83, 84–86, 87
inflammatory arthritis and, 233, 234
multiple sclerosis and, 206
vaccines and, 101, 102, 103, 106
Cefotaxime, 189
Ceftriaxone, 49
Cell and molecular biology, 29–40
physical localization of components and, 31–32
structural appearance and, 30–31
of vesicle trafficking, 39
Cell-mediated immunity, 81, 83, 84–88
in animals, 84–86
in humans, 86–88
vaccines and, 104
Cephalosporin, 49, 176
Cethromycin, 47, 50
Chickens, 154, 156
Children. *See* Pediatric patients
Chlamydial anomaly, 46
Chlamydia pecorum, 13, 15, 45
Chlamydia pneumoniae and Myocardial Infarction (CLAINF) study, 128
Chlamydia pneumoniae AR39 strain, 32, 33, 35, 114, 136, 145, 146, 148, 159
Chlamydia pneumoniae A-03 strain, 121
Chlamydia pneumoniae CWL029 strain, 33, 35, 146, 148
Chlamydia pneumoniae IOL-207 strain, 15, 148
Chlamydia pneumoniae J138 strain, 32, 33
Chlamydia pneumoniae Kajaani-6 strain, 15
Chlamydia pneumoniae KKpn strain, 15, 18f, 19–20
Chlamydia pneumoniae outer membrane complex (COMC), 30–32, 35, 36, 37
Chlamydia pneumoniae polymorphic proteins. *See* Ppps
Chlamydia pneumoniae TWAR strain, 15, 16, 30, 61, 114–115, 248
Chlamydia pneumoniae TW-183 strain, 35, 114, 148
Chlamydia psittaci, 25, 36, 114, 115
antibiotics and, 45
developmental cycle of, 13, 15
endocarditis and, 191
morphology of, 16
myocarditis and, 187, 190
outer membrane structure of, 18–19
secreted proteins and, 37
surface projections of, 20, 21, 23, 26
type III secretions and, 37
Chlamydia psittaci california 10 strain, 16
Chlamydia suis, 47
Chlamydia trachomatis, 2, 3, 115, 144, 173
antibiotics and, 45–46, 47
antibodies to, 90
developmental cycle of, 14, 15
endocarditis and, 191
immune response and, 83, 87, 88
lung infections and, 157
morphology of, 16
multiple sclerosis and, 206
myocarditis and, 187, 190, 193
PCR and, 68
physical localization of components, 31
reactive arthritis and, 227, 229–231, 232, 233, 234, 236
secreted proteins and, 39
surface projections of, 20, 23
type III secretions and, 36–37
vaccines and, 101, 102, 103, 104, 105, 108
Chlamydia trachomatis A, 35, 39
Chlamydia trachomatis D, 32, 33, 35, 39
Chlamydia trachomatis L2, 35, 39
Chlamydia trachomatis MoPn, 32
Cholesterol, 120, 154–155, 156, 158–159, 162, 163, 166, 174, 175, 212. *See also* High density lipoprotein; Hypercholesterolemia; Low density lipoprotein; Very low density lipoprotein
Chronic bronchitis, 4, 57, 100, 240, 243, 249, 253–254
Chronic diseases, 6–7
Chronic nonspecific lung disease (CNSLD), 240, 253–254
Chronic obstructive pulmonary disease (COPD), 23, 100, 240, 241, 251–252, 253, 254
Chronic sequelae, 1, 2t, 5
Ciprofloxacin, 24, 47
CLAINF study, 128
CLARICOR study, 128

CLARIFY. *See* Clarithromycin in Acute Coronary Syndrome Patients in Finland
Clarithromycin, 49
  asthma and, 246
  atherosclerosis and, 127, 175
  coronary artery disease and, 182
  persistent infections and, 48
  respiratory infections and, 50–51, 270
  susceptibility to, 46
Clarithromycin in Acute Coronary Syndrome Patients in Finland (CLARIFY) study, 127, 182
Co-amoxiclav, 49
Colony stimulating factor, 121
COMC. *See Chlamydia pneumoniae* outer membrane complex
Common cold, 264
Community-acquired pneumonia (CAP), 4, 57–63, 116, 173, 267
  antibiotics and, 50, 59
  epidemiology of, 58
  in immunocompromised patients, 62
  nursing home outbreaks of, 61–62
  representative studies of, 268t
Complement fixation (CF) test, 187, 191
Coronary artery disease (CAD). *See also* Atherosclerosis
  antibiotics and, 51–52, 173–185
    in observational studies, 175–176
    in ongoing clinical trials, 183–184
    in randomized secondary prevention trials, 178–183
  blood *C. pneumoniae* and, 70, 74, 75–76
Coronary heart disease (CHD), 87, 113, 116, 117, 123, 126, 129. *See also* Atherosclerosis; Cardiovascular disease
Corticosteroids. *See* Inhaled corticosteroids
CPAF, 35, 38f, 39
Cpn058 gene, 23
C-reactive protein (CRP)
  atherosclerosis and, 123, 126, 127, 129, 130
  CAP and, 59
  cardiovascular disease and, 139
  coronary artery disease and, 180
  dilated cardiomyopathy and, 192
  peripheral vascular disease and, 177, 178
Croatian Azithromycin in Atherosclerosis Study (CROAATS), 128
Cryptococcal meningitis, 203
CTLA4, 108
Cyclooxygenase-2, 252
Cycloserine, 24
Cystic fibrosis, 245–246
Cytochrome c, 123, 193
Cytokines. *See also* specific types
  antiinflammatory, 83, 84
  asthma and, 252–253
  atherosclerosis and, 121, 154, 159
  inflammatory arthritis and, 233–234
  proinflammatory, 4, 83, 84, 217
Cytomegalovirus (CMV), 114, 118, 136
Cytoplasmic body, 20–21, 25

Def gene, 47
Dendritic cells, 83, 84, 86, 102, 104, 107
Dental (periodontal) pathogens, 114, 118
Deoxyuridine triphosphate (dUTP), 66
Developmental cycle, chlamydial, 11–15
Diabetes, 113, 130
Diffuse panbronchiolitis (DPB), 245
Digoxigenin (DIG)-PCR enzyme-linked immunosorbent assay (ELISA), 70
Dilated cardiomyopathy (DCM), 192
DNA A gene, 230
DNA amplification, 66, 68, 268
DNA extraction, 66, 67, 71, 74
DNA vectors, 3
Dogs, 120, 154, 156
Doxycycline, 46
  atherosclerosis and, 128
  coronary artery disease and, 176
  endocarditis and, 191
  myocarditis and, 189
  peripheral vascular disease and, 177–178
  respiratory infections and, 49
Dutch Hypothesis, 240, 253–254
Dynein, 39

EBV. *See* Epstein-Barr virus
Elementary body (EB), 1–3, 57, 115, 137–138, 143, 173
  Alzheimer's disease and, 213
  antibodies and, 89, 90
  basic biology of, 2–3
  cardiovascular disease and, 148
  cell and molecular biology of, 29
  developmental cycle and, 11–12
  immune response and, 81, 82, 84, 88
  morphology of, 14t, 15–16
  multiple sclerosis and, 205
  outer membrane structure, 16–20
  PCR and, 67
  persistence and, 23–25

Elementary body (*cont.*)
  physical localization of components, 31
  protein composition of, 33–36
  reactive arthritis and, 229, 230
  surface projections of, 20–23
  type III secretions in, 37
  ultrastructure of, 30
Emphysema, 4, 240, 251, 253, 254
Endocarditis, 5, 187, 188t, 190–192, 194
Endothelial cells, 86
  Alzheimer's disease and, 215, 216, 222
  atherosclerosis and, 121, 154, 158, 161, 166
  cardiovascular disease and, 139, 140, 141, 142, 147
  human brain microvascular, 214–215, 216, 218
  myocarditis and, 193
  peripheral vascular disease and, 178
  vaccines and, 103
Energy parasitism, 14, 33
Enzyme-linked immunosorbent assay (ELISA), 70, 117
Eosinophil cationic protein, 243
Eosinophils, 253
Epidemiology
  of *C. pneumoniae*, 115–116
  of CAP, 58
Epithelial adhesion molecule-1, 82
Epithelial cells, 3–4, 12, 29, 82, 84, 136, 147, 252–253
Epstein-Barr virus (EBV), 204–205
Erm genes, 48
Erythromycin
  asthma and, 252
  atherosclerosis and, 128
  CAP and, 59
  coronary artery disease and, 176
  myocarditis and, 189
  resistance to, 48
  respiratory infections and, 49, 50–51, 270
  susceptibility to, 45, 46
E-selectin, 121, 129, 142, 178
Ethidium bromide detection method, 70
European Respiratory Science Meeting, 242
Experimental allergic encephalitis (EAE), 200, 206–207

Familial Alzheimer's disease, 212
Fibrinogen, 123, 127, 163, 192
Fluorescent-probe-based polymerase chain reaction (PCR), 68
Fluoroquinolones, 63, 176, 269
Fluorphenicol, 206
Foam cells, 122, 136, 143, 193
FtsK gene, 230
FtsW gene, 230

Garenoxacin (BMS-284756), 47
Gatifloxacin, 63, 184
Gemifloxacin, 48
Gender
  antibody prevalence and, 89–90
  atherosclerosis and, 130
  cardiovascular disease and, 76
  multiple sclerosis and, 199
Genes. *See also* specific genes
  PCR and, 67–68
  of Pmp family, 35–36
Genital chlamydial disease, 45–46, 101, 139, 254
Genitourinary tract infections, 3
Genomes, 32–33
Growth phase, obstacles to, 3, 4
Guinea pigs, 154, 155, 167

Hamsters, 154, 155
Hc1, 29
Heat shock protein (HSP), 115, 217
Heat shock protein 60 (Hsp60), 139
  antibodies and, 89
  asthma and, 243
  atherosclerosis and, 122, 123
  cardiovascular disease and, 5, 6, 136, 141, 142–143, 144
  myocarditis and, 193
  reactive arthritis and, 230
  vaccines and, 101, 102, 105
HeLa cells, 16, 229, 230
Helicobacter pylori, 114, 118, 127, 130, 182
Hepatitis, 206
HEp-2 cells, 13, 35, 48, 232
Herpes simplex encephalitis, 203
Herpes simplex virus (HSV), 118
Herpes simplex virus-1 (HSV-1), 206, 212
Hexagonally arrayed structures (HAS), 17–19, 30
HHV. *See* Human herpes virus
High density lipoprotein (HDL), 155, 163, 166, 212
HIV. *See* Human immunodeficiency virus
HLA B27 allele, 228
HMG-CoA inhibitors, 142
Host attack, 39–40
Host cell signaling, 139–142
Host tropism, 147–148

HSP. *See* Heat shock protein
HSV. *See* Herpes simplex virus
HTLV. *See* Human T-cell lymphotropic virus
Human brain microvascular endothelial cells (HBMECs), 214–215, 216, 218
Human herpes virus-6 (HHV-6), 204
Human immunodeficiency virus-1 (HIV-1), 62, 202, 204, 217, 221
Human T-cell lymphotropic virus (HTLV), 204
Human T-cell lymphotropic virus-1 (HTLV-1) myelitis, 203
Humoral immunity, 83, 104
Hypercholesterolemia, 5, 154, 155, 156, 162, 166. *See also* Cholesterol
Hypertension, 113

IκBα, 140
Immune escape, 88, 103
Immune response, 3, 39–40, 81–90
  acquired, 83, 84–88
  antigen-specific, 81, 82, 83
  atherosclerosis and, 121–123
  cell-mediated (*see* Cell-mediated immunity)
  humoral, 83, 104
  innate mechanisms in, 82–83
  vaccines and, 101–102
Immunocompromised patients, 62, 100
Immunofluorescence microscopy, 35, 36, 37
Immunoglobulin gamma A (IgA), 89, 188
  antibiotics and, 51, 128, 176
  asthma and, 242, 243
  cell-mediated immunity and, 87–88
  dilated cardiomyopathy and, 192
  endocarditis and, 191
  myocarditis and, 189, 190
Immunoglobulin gamma G (IgG), 89, 90, 188
  antibiotics and, 49, 51, 125–126, 176
  asthma and, 242
  cardiovascular disease and, 6
  cell-mediated immunity and, 87
  endocarditis and, 191, 192
  HHV-6 and, 204
  multiple sclerosis and, 203
  myocarditis and, 189, 190
  respiratory infections and, 268, 269
Immunoglobulin gamma M (IgM), 89, 188
  antibiotics and, 49, 51
  asthma and, 249
  endocarditis and, 191
  HHV-6 and, 204
Immunoglobulin gamma M (*cont.*)
  myocarditis and, 189, 190
  respiratory infections and, 268, 269
Immunostaining, 66, 75, 76–77
IncA, 37–39
IncB, 39
IncC, 39
Inclusion, 2, 23, 37
  antibiotics and, 48
  cell and molecular biology of, 29, 30
  morphology of, 12, 13–15
  proteome and, 35
Inclusion body, 115
Inclusion-forming units (IFUs), 156, 158, 159
Inclusion membrane, 2, 3, 15, 21, 26, 29, 30, 37
Inc proteins, 3, 15, 33
Indoleamine-2,3-dioxygenase (IDO), 102, 144–145, 234
Inducible nitric oxide synthase (iNOS), 102, 107, 139
Infectious mononucleosis (IM), 205
Inflammatory arthritis, 227–236. *See also* Reactive arthritis
Inhaled corticosteroids (ICS), 241–242, 246, 248t, 265
Interferon-α (IFN-α), 234
Interferon-β (IFN-β), 88, 234, 253
Interferon-γ (IFN-γ), 40, 81, 83, 84, 86, 90, 115
  Alzheimer's disease and, 217
  antibiotics and, 48
  atherosclerosis and, 121, 123
  cardiovascular disease and, 144–145
  multiple sclerosis and, 201
  persistence and, 23
  reactive arthritis and, 230, 232, 233, 234
  respiratory infections and, 270
  vaccines and, 101–102, 103, 106, 107–108
Interferon (IFN) type 1. *See* Type 1 interferons
Interleukin-1 (IL-1), 83, 107, 121, 126, 233
Interleukin-1-β (IL-1β), 136, 217, 218, 234, 252
Interleukin-2 (IL-2), 121, 233
Interleukin-4 (IL-4), 83, 233
Interleukin-5 (IL-5), 233, 246
Interleukin-6 (IL-6), 83, 139
  Alzheimer's disease and, 217
  asthma and, 253
  atherosclerosis and, 121, 123, 126
  cardiovascular disease and, 140
  coronary artery disease and, 180
  inflammatory arthritis and, 233

Interleukin-8 (IL-8), 82
asthma and, 252, 253
atherosclerosis and, 121
cardiovascular disease and, 141, 142
inflammatory arthritis and, 234
Interleukin-10 (IL-10), 83–84, 88, 193, 233, 234
Interleukin-12 (IL-12), 83, 84, 88
Alzheimer's disease and, 217
inflammatory arthritis and, 233, 234
vaccines and, 107, 108
Interleukin-13 (IL-13), 83
Interleukin-15 (IL-15), 234
Interleukin-18 (IL-18), 83, 84
Intervention with Clairthomycin in patients with stable Coronary heart disease (CLARICOR) study, 128
Intracellular adhesion molecule-1 (ICAM-1), 121, 142, 215, 252
Intracoronary Stenting and Antibiotic Regime (ISAR-3) trial, 126, 180
In vitro studies
of Alzheimer's disease, 216, 218–219
of antibiotic susceptibility, 45–48, 50
of antibodies, 90
of atherosclerosis, 153
of cardiovascular disease, 141
of cell-mediated immunity, 87
of the developmental cycle, 12
of persistent infections, 23
of reactive arthritis, 230
of vaccines, 103
In vivo studies
of antibodies, 90
of cardiovascular disease, 136
of the developmental cycle, 12
ISAR-3. *See* Intracoronary Stenting and Antibiotic Regime
Ischemic stroke, 113, 176

Ketolides, 46–47
Koalas, 167
Koch's postulate, 130
Kynurenine, 145

Leukocytes
atherosclerosis and, 121, 122
cardiovascular disease and, 136, 139, 141, 142, 143
PCR and, 75
reactive arthritis and, 228
Levofloxacin, 47, 48, 49, 50, 63
Life cycle, chlamydial, 137–139
Lipid X, 143, 147
Lipopolysaccharide (LPS), 5, 29, 40, 71, 115, 142
Alzheimer's disease and, 217
antibodies and, 89
atherosclerosis and, 122
cardiovascular disease and, 136, 147
inflammatory arthritis and, 234
myocarditis and, 187
physical localization of, 31, 32
vaccines and, 101
Listeria monocytogenes, 107
Low density lipoprotein (LDL), 120, 121–122, 136, 139
in animal models, 155, 156, 157, 159, 162, 163, 166, 167
macrophages and, 143
myocarditis and, 193
Lower respiratory tract infections, 4, 57, 265–267
Lung infections, 156–157
Lung remodeling, 251–253
Lymphocyte proliferation, 86–88
Lymphocytes, 75, 136, 194
Lysosome fusion, 3, 12, 29, 37, 138

Macrolides
asthma and, 243–246, 252, 265
atherosclerosis and, 128, 129, 163
CAP and, 59, 62, 63
coronary artery disease and, 176, 184
resistance to, 47, 48
respiratory infections and, 51, 269–270
susceptibility to, 45, 46–47
Macrophages, 4, 30
Alzheimer's disease and, 217
asthma and, 252, 253
atherosclerosis and, 121–122, 123, 154, 156
cardiovascular disease and, 136, 139, 141, 142–144, 147
Chlamydiae interactions with, 142–144
inflammatory arthritis and, 234
myocarditis and, 193
role in *C. pneumoniae* immunity, 82, 83–84, 86
vaccines and, 102, 106
Major histocompatibility complex (MHC) class I, 39, 83, 86, 88, 102, 103, 104, 106
Major histocompatibility complex (MHC) class II, 39, 83, 84, 86, 88, 103, 104, 106

Major outer membrane protein (MOMP), 29–30, 40, 115, 137
  cardiovascular disease and, 147, 148
  multiple sclerosis and, 205
  PCR and, 71, 74
  persistence and, 23, 24
  physical localization of, 31, 32
  respiratory infections and, 269
  vaccines and, 103, 104–105, 108
Malaria, 202
Mast cells, 253
Matrix metalloproteinase (MMP), 123, 136, 143–144, 177, 178, 253
Mecillinam, 45
Meeting of the European Society for Chlamydia Research, 242
MefA gene, 48
Metabolic syndrome, 130
Metronidazole, 127, 181
MHC. *See* Major histocompatibility complex
Mice, 4
  in Alzheimer's disease studies, 219
  in atherosclerosis studies, 120, 154, 155, 157–159, 160t, 163, 166, 167, 168
  in cardiovascular disease studies, 5, 135–136
  in genital infection studies, 139
  immune response in, 82, 84–86
  in lung infection studies, 156–157
  in multiple sclerosis studies, 206
  in myocarditis studies, 194
  in respiratory infection studies, 30
  in vaccine studies, 100, 101–102, 103, 104, 108
Microbiology of *C. pneumoniae*, 115–116
Microimmunofluorescence (MIF), 89, 116–117, 188
  endocarditis and, 191
  myocarditis and, 189, 190
  respiratory infections and, 49, 268, 269
Minocycline, 46
Mitogen-activated protein kinase (MAPK), 123, 142
Molecular biology. *See* Cell and molecular biology
MOMP. *See* Major outer membrane protein
Monkeys, 156, 157, 200. *See also* Primates, nonhuman
Monocyte chemotactic protein-1 (MCP-1), 121, 141, 142, 234
Monocyte integrins, 121
Monocytes, 6, 30, 83, 86, 88, 184
  Alzheimer's disease and, 213, 214, 215, 216, 217, 218–219, 222
  atherosclerosis and, 121–122, 123
  cardiovascular disease and, 139, 141, 147
  inflammatory arthritis and, 234
  myocarditis and, 193
  PCR and, 75
Moraxella osloensis, 228
Morphology, 11–26
  developmental cycle, 11–15
  of the elementary body, 14t, 15–16
  outer membrane structure, 16–20
Moxifloxacin, 49, 50, 63
Multiple sclerosis, 6, 199–208, 235
  autoimmune basis for, 199–201
  C. pneumoniae mediation of autoimmunity, 205–207
  candidate pathogens in, 203–205
  epidemics of, 202, 203
  infection hypothesis for, 201–203
  prevalence of, 199
Multiple sclerosis-associated retrovirus (MSRV), 204–205
MutS gene, 230
Myasthenia gravis, 201
Mycoplasma, 228
Mycoplasma pneumoniae, 51, 59, 61, 118, 120, 161, 239, 246, 253, 263, 265, 267
Myelin antigens, 200
Myelin basic protein (MBP), 200, 206
Myelin oligodendrocyte glycoprotein (MOC), 200, 206
Myocardial infarction (MI), 141
  animal models of, 156, 168
  antibiotics and, 125–126, 127, 128, 175–176, 178–180, 182–183
  non-Q-wave, 126, 127, 179, 182
Myocarditis, 5, 187–190, 192–194
  animal models of, 192–193
  possible mechanisms in, 193–194

NADPH oxidase, 102
Nasopharyngitis, 264
Natural killer (NK) cells, 84, 102
Neopterin, 123
Nested polymerase chain reaction (PCR), 68, 74
Netilmicin, 189
Neural antigens, 200
Neuritic senile plaques (NSPs), 211
Neuroborreliosis, 203
Neurofibrillary tangles (NFTs), 211, 219, 222

Neuropil threads, 211
Neutrophil cationic protein, 243
Neutrophils, 122, 141, 193, 253
NF-κB, 123, 140, 142, 252
Nitric oxide, 83, 123, 139, 234
Nitric oxide synthase, 83. *See also* Inducible nitric oxide synthase
NlpD gene, 23
Non-Q-wave myocardial infarction, 126, 127, 179, 182
Nonsteroidal antiinflammatory drugs, 217
NVP-PDF386, 47

Observational studies, 175–176
Occludin, 216
Ocular chlamydial infections, 101, 254
Ofloxacin, 24, 47, 48, 49
Olfactory bulbs, 216, 219, 222
Oligoclonal bands (OCBs), 203, 205
Oligodendrocytes, 201
OmcA, 71
OmcB, 71
Omeprazole, 127, 181
OmpA gene, 23, 71
OmpB gene, 23
Omp1 gene, 68, 71, 74, 230, 234
Omp2 gene, 29, 31, 101, 105, 108
Omp3 gene, 29, 31
Otitis media, 264
Outer membrane, structure of, 16–20
Outer membrane protein (OMP), 213, 217. *See also* Major outer membrane protein; Pmps

Pathogen-associated molecular patterns (PAMPs), 107
Peak expiratory flow rate (PEFR), 246
Pediatric patients
  antibodies in, 89
  asthma in, 239, 240, 241, 242, 248, 249, 265–267
  CAP in, 59, 60t, 62, 267, 268t
  ocular infections in, 254
  respiratory infections in, 4, 50–51, 263–271
  vaccines for, 100
Pelvic inflammatory disease, 1, 142, 243, 251, 254
Penicillin, 24
  cardiovascular disease and, 144
  coronary artery disease and, 176
  myocarditis and, 189, 190
  reactive arthritis and, 230
  susceptibility to, 45, 46
Penicillin-binding proteins (PBPs), 46
Peptide deformylase (PDF) inhibitors, 47
Peptidoglycan (PG), 16
Pericarditis, 190
Perimyocarditis, 5, 187, 188t, 190, 194
Periodicity, 18, 19t, 30
Peripheral arterial disease (PAD), 113, 116, 123, 128
Peripheral blood mononuclear cells (PBMC), 75, 147, 157, 204
Peripheral vascular disease, 176–178
Persistence, 2, 23–25, 137–139
  in animal models, 157
  antibiotic treatment and, 48
  atherosclerosis and, 123
  reactive arthritis and, 229–231
Persistent body (PB), 115–116, 123, 130, 173–174
Persistent symptomatic reversible airway obstruction, 240
Phagocytes, 4, 82, 147
Phagocytosis, 12
Phagosomes, 3, 29
Pharyngitis, 264
Pigeons, 154, 156
Pigs, 120, 156, 167
Plasminogen activator inhibitor 1 (PAI-1), 140
Platelet-derived growth factor (PDGF), 121, 161, 166
*Pmp*1 gene, 35
*Pmp*2 gene, 35
*Pmp*6 gene, 35, 36
*Pmp*7 gene, 35
*Pmp*8 gene, 35, 36
*Pmp*10 gene, 35
*Pmp*11 gene, 35
*Pmp*13 gene, 35
*Pmp*14 gene, 35
*Pmp*20 gene, 35, 36
*Pmp*21 gene, 35, 36
Pmps, 30, 31, 33, 34, 40
  cardiovascular disease and, 147–148
  gene family of, 35–36
  structure of, 36
Pneumonia, 1, 30, 188, 189, 191
  antibiotics and, 49, 52
  asthma and, 248
  atypical, 4
  community-acquired (*see* Community-acquired pneumonia)
  in pediatric patients, 267, 269
  signs and symptoms of, 58t
  vaccines and, 101

PolA gene, 230
Polymerase chain reaction (PCR), 6, 30, 65–77, 89
  in Alzheimer's disease studies, 213, 216, 219
  in asthma studies, 242, 246, 249, 251, 253
  in atherosclerosis studies, 117, 118, 120, 129, 157
  in cardiovascular disease studies, 135, 140
  Chlamydia detection in blood, 65–66, 70–76, 77
  contamination in, 66–67, 74
  detection of products, 70
  DNA extraction in, 66, 67, 71, 74
  in endocarditis studies, 192
  endpoint readings, 68
  fluorescent-probe-based, 68
  immunoassay detection method, 70
  methodological aspects of, 66–70
  in multiple sclerosis studies, 205
  in myocarditis studies, 188, 189
  nested, 68, 74
  in pneumonia studies, 267
  protocols in, 68–69
  in reactive arthritis studies, 227, 231–232
  real-time readings, 68–69, 74
  in respiratory infection studies, 264, 268–269
  reverse-transcriptase, 67–68, 161, 213, 216, 232
  target genes in, 67–68
  touchdown, 68
  vaccines and, 106
Polymorphic outer membrane proteins. *See* Pmps
Polymorphisms
  single-nucleotide, 146
  tyrP gene, 144–146
Polymorphonuclear neutrophils (PMN), 82
Pomp genes, 35, 36
Ppp2 gene, 148
Ppp6 gene, 148
Ppps, 148
Pravastatin, 184
Pravastatin or Atorvastatin Evaulation and Infection Therapy (PROVE-IT) study, 128, 184
Presenilin 1 (PS1) protein, 212
Presenilin 2 (PS2) protein, 212
Primary MS affection (PMSA), 202
Primates, nonhuman, 102, 154, 156, 157, 167
Proinflammatory cytokines, 4, 83, 84, 217
Prostacyclin, 123
Prostaglandin-E2, 82, 252
Proteolipid protein (PLP), 200
Proteome, 33–35, 37
PROVE-IT. *See* Pravastatin or Atorvastatin Evaulation and Infection Therapy
P-selectin, 121
Pseudomonas, 228
Pyk gene, 23

Quail, 154, 156
Quinolones, 45, 47, 48, 128, 176, 184

Rabbits, 120, 154–155, 156, 157, 159–163, 164–165t, 166, 167, 168, 174–175
Race, multiple sclerosis and, 202
Randomized controlled trials, 246–248
Randomized secondary prevention trials
  for coronary artery disease, 178–183
  for peripheral vascular disease, 176–178
RANTES, 234
Rats, 154, 155, 168
Reactive airway disease, 4, 50, 51
Reactive arthritis, 1, 6, 189, 216–217, 227–236
  C. pneumoniae and, 231–232
  overview of, 228–229
  persistence and, 229–231
Reiter's syndrome, 228–229
Respiratory system, 3–4, 12, 29
Respiratory tract infections, 23. *See also* Lower respiratory tract infections; Upper respiratory tract infections
  antibiotics and, 49–51, 269–271
  asthma following, 248
  carrier status, 263–264
  diagnosis of, 267–269
  insights into, 4
  in pediatric patients, 4, 50–51, 263–271
  treatment of, 269–270
  with wheezing, 265–267, 270
Response to injury hypothesis, 121
Restenosis, 126, 180
Reticulate body (RB), 1, 11, 26, 36, 57, 115, 137–138, 173
  Alzheimer's disease and, 213
  basic biology of, 2–3
  cell and molecular biology of, 29
  developmental cycle and, 12
  immune response and, 81
  outer membrane structure of, 19–20
  persistence and, 23–24
  physical localization of components, 31
  protein composition of, 33, 35

Reticulate body (*cont.*)
reactive arthritis and, 229
surface projections and, 21–23
type III secretions and, 37
ultrastructure of, 30
vesicle trafficking and, 39
Retroviruses, 204–205
Reverse-transcriptase polymerase chain reaction (RT-PCR), 67–68, 161, 213, 216, 232
Rheumatoid arthritis, 228
Rifalazil, 47
Rifampin, 6
Rifamycin, 47
Rodents, 154, 155. *See also* specific types
Roxithromycin
asthma and, 246
atherosclerosis and, 126, 127, 128–129
coronary artery disease and, 52, 179–180
peripheral vascular disease and, 176–177
for respiratory infections, 51
Roxithromycin in Ischemic Syndromes (ROXIS) study, 126

Salmonella, 227
Secreted proteins, 37–39
Seroepidemiology
of asthma, 242–243
of atherosclerosis, 116–117
Shigella, 227
Sickle cell disease, 202
Single-nucleotide polymorphisms (SNPs), 146
Sinusitis, 264
Smoking, 88, 90, 113, 130, 240, 249, 251
Smooth muscle cells, 121, 122–123, 139, 141, 147, 154, 166, 193
South Thames Trial of Antibiotics in Myocardial Infarction and Unstable Angina (STAMINA) study, 127, 181
Sparfloxacin, 47
Sporadic Alzheimer's disease. *See* Alzheimer's disease
STAMINA. *See* South Thames Trial of Antibiotics in Myocardial Infarction and Unstable Angina
Stenotrophomonas maltophilia, 228
Streptococcus pneumoniae, 48, 59, 61
Streptococcus pyrogens, 264
Stroke, 113, 116, 123, 176
Subacute sclerosing panencephalitis (SSPE), 203, 205
Sulfonamides, 45, 176
Surface adhesion molecules, 215
Surface projections, 20–23, 25–26, 30, 36
Swine, 154
Syphilis, 203

T cells, 4, 7, 81, 83, 84–86, 88, 90
inflammatory arthritis and, 234
multiple sclerosis and, 200, 206
myocarditis and, 193
vaccines and, 102, 106, 107–108
Telithromycin, 47, 50
Tetracyclines
atherosclerosis and, 128, 129
coronary artery disease and, 176
endocarditis and, 191
resistance to, 47
respiratory infections and, 49, 269
susceptibility to, 45, 46, 47
TGF. *See* Transforming growth factor
Th1 cells, 3, 101, 102, 108, 200, 217, 233, 234
Th2 cells, 101, 217, 233, 234
Theophylline, 243
Tigecycline (GAR-936), 46
Tissue factor, 121
T lymphocytes, 122, 123, 139, 147, 200
TNF. *See* Tumor necrosis factor
Toll/interleukin-1 receptor domain (TIR), 107
Toll-like receptors (TLR), 107–108
Touchdown polymerase chain reaction (PCR), 68
Trachoma, 1, 142, 207, 243, 251, 254
Transforming growth factor-β (TGF-β), 83, 161, 166, 233
Triacetyloleandomycin (TAO), 243
Tryptophan, 83, 102, 144–145, 234, 270
Tubal infertility, 1, 84, 87, 243, 251, 254
Tuberculosis, 202, 203
Tumor necrosis factor (TNF), 201
Tumor necrosis factor-α (TNF-α), 83
Alzheimer's disease and, 217, 218
asthma and, 252
atherosclerosis and, 121, 123, 126
cardiovascular disease and, 136, 142, 143
inflammatory arthritis and, 233, 234
Tumor necrosis factor-γ (TNF-γ), 107
Type 1 immune response, 81, 83, 84, 86, 90
Type 2 immune response, 83–84, 88, 90
Type 1 interferons (IFNs), 104, 107, 108
Type III secretion, 2–3, 23, 34, 39, 146
description of system, 36–37
genes encoding, 33

Type III secretion (*cont.*)
immune response and, 83
vaccines and, 102, 105
TyrP gene polymorphism, 144–146

Unstable angina, 126, 127, 179, 182, 183
Upper respiratory tract infections, 4, 30, 57, 205, 264
Uracil N-glycosylase (UNG), 66
Uridine kinase, 146

Vaccines, 4, 7, 88, 99–108
adjuvants for, 107–108
antigens and, 101, 104–106
delivery systems for, 108
development of, 100–103
populations requiring, 100
Vascular cell adhesion molecule-1 (VCAM-1), 121, 142, 215
Vascular smooth muscle cells (VSMC), 140
Very low density lipoprotein (VLDL), 155, 163, 166, 212
Vesicle trafficking, 39
VF-cadherin, 216
Vitamin E, 143
VLA-4, 215
Von Willebrand factor, 129, 178
VRC4887, 47

Weekly Intervention with Zithromax for Atherosclerosis and its Related Disorders (WIZARD) study, 127, 166, 182–183
Wheezing, 265–267, 270–271

Yersinia, 227

Zonula adherens junction, 215–216
Zonula occludens junction, 215–216